Communications Handbook

Volume I

Communications Handbook
Volume I

Edited by **Akira Hanako**

CLANRYE
INTERNATIONAL

New Jersey

Published by Clanrye International,
55 Van Reypen Street,
Jersey City, NJ 07306, USA
www.clanryeinternational.com

Communications Handbook: Volume I
Edited by Akira Hanako

International Standard Book Number: 978-1-63240-108-3 (Hardback)

Printed in the United States of America.

Contents

Preface

Communication is the activity of conveying information, ideas, attitudes, perceptions or feelings through various means of expression like writings, gestures, speech, behavior as well as by other possible means which can include physical, chemical or electromagnetic phenomena. It can also be explained as a meaningful exchange of information and ideas between two or more participants, either of which can be machines or organisms. This arena can also be seen as a field of study and research that encompasses a variety of topics, from face-to-face conversation to mass media outlets such as television broadcasting. Through this discipline, one is also able to examine how messages are interpreted through the cultural, political, economic, social, hermeneutic and semiotic dimensions of their milieu. One can also say that from a humanistic perspective, communication is related to rhetoric and persuasion. It is thus a diverse field that includes examination by humanists, social scientists as well as scholars focusing on critical and cultural studies. The discipline of communication concentrates on how people use messages to create meanings within and across various perspectives, media, cultures and channels. The discipline is thus effective in promoting the practice of human communication. More effective means of communication and communication research are being discovered at a rapid pace that highlights the importance of communication in our lives.

This book is an attempt to compile and collate all available research on communication under one umbrella. I am grateful to those who put their hard work, effort and expertise into these researches as well as those who were supportive in this endeavor.

Editor

Nonlinear Demodulation and Channel Coding in EBPSK Scheme

Xianqing Chen and Lenan Wu

School of Information Science and Engineering, University of Southeast, 2 Sipailou, Nanjing 210096, China

Correspondence should be addressed to Xianqing Chen, xqchen213@126.com

Academic Editors: P. Colantonio, J. Dauwels, A. Ruano, and S. Sinha

The extended binary phase shift keying (EBPSK) is an efficient modulation technique, and a special impacting filter (SIF) is used in its demodulator to improve the bit error rate (BER) performance. However, the conventional threshold decision cannot achieve the optimum performance, and the SIF brings more difficulty in obtaining the posterior probability for LDPC decoding. In this paper, we concentrate not only on reducing the BER of demodulation, but also on providing accurate posterior probability estimates (PPEs). A new approach for the nonlinear demodulation based on the support vector machine (SVM) classifier is introduced. The SVM method which selects only a few sampling points from the filter output was used for getting PPEs. The simulation results show that the accurate posterior probability can be obtained with this method and the BER performance can be improved significantly by applying LDPC codes. Moreover, we analyzed the effect of getting the posterior probability with different methods and different sampling rates. We show that there are more advantages of the SVM method under bad condition and it is less sensitive to the sampling rate than other methods. Thus, SVM is an effective method for EBPSK demodulation and getting posterior probability for LDPC decoding.

1. Introduction

Nowadays, wireless communication is playing a very important role in our daily life. The growing demands on wireless multimedia services and products lead to increasing needs for radio spectrum and data rates. Thereby, the research on modulations with high bandwidth efficiency is on focus [1]. In order to satisfy the higher and higher demand for communication systems, an extended binary phase shift keying (EBPSK) system with very high spectra efficiency is introduced in [2]. A special impacting filter (SIF) which can produce high impact at the phase jumping point, narrow in bandwidth, and great improvement in SNR, was applied at the demodulator [3]. Therefore, a simple amplitude detector followed would perform the demodulation of EBPSK signals [4]. However, the conventional threshold decision may not be best to achieve the optimum performance, and the SIF used in EBPSK demodulator brings more difficulty in obtaining posterior probability for low-density parity check (LDPC) codes decoding. A simple and general bit metric generation method is proposed by Hyun and Yoon [5] for the soft information to initial channel decoding. We modify the scheme to suit our system and the method is referred

to as MHY in this paper. Meanwhile, nonlinear detectors are specifically designed to get the optimum performance of a blind multiuser detector [6, 7] and nonlinear channel equalization [8–10] and providing accurate posterior probability estimates (PPEs) for LDPC decoding [11, 12]. All results have shown that a nonlinear demodulator performs similar to an optimum receiver. One of the goals of this paper is the analysis of nonlinear demodulation with the channel decoder. We make use of the fact that the demodulator performance should not only be measured by low BER, but also in its ability to provide accurate PPEs that can be exploited by a soft-input channel decoder to achieve capacity. In this paper, we will introduce a nonlinear demodulation technique called the support vector machine (SVM) classifier [13]. The design approach is completely novel, where we select only a few samples of the SIF output for SVM training and testing at intermediate frequency (IF) without downconversion. We propose to measure the performance of this demodulator after an LDPC channel decoder, and the ability of SVM to provide accurate posterior probability predictions boosts the demodulator performance compared to the MHY method.

The rest of the paper is organized as follows. Section 2 is devoted to introducing SVM. We present the receiver scheme in Section 3 and briefly describe the EBPSK modulation and LDPC decoding. In Section 4, we include illustrative experiments to compare the performance of the proposed demodulators. We conclude in Section 5 with some final comments.

2. Support Vector Machine

The SVM is a classifier introduced by Cortes and Vapnik [14], which can realize the same performance as the so-called artificial neural networks (ANNs) for classification. Generally, ANN has the problem of a local minimum. On the other hand, the SVM is mathematically transparent and can provide global and unique solutions.

2.1. Binary Classification of SVM. For the binary classification problem, the training set consists of vectors from the pattern space $x_i \in \mathbf{R}^n$, $i = 1, 2, \ldots, L$ and to each vector a classification $y_i \in \{1, -1\}$. During the initial training stage, a decision function is constructed via

$$f(x) = \sum_{i=1}^{L} \alpha_i y_i K(\mathbf{x}, \mathbf{x}_i) + b, \tag{1}$$

where α_i is a Lagrangian constant, $K(\mathbf{x}, \mathbf{x}_i) = \Psi(\mathbf{x}_i)^T \cdot \Psi(\mathbf{x})$ is a kernel function, $\Psi(\mathbf{x})$ maps the training data vector $\mathbf{x}_i$ into the high-dimensional feature space, and b is a bias term.

Define a coefficient vector $\mathbf{w}$, such that

$$\mathbf{w} = \sum_{i=1}^{L} \alpha_i y_i \Psi(\mathbf{x}_i), \tag{2}$$

then the training is completed by solving the following optimization problem:

$$\min_{\mathbf{w} \in \mathbf{H}, b \in \mathbf{R}, \xi \in \mathbf{R}^L} \frac{1}{2} \|\mathbf{w}\|^2 + C \sum_{i=1}^{L} \xi_i y_i((\mathbf{w} \cdot \mathbf{x}_i) + b) \geq 1 - \xi_i, \tag{3}$$

$$\xi_i \geq 0, \quad i = 1, 2, \ldots, L,$$

where C is the tradeoff parameter between the training error and the margin of the decision function, and ξ_i is a slack variable to compensate for any nonlinearly separable training points.

In this paper, the SVM demodulator uses two types of kernel functions to compare the performance with each other. The first is the simplest linear kernel, shown as

$$K(x_i, x_j) = x_i^T x_j. \tag{4}$$

The second is a more popular radial basis function (RBF) kernel, shown as

$$K(x_i, x_j) = \exp\left(-\gamma \|x_i - x_j\|^2\right), \quad \gamma > 0, \tag{5}$$

where γ controls the width of the function.

2.2. Complexity Analysis. The complexity of training an SVM for binary classification is $O(n^2)$, using the sequential minimal optimization [15], and Platt's method adds a computational complexity of $O(n^2)$. However, the SVM demodulator should be analyzed for the testing stage only because the training time is very small compared with the actual testing time. The main focus thus becomes analyzing the complexity required for the computing decision function in (1), which is using the simplest kernel. This issue will be discussed in detail later. A great amount of complexity can be reduced further in (1) if the expression is simplified as follows:

$$\begin{aligned} f(x) &= \sum_{i=1}^{L} \alpha_i y_i K(\mathbf{x}, \mathbf{x}_i) + b = \sum_{i=1}^{L} \alpha_i y_i \left(x_i^T x\right) + b \\ &= \left[\sum_{j=1}^{N} y_j \alpha_j \left(\sum_{i=1}^{n} x_{j,i} x_i\right)\right] + b = \sum_{i=1}^{n} x_i \left(\sum_{j=1}^{N} y_i \alpha_i x_{j,i}\right) + b \\ &= \sum_{i=1}^{n} A_i x_i + b, \end{aligned} \tag{6}$$

where N is the number of support vectors, and the constants $A_i = \sum_{j=1}^{N} y_i \alpha_i x_{j,i}$ and b can be precomputed before the testing stage to save the computation time. Therefore, the complexity of the SVM demodulator is $O(n)$.

2.3. Probabilistic Outputs of SVM. Instead of predicting the label, many applications require a posterior class probability $P(y = 1 \mid x)$. The transformation of SVM output into posterior probabilities has been proposed by Platt in [16]. Platt's method squashes the SVM soft output through a trained sigmoid function to predict posterior probabilities:

$$p(y = 1 \mid \mathbf{x}) \approx P_{A,B}(f) = \frac{1}{1 + \exp(Af + B)}, \tag{7}$$

where $f = f(\mathbf{x})$, let each f_i be an estimate of $f(\mathbf{x}_i)$. The best parameter setting $z^* = (A^*, B^*)$ is determined by solving the following regularized maximum likelihood problem:

$$\min_{z=(A,B)} F(z) = -\sum_{i=1}^{l} (t_i \log(p_i) + (1 - t_i) \log(1 - p_i)), \tag{8}$$

where $p_i = P_{A,B}(f_i)$, $t_i = (y_i + 1)/2$.

Furthermore, log and exp could easily cause an overflow, if $Af_i + B$ is large, $\exp(Af_i + B) \to \infty$ and $1 - p_i = 1 - 1/(1 + \exp(Af_i + B))$ is a "catastrophic cancellation" when p_i is close to one. The problem can usually be resolved by reformulation [17]:

$$-(t_i \log p_i + (1 - t_i) \log(1 - p_i)) \tag{9}$$

$$= (t_i - 1)(Af_i + B) + \log(1 + \exp(Af_i + B)) \tag{10}$$

$$= t_i(Af_i + B) + \log(1 + \exp(-Af_i - B)). \tag{11}$$

If $Af_i + B \geq 0$ then use (11), else use (10). Then (7) can be rewritten as follows:

$$p(y = 1 \mid \mathbf{x}) \approx \begin{cases} \dfrac{1}{1 + \exp(Af + B)}, & Af + B < 0, \\[2mm] \dfrac{\exp(-Af - B)}{1 + \exp(-Af - B)}, & Af + B \geq 0. \end{cases} \quad (12)$$

From (12), we can see that SVM does not provide PPE and its output needs to be transformed, before it can be interpreted as posterior probabilities; therefore, the posterior probability is an approximate one.

3. Communication System

3.1. EBPSK Modulation. EBPSK is a modulation method with high frequency spectra efficiency, which is defined as follows:

$$f_0(t) = A \sin 2\pi f_c t, \quad 0 \leq t < T,$$

$$f_1(t) = \begin{cases} B \sin(2\pi f_c t + \theta), & 0 \leq t < \tau, 0 \leq \theta \leq \pi, \\ A \sin(2\pi f_c t), & \tau \leq t < T, \end{cases} \quad (13)$$

where f_0 and f_1 are modulation waveforms corresponding to bit "0" and bit "1," respectively, $T = N/f_c$ is the bit duration, $\tau = K/f_c$ is the phase modulation duration, and θ is the modulating angle. Obviously, if $\tau = T$ and $\theta = \pi$, (13) degenerates to the classical binary phase shift keying (BPSK) modulation.

3.2. LDPC Decoding. LDPC codes can be decoded by an iterative message-passing (MP) algorithm which passes messages between the variable nodes and check nodes iteratively. If the messages passed along the edges are probabilities, then the algorithm is also called belief propagation (BP) decoding, which is the optimal if there are no cycles or cycles are ignored. Moreover, with BP decoding, complicated calculations are distributed among simple node processors, and after several iterations, the solution of the global problem is available. The steps of BP decoding are as follows.

(1) Initialization: $p_n^0(x) = q_{nm}^0 = p(x_n = x \mid y_n)$, where $p(x_n = x \mid y_n)$ is the soft information of channel outputs.

(2) Horizontal step: the MAP output from c_m to v_n:

$$r_{mn}^k(0) = p(v_n = 0 \mid c_m = 0, y_{i \in B(m) \backslash n}),$$

$$r_{mn}^k(0) = \frac{1}{2} + \frac{1}{2} \prod_{i \in B(m) \backslash n} \left(1 - 2q_{im}^k(1)\right), \quad (14)$$

$$r_{mn}^k(1) = 1 - r_{mn}^k(0).$$

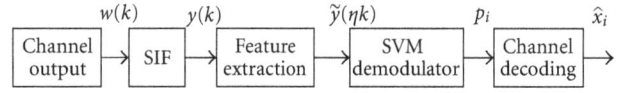

FIGURE 1: The block diagram of EBPSK receiver.

(3) Vertical step: updating the message from v_n to c_m:

$$q_{nm}^{k+1}(0) = \theta p_n^0(0) \prod_{j \in A(n) \backslash m} r_{jn}(0),$$

$$q_{nm}^{k+1}(1) = \theta p_n^0(1) \prod_{j \in A(n) \backslash m} r_{jn}(1), \quad \theta \text{ is chosen to ensure}$$

$$q_{nm}^{k+1}(0) + q_{nm}^{k+1}(1) = 1, \quad \text{Compute } p_n^k(x),$$

$$p_n^{k+1}(0) = \theta p_n^0(0) \prod_{j \in A(n)} r_{jn}(0),$$

$$p_n^{k+1}(1) = \theta p_n^0(1) \prod_{j \in A(n)} r_{jn}(1). \quad (15)$$

(4) Tentative output:

$$v_n^{k+1} = \begin{cases} 1, & p_n^{k+1}(1) \geq 0.5, \\ 0, & p_n^{k+1}(1) < 0.5, \end{cases} \quad (16)$$

if all parity check equations are satisfied or the maximum iteration number is reached, stop iteration, else return to Step (2).

In this paper, we focus on the initialization step for the posterior probabilities obtained by the nonlinear demodulator.

3.3. System Model. Figure 1 shows the receiver of EBPSK system. Suppose the system is synchronized, the signal of the channel output can be expressed as $w(k) = z(k) + n(k)$, where $n(k)$ is Gaussian white noise with zero mean. Input $w(k)$ into a SIF, and then the output signal can be expressed as $y(k) = w(k) * h(k)$, where $h(k)$ is the impulse response of SIF. In order to reduce the demodulation complexity, we select a few sample points as the features for SVM training and testing. Then, using the decision function (1), we can get the binary output as follows:

$$\text{sign}(f(\widetilde{\mathbf{y}}(\eta k))) = \text{sign}\left(\sum_{i=1}^{l} \alpha_i c_i K(\widetilde{\mathbf{y}}_i(\eta k), \widetilde{\mathbf{y}}(\eta k)) + b\right). \quad (17)$$

Then, we can get the posterior probability $p(\widetilde{x}_i = 1 \mid \widetilde{\mathbf{y}}_i(\eta k))$ and $p(\widetilde{x}_i = 0 \mid \widetilde{\mathbf{y}}_i(\eta k)) = 1 - p(\widetilde{x}_i = 1 \mid \widetilde{\mathbf{y}}_i(\eta k))$ through (12): finally, we use $p(x_i = x \mid y_i)$ to initiate the LDPC decoder.

4. Simulation Results and Discussions

In this section, we illustrate the performance of the proposed SVM demodulation and its soft output for LDPC decoding.

TABLE 1: Comparison of SVM models.

	Selected kernel	
	RBF	Linear
C	4	2
γ	8	—
SVs	271	210

Unless specified otherwise, all simulations assume that the system had 3000 random symbols for training and the reported BER is computed using 10^5 symbols and we average the results over 1000 independent trials with random training and test data. We choose $K = 2$, $N = 20$, $A = B = 1$, $\theta = \pi$ as the parameters of EBPSK modulation. LDPC codes are also applied to measure the BER performance of the communication system and the accurate posterior probability obtained by the SVM method. During simulations, we use a 1/2 rate regular LDPC code with 1000 bits per codeword and 3 ones per column. The whole system was simulated under MATLAB.

4.1. Kernel Selection and Demodulation. In this subsection, the performance of the SVM demodulator, using the kernel functions (4) and (5), introduced in Section 2, is compared. For the RBF kernel, a 10-fold cross-validation sweep from the training samples was used to find the optimum parameters of C and γ. A similar search was conducted for the linear kernel, but it only has the C parameter to adjust. Table 1 summaries the optimum SVM model obtained after the parameter search.

The linear kernel has less support vectors than the RBF one; therefore, it has a less computational complexity and thus would perform faster. In order to compare the BER performance fairly, both kernels used by the SVM receiver were classifying exactly the same received signals.

Figure 2 shows the BER performance of the SVM demodulator when employing different kernels; also, the performance of conventional threshold decision is analyzed. Evidently, the linear kernel, though much simpler, has slightly better performance than the RBF kernel. Moreover, the SNR gain between the SVM method and the threshold decision is around 1.8 dB; therefore, a linear SVM is chosen for the task. Training on a "worse-case" scenario works well (SNR = −7 dB in this case), proving that the SVM receiver needs not frequently retraining in different SNRs.

4.2. Kernel Optimization. To optimize the linear kernel, the only controlling parameter is C, which restrains the maximum size of the Lagrangian dual variable. The SVM detector is tested on the 20 sets of 20000 noisy sequences at SNR = 2 dB for various C values. The results are shown in Figure 3. While the error performance for various C is very similar, it is still ideal to choose a model with the least number of support vector (SV) in order to reduce the complexity. In this case, when C is beyond 6, the model gives the same number of SV because variable α_i is no longer

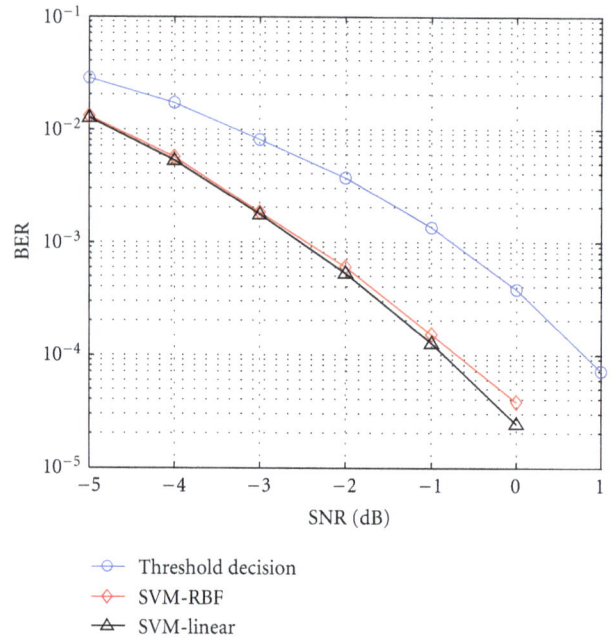

FIGURE 2: Demodulation with SVM-RBF, SVM-linear, and threshold decision.

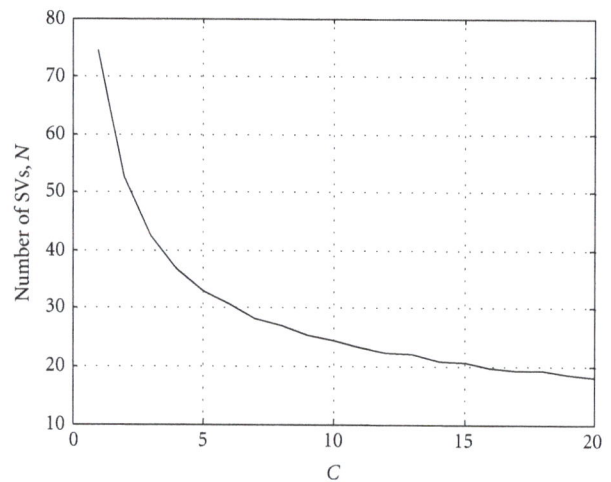

FIGURE 3: Number of support vectors from the SVM model for different C parameters, $n = 5$.

constrained by C. The correct rate remains around 99.47%, as shown in Figure 4.

The training size for the SVM detector is another parameter that the designer needs to control. In general, for any machine learning algorithms, the training size should be as large as possible to improve the prediction of the unknown testing data. The tradeoff in this application is the increased time required to produce and collect the training data. Figures 5 and 6, respectively, show the SVM demodulator's error performance and the number of SVs required on the same system as stated above with different training sizes. When the C parameter is fixed at 2, and with a training size of about 200, the performance of the SVM detector would reach

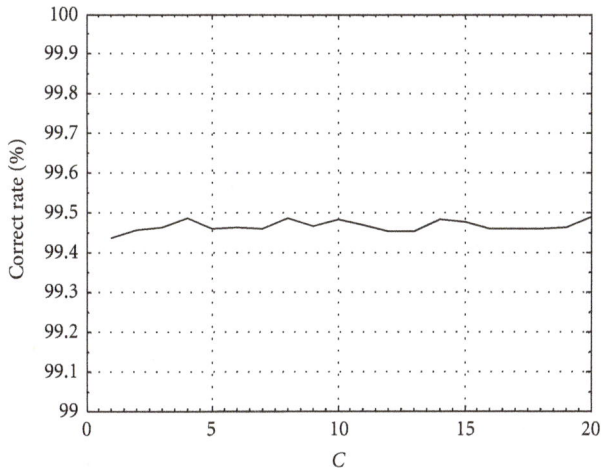

FIGURE 4: Correct rate of the SVM model with linear kernel for different C parameters, SNR = -4 dB, $n = 5$.

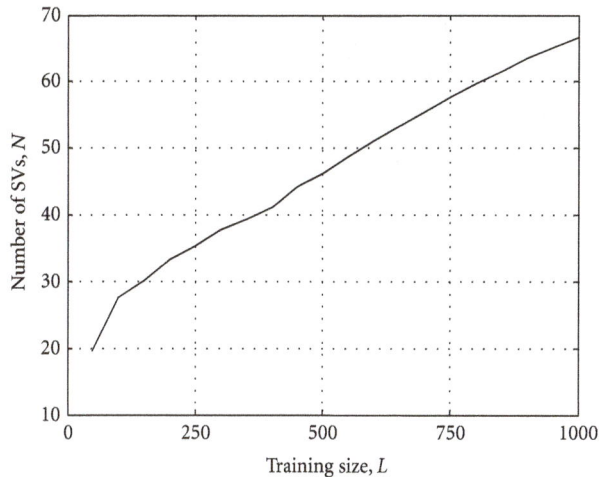

FIGURE 6: Number of support vectors from the SVM model for different training sizes, $n = 5$.

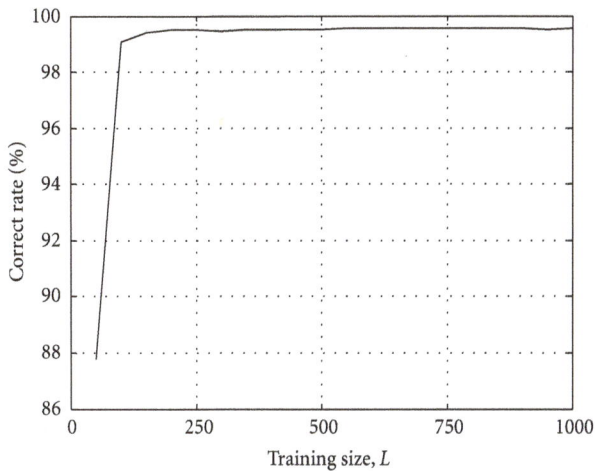

FIGURE 5: Correct rate of the SVM model with linear kernel for different training sizes, $n = 5$.

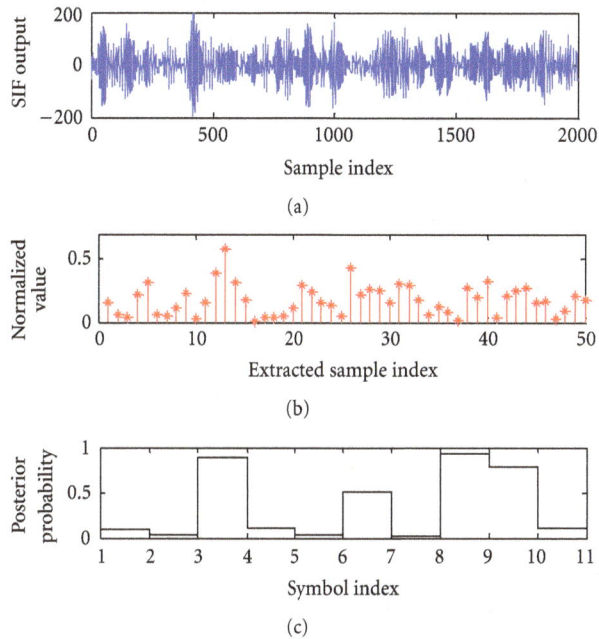

(a)

(b)

(c)

FIGURE 7: The waveform of SIF output and the posterior probability output obtained by SVM at SNR = -9 dB.

to its limit where the increase of SVs would not improve its accuracy.

4.3. Posterior Probability Estimates. In order to reduce the complexity of the SVM analyzed in Section 2, we select only a few samples from the filter output as the features for training and testing (i.e., $n = 5$ in this case). We depict the probabilities obtained by the SVM output of SNR = -9 dB in Figure 7. The signal in Figure 7 is submerged in noise, so the optimal performance cannot be achieved by using a conventional threshold decision. Yet, the probability which the demodulator output by SVM technique is accurate while a source symbol sequence $[0, 0, 0, 0, 1, 1, 0, 1, 0, 1]$ is transmitted, and the noise from the part which did not carry any information of the waveform of symbol "1" is almost removed.

To understand the difference in PPEs, we have plotted the curves for the SVM and the MHY in Figures 8(a) and 8(b), respectively, with SNR = -5 dB. We depict the

estimated probabilities $P(y = 1 \mid x)$ versus the ones when a source symbol sequences with all ones are transmitted. We can appreciate that the SVM PPEs are closer to "1" and less spread, most of the values of demodulation output are between 0.9 and 1. Thereby, SVM estimates are closer to the true posterior probability, which explains its improved performance with respect to the MHY, when we measure the BER after the LDPC decoder.

In a previous subsection, we have shown that the demodulator is based on an SIF and SVM classifier, when we compare performances at a low BER. In this section, we focus on the performance after the sequence has been corrected by an LDPC decoder. The ability of SVM to provide accurate

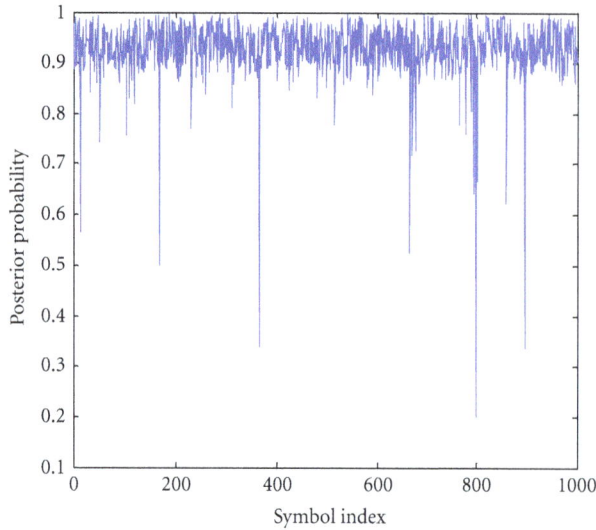

(a)

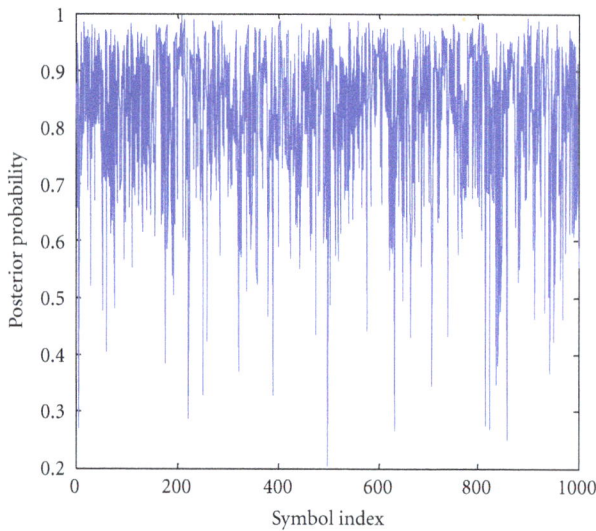

(b)

FIGURE 8: The posterior probability $P(y = 1 \mid x)$ obtained by SVM and MHY method, in (a) and (b), respectively, where source symbols with all ones are transmitted.

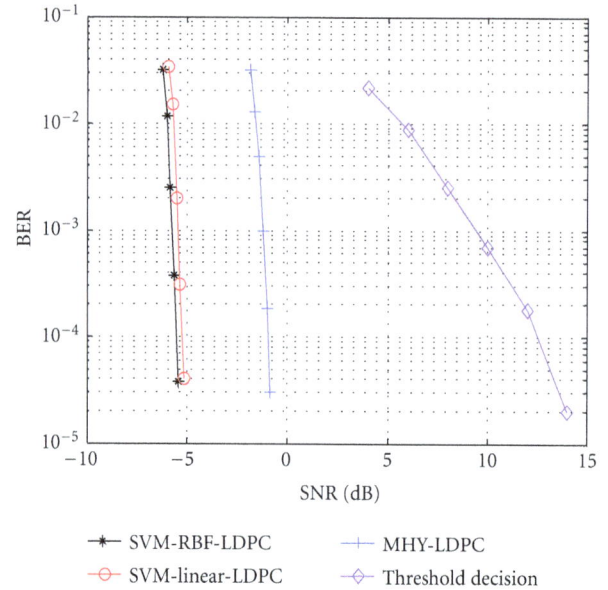

FIGURE 9: Performance at the output of the LDPC decoder with the soft-input and threshold decision.

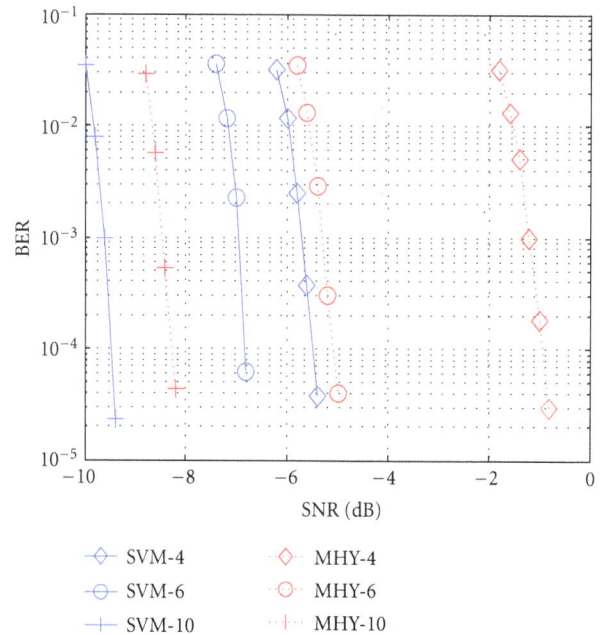

FIGURE 10: BER performance comparisons of the SVM with MHY method at the output of the LDPC decoder with different sampling rates. Using SVM-4, SVM-6, and SVM-10 for the SVM method (solid lines) and MHY-4, MHY-6, and MHY-10 for the MHY method (dashed lines) with $f_s = 4f_c$, $f_s = 6f_c$, and $f_s = 10f_c$, respectively.

posterior probability predictions boosts the demodulator performance compared to the MHY method.

From Figure 8, we can understand that the improved performance of the SVM with respect to the MHY is based on its ability to provide accurate PPEs. In Figure 9, we can appreciate that the SVM-LDPC significantly reduces the BER at lower SNR, because the PPEs are more accurate and the LDPC decoder can rely on these trustworthy predictions. Also, Figure 9 shows that the performance of SVM-RBF-LDPC is a little more superior to SVM-linear-LDPC, it is not the same as the results in Section 4.1 which are analyzed without channel coding. Moreover, the SVM-linear-LDPC decoding outperforms the MHY-LDPC decoding by 4.5 dB and by 18 dB without channel coding when BER = 10^{-4} and sampling rate $f_s = 4f_c$. In Figure 10, we compare the

BER performance of the SVM-LDPC with MHY-LDPC by a different sampling rate. Compared to the MHY-LDPC, the SVM-LDPC can upgrade more than 4.6 dB, 1.7 dB, and 1.2 dB for $f_s = 4f_c$, $f_s = 6f_c$, and $f_s = 10f_c$, respectively. This means that the performance of SVM-LDPC improved significantly while the sampling rate is low, and it is not sensitive to the sampling rate for SVM-LDPC. Also, Figure 10 illustrates that it is more superior for the SVM demodulator than MHY in a bad condition.

We have shown that SVM-LDPC is far superior to the MHY method. This result shows that using a method that can predict accurately the PPEs allows the LDPC decoding algorithm to perform to its fullest.

5. Conclusions

In this paper, we introduce a nonlinear demodulator which is a novel solution for the EBPSK scheme. We have shown that the performance can be significantly improved by using a linear kernel for demodulation, which has a less computational complexity thus saves the computation time.

SVM is a nonlinear probabilistic classifier that produces accurate PPEs. The performance comparisons of different probabilistic demodulators at the output of an LDPC channel decoder are made, which has shown that the SVM outperforms the MHY with probabilistic output.

The SVM probability output method does not need to estimate the channel noise power σ, and uses only a few samples as the features of SVM for training and testing, which reduces the complexity significantly.

A simulator of the system was designed and the BER performance was significantly improved for the SVM-LDPC comparing with the MHY-LDPC approach. Moreover, the SVM method is more robust to sampling rate than MHY method.

Yet, the performance of the system can be improved significantly at the cost of complexity, and the probability is still approximate. More investigations are undertaken to reduce the computational complexity of this approach and test its performance under more severe channel conditions, such as the fading channel.

Acknowledgment

The authors would like to thank the support of the National Natural Science Foundation of China (NSFC) under Grant 61271204.

References

[1] H. R. Walker, "VPSK and VMSK modulation transmit digital audio and video at 15 Bits/Sec/Hz," *IEEE Transactions on Broadcasting*, vol. 43, no. 1, pp. 96–103, 1997.

[2] M. Feng and L. Wu, "Special non-linear filter and extension to Shannon's channel capacity," *Digital Signal Processing*, vol. 19, no. 5, pp. 861–873, 2009.

[3] L. Wu and M. Feng, "On BER performance of EBPSK-MODEM in AWGN channel," *Sensors*, vol. 10, no. 4, pp. 3824–3834, 2010.

[4] M. Feng, L. Wu, J. Ding, and C. Qi, "BER analysis and verification of EBPSK system in AWGN channe," *IEICE Transactions on Communications*, vol. 94, no. 3, pp. 806–809, 2011.

[5] K. Hyun and D. Yoon, "Bit metric generation for Gray coded QAM signals," *IEE Proceedings Communications*, vol. 152, no. 6, pp. 1134–1138, 2005.

[6] J. W. H. Kao, S. M. Berber, and V. Kecman, "Blind multiuser detector for chaos-based CDMA using support vector machine," *IEEE Transactions on Neural Networks*, vol. 21, no. 8, pp. 1221–1231, 2010.

[7] K. C. Ho, X. Lu, and V. Mehta, "Adaptive blind narrowband interference cancellation for multi-user detection," *IEEE Transactions on Wireless Communications*, vol. 6, no. 3, pp. 1024–1033, 2007.

[8] F. Perez-Cruz, J. J. Murillo-Fuentes, and S. Caro, "Nonlinear channel equalization with Gaussian processes for regression," *IEEE Transactions on Signal Processing*, vol. 56, no. 10, pp. 5283–5286, 2008.

[9] H. Zhao and J. Zhang, "Functional link neural network cascaded with Chebyshev orthogonal polynomial for nonlinear channel equalization," *Signal Processing*, vol. 88, no. 8, pp. 1946–1957, 2008.

[10] J. C. Patra, P. K. Meher, and G. Chakraborty, "Nonlinear channel equalization for wireless communication systems using Legendre neural networks," *Signal Processing*, vol. 89, no. 11, pp. 2251–2262, 2009.

[11] P. M. Olmos, J. J. Murillo-Fuentes, and F. Pérez-Cruz, "Joint nonlinear channel equalization and soft LDPC decoding with Gaussian processes," *IEEE Transactions on Signal Processing*, vol. 58, pp. 1183–1192, 2010.

[12] N. Singla and J. A. O'Sullivan, "Joint equalization and decoding for nonlinear two-dimensional intersymbol interference channels," in *Proceedings of the IEEE International Symposium on Information Theory (ISIT '05)*, pp. 1353–1357, Adelaide, Australia, September 2005.

[13] W. J. Park and R. M. Kil, "Pattern classification with class probability output network," *IEEE Transactions on Neural Networks*, vol. 20, no. 10, pp. 1659–1673, 2009.

[14] C. Cortes and V. Vapnik, "Support-vector networks," *Machine Learning*, vol. 20, no. 3, pp. 273–297, 1995.

[15] J. Platt, "Sequential minimal optimization: a fast algorithm for training support vector machines," in *Advances in Kernel Method: Support Vector Learning*, B. Scholkopf, Ed., pp. 185–208, The MIT Press, Cambridge, Mass, USA, 1998.

[16] J. Platt, "Probabilities for SV machines," in *Advances in Large Margin Classifiers*, A. J. Smola and P. L. Bartlett, Eds., pp. 61–73, The MIT Press, Cambridge, Mass, USA, 2000.

[17] H. T. Lin, C. J. Lin, and R. C. Weng, "A note on Platt's probabilistic outputs for support vector machines," *Machine Learning*, vol. 68, no. 3, pp. 267–276, 2007.

2

Energy-Efficient Data Gathering Scheme Based on Broadcast Transmissions in Wireless Sensor Networks

Soobin Lee[1] and Howon Lee[2]

[1] Institute for IT Convergence, KAIST, Yuseong-gu, Daejeon 305-701, Republic of Korea
[2] Department of Electrical, Electronic and Control Engineering, Hankyong National University, Anseong, Gyeonggi 456-749, Republic of Korea

Correspondence should be addressed to Howon Lee; hwlee@hknu.ac.kr

Academic Editors: A. Kanatas and J. Li

Improving energy efficiency is the most important challenge in wireless sensor networks. Because sensing information is correlated in many sensor network applications, some previous works have proposed ideas that reduce the energy consumption of the network by exploiting the spatial correlation between sensed information. In this paper, we propose a distributed data compression framework that exploits the broadcasting characteristic of the wireless medium to improve energy efficiency. We analyze the performance of the proposed framework numerically and compare it with the performance of previous works using simulation. The proposed scheme performs better when the sensing information is correlated.

1. Introduction

Energy efficiency is the most important issue in wireless sensor networks because sensor nodes are usually battery-powered, and in many sensor network applications, it is not possible to replenish the energy of sensor nodes. To improve energy efficiency and extend the network lifetime of sensor networks, many proposals have been made for preventing redundant information from being transmitted and received by exploiting the spatial correlation between the sensing information gathered by sensor nodes. Typically, sensor nodes are deployed densely to achieve satisfactory coverage; hence, the sensing information gathered by sensor nodes is highly correlated [1–3].

Depending on the sensor network applications that are used, there exist two approaches to exploiting the correlation between sensing information. The first approach is used when the goal of the sensor network application is to estimate an event from the sensor field with a certain reliability at a sink node [4–6]. In this approach, only some of the nodes transmit their sensed information to the sink node. The second approach is used in which the sensed information at each location is equally important, such as an environmental monitoring application or video surveillance system. This

approach typically uses joint coding of correlated information to compress the data. The distributed source coding (DSC) technique [7] allows sensor nodes to use joint coding without explicit communication between the nodes. This technique makes it easy to find the optimal transmission structure. However, the use of DSC in large-scale networks encounters practical problems because this technique requires complex encoders and global knowledge of the network.

In view of these practical difficulties, [8, 9] proposed ideas for compressing correlated sensed information using joint coding with explicit communications between nodes, which is called the explicit communication approach. In [8], it is claimed that there is no need to impose the constraint that the encoding should be performed without sharing information from the other nodes because nodes receive data from the other nodes in each data gathering path to the sink node. Reference [9] addressed an optimization problem for the minimum cost correlated data gathering tree (MCCDGT) and proposed distributed heuristic approximation algorithms to solve this problem.

In the previous works that use the explicit communication approach, a communication channel is abstracted as a point-to-point link, which ignores the fact that wireless channels transmit information by broadcasting it, which makes it available to any receiver of the right type. In wireless networks, when one node transmits data to its destination node, other nodes within the transmission range of the transmitting node can also receive the data and may use it to compress their own information. Herein, we propose a framework for data compression that exploits the broadcasting characteristic of the wireless medium, thereby achieving greater energy efficiency.

The remainder of this paper is organized as follows. In Section 2, we review related work. In Section 3, we explain the proposed scheme and analyze its performance numerically. In Section 4, we evaluate the performance of the proposed scheme using simulations. In Section 5, we consider MAC protocols for the proposed scheme. Section 6 concludes.

2. Related Work

References [4–6] exploit the broadcasting characteristic of the wireless medium to reduce the energy consumption. In these works, if one node transmits its sensing information, the other nodes within transmission range overhear the information. They then determine whether or not their own information is redundant, either by (i) checking that the distance from the transmitting node is less than the correlation radius [4] or the influential range [5] or by (ii) spatial interpolation using the data that they overheard [6]. The distance between sensor nodes is assumed to be known from the exchange of messages during network initialization or estimated using the received signal strength. If a node determines that its sensing information is redundant, it does not try to transmit the information. Consequently, only some of the total nodes transmit their sensing information to the sink node, while satisfying a certain distortion constraint; hence, the amount of energy that is consumed by transmitting redundant information is reduced. However, the schemes that are proposed in these works cannot be used for the wireless sensor network applications in which all information that is collected from the sensor field is equally important and should be sent to the user of the application, such as environmental monitoring systems or video surveillance systems.

In [9], the authors considered a sensor network as a connectivity graph with point-to-point communication links instead of a full wireless multipoint communication structure. They formulated the optimization problem for minimizing the network's energy consumption by jointly optimizing the data gathering structure and data compression. This problem is called the MCCDGT problem. Due to the fact that this optimization problem is NP-hard, the authors proposed a number of distributed heuristic approximation algorithms for solving it.

Herein, we compare the performance of our proposed scheme to the performance of MCCDGT whereas in [9], nodes compress their information only using data that they have received from their child nodes in the data gathering structure; in the proposed scheme, nodes can also compress their information using the data that they have received from neighbor nodes that are not their child nodes. We separate the data gathering structure and the data compression scheme

to fully exploit the broadcasting characteristic of the wireless medium. Furthermore, we use a widely used spatial correlation model to consider more realistic environments though a simplified version of the correlation model is used in [9].

3. Energy-Efficient Data Gathering Based on Broadcast Transmissions

3.1. Assumptions. We consider sensor network applications in which sensing information at each sensor node is equally important; hence, all sensor nodes measure the environmental variables and transmit the sensing information to the sink node periodically. In addition, we assume that N sensor nodes are deployed randomly according to a Poisson point process.

3.1.1. Model of Energy Behavior of Nodes. A simple model of the energy behavior of sensor nodes is introduced in [10]. This model assumes a path loss of $(1/d)^n$, where d is the distance between a sender and a receiver, and uses the following definitions: α_{11} is the energy per bit consumed by the transmitter electronics, α_2 is the energy per bit per m^n used in the transmitter amplifier, α_{12} is the energy per bit consumed by the receiver electronics, and α_3 is the energy used to sense a bit. Then, the energy consumed by a transmitter and a receiver that are separated by a distance of d to transport one bit is represented as

$$P_{\text{relay}}(d) = \alpha_{11} + \alpha_2 d^n + \alpha_{12} = \alpha_1 + \alpha_2 d^n, \qquad (1)$$

where $\alpha_1 = \alpha_{11} + \alpha_{12}$. Using this model as a basis, the author demonstrates that the energy required to relay a bit over the distance L is bounded as

$$E(L) \geq \alpha_1 \frac{n}{n-1} \frac{L}{d_m} - \alpha_{12} \qquad (2)$$

with equality if and only if L is an integral multiple of d_m, where d_m is the optimal distance between intervening nodes, which is given by

$$d_m = \sqrt[n]{\frac{\alpha_1}{\alpha_2(n-1)}}. \qquad (3)$$

3.1.2. Data Compression Model. There are several correlation models that describe the spatial correlation of sensed information or methods that estimate the correlation in the wireless sensor networks [1–3]. Herein, we use the power exponential model for simplicity. The model is represented as

$$K_{ji} = e^{-\gamma d_{ji}^\theta}, \quad \theta \in \{1, 2\}, \qquad (4)$$

where K_{ji} and d_{ji} are the correlation coefficient and the distance between node j and node i, respectively. γ indicates the degree of spatial correlation; its value depends on the sensor network application that is used.

To evaluate the amount of compressed data, we consider the relation between spatial correlation and data compression. Let random variables $X_1, X_2, \ldots, X_N$ be the sensing information from each sensor node. As in [9], $X_1, X_2, \ldots, X_N$ are entropy coded with $h_X = h(X_1) = h(X_2) \cdots = h(X_N)$ if there is no explicit information from other nodes. For the maximum possible lossless compression, each node's information is compressed by conditional entropy coding, given the data that is received from other nodes [9, 11]. Given our assumption that the conditional entropy is characterized by the correlation model in (4), the conditional entropy $h(X_j \mid X_i)$ is given by

$$h\left(X_j \mid X_i\right)\left(d_{ji}\right) = \left(1 - K_{ji}\right)h_X = \left(1 - e^{-\gamma d_{ji}^\theta}\right)h_X, \quad (5)$$

where $1 - e^{-\gamma d_{ji}^\theta}$ is the compression ratio. After it has received the compressed data, the sink node can decode it if the information for identifying the source, which is used in joint coding as the explicit information, is included.

3.2. Proposed Data Gathering Scheme. We propose a distributed data compression scheme based on wireless point-to-multipoint communication. We use the restriction that each sensor node is allowed to compress its own sensing information, using only the data that is not compressed. In addition, we assume that the data gathering structure is already constructed using the optimal algorithm. In the proposed scheme, raw data transmitter (RDT) nodes transmit their sensing information without compression. Nodes that are not chosen as RDT nodes compress their information using only the data received from the RDT nodes.

The proposed scheme has two phases. In the first phase, which is called the *RDT selection phase*, RDT nodes are chosen among N sensor nodes. Each sensor node becomes a RDT node with probability β, where $0 \leq \beta \leq 1$. After the RDT nodes are selected, each non-RDT node selects its RDT node. RDT nodes broadcast advertising messages to their neighbor nodes. Non-RDT nodes that receive those advertising messages estimate the distance to the RDT nodes on the basis of the received signal strength of the advertising messages. Let S be the set of all sensor nodes and S_R the set of nodes that are chosen as RDT nodes. Then, the set of non-RDT nodes is given by $S_{NR} = S \setminus S_R$. Let C_j be the set of RDT nodes from which non-RDT node $j \in S_{NR}$ receives advertising messages. Node j chooses its RDT node r_j that satisfies $r_j = \arg\min_{i \in R_j} d_{ji}$ from the nodes in

$$R_j = \left\{ i \in C_j \mid d_{ji} \leq R_{th} \right\}, \quad (6)$$

where d_{ji} is the estimated distance between node j and node i and R_{th} is the predefined threshold distance. Because the correlation between sensing information that is generated at two sensor nodes increases when the distance between those two nodes decreases, sensing information is mostly compressed using the data from the nearest RDT node, which minimizes the energy consumed to send the compressed data to the sink node.

In the second phase, which is called the *data gathering phase*, all sensor nodes send their sensing information to the sink node through the data gathering structure. As mentioned above, RDT nodes transmit their sensing information without compression. However, not all the data that the RDT nodes transmit is sensing information generated at the RDT nodes because sensor nodes also relay the data that is received from child nodes in their data gathering paths. Non-RDT nodes need to distinguish the data that is generated at their RDT nodes from the data that is just relayed by the RDT nodes. To do this, additional information that indicates whether or not the corresponding sensing information is generated at the transmitting node is required; for example, we can add a new field to RTS frames in CSMA-based MAC protocols, such as IEEE 802.11 or S-MAC [12]. After receiving a RTS frame, neighbor nodes can detect whether the information in received data frames following the RTS frame is generated at the transmitting node or not.

Further, because non-RDT nodes should transmit their own information after receiving data from their RDT nodes and compressing the information using the received data, the transmission of data that is generated at RDT nodes should be given priority over the transmission of data that is generated at non-RDT nodes and has not yet been compressed. For example, in TDMA-based MAC protocols, the priority scheme can be set up by scheduling the transmissions of RDT nodes first. In CSMA-based MAC protocols, we can use a priority-based transmission policy as follows. RDT nodes or non-RDT nodes that do not have RDT nodes participate in contention to acquire a channel when they have data to transmit. Non-RDT nodes that have RDT nodes participate in contention only when they have either (i) data that is generated by them and that they have already compressed using the data received from their RDT nodes or (ii) data that needs to be relayed.

Let $P(k)$ be a parent node of node $k \in S$ in the data gathering structure. During the data gathering phase, node $m \in S$ always receives the data from node $k \in S$ if $m = P(k)$. For nodes $j \in S_{NR}$ and $j \neq P(r_j)$, when j detects r_j's transmission, j checks whether or not r_j has generated its data itself. If r_j's data is locally generated, j receives it. If not, j turns off the radio and enters the idle state until the current transmission is complete. After receiving r_j's data, j compresses its own information using joint coding, given the r_j data. If j does not have r_j or it does not receive the data from r_j correctly within the prespecified duration, j transmits its sensing information without compression. The proposed framework is illustrated in Figure 1.

Several different definitions are in use for the network lifetime; for example, the interval from the time at which the sensor network starts its operation to the time at which the first sensor dies, when the number of active nodes falls below a prespecified threshold, or when the sensing coverage falls below a prespecified threshold [13, 14]. In general, the definition of network lifetime that is appropriate to use in any given situation may depend on the wireless sensor network application in question. However, whatever definition is used, balancing the energy usage across sensor nodes extends the network lifetime. In the proposed scheme, RDT nodes consume more energy than non-RDT nodes because they transmit uncompressed data. To prevent some sensor nodes from dying much earlier than the rest of nodes and to extend the network lifetime, RDT nodes can be selected periodically.

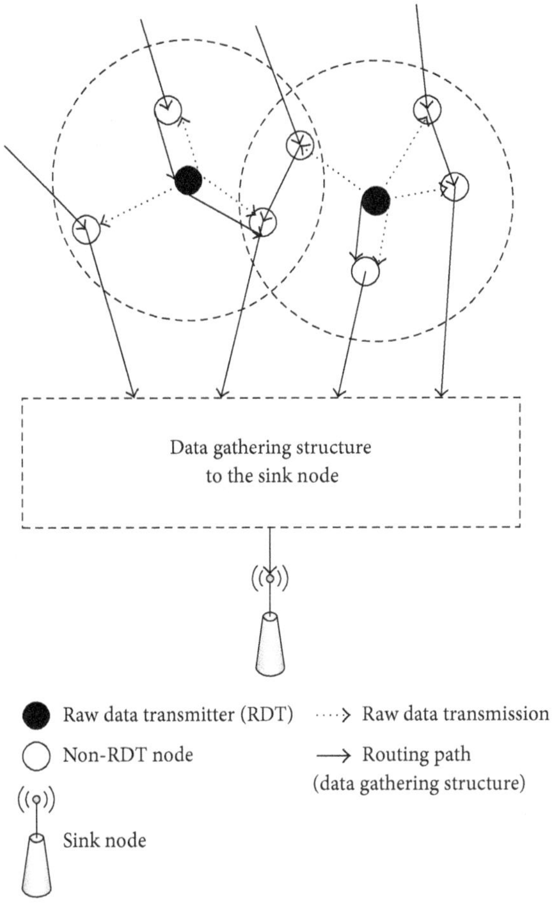

FIGURE 1: Proposed data compression framework.

Reference [9] does not attempt to balance the energy of sensor nodes.

3.3. Numerical Analysis. Let D be the deployment area of the sensor network. Each non-RDT node chooses its RDT node from the RDT nodes that lie within R_{th}. Thus, the probability of a non-RDT node in D being covered by its RDT node, $P_{\text{NR,cov}}$, can be given by

$$P_{\text{NR,cov}} = 1 - e^{-(\beta N/D)\pi R_{\text{th}}^2} \qquad (7)$$

using a theorem in [15, 16]. Then, the number of non-RDT nodes that are covered by the RDT nodes is

$$N_{\text{NR}} = (1 - \beta) N P_{\text{NR,cov}}. \qquad (8)$$

Let a random variable R_{min} be the distance between any arbitrary non-RDT node and the nearest RDT node. The distribution density function of R_{min} is given by

$$f_{R_{\text{min}}}(d) = \frac{\beta N}{D} 2\pi d e^{-(\beta N/D)\pi d^2} \qquad (9)$$

as in [15]. Then, the average number of bits to be transmitted by each non-RDT node that is covered by its RDT node after compression is

$$
\begin{aligned}
h_{\text{avg,NR}} &= \int_0^{R_t} f_{R_{\text{min}}}(x) h\left(X_j \mid X_i\right)(x)\, dx \\
&= \frac{2\pi\beta N h_X}{D} \int_0^{R_t} x e^{-(\beta N/D)\pi x^2} \left(1 - e^{-\gamma x^\theta}\right) dx.
\end{aligned}
\qquad (10)
$$

If $\theta = 2$, that is, if the correlation model is squared exponential, $h_{\text{avg,NR}}$ is given by

$$
\begin{aligned}
h_{\text{avg,NR}} = \pi\beta N h_X \\
\times \left(\frac{1 - e^{-(\pi\beta N/D)R_{\text{th}}^2}}{\pi\beta N} + \frac{-1 + e^{-(\pi\beta N/D+\gamma)R_{\text{th}}^2}}{\pi\beta N + \gamma D} \right).
\end{aligned}
\qquad (11)
$$

On the other hand, RDT nodes or non-RDT nodes that are not covered by RDT nodes transmit their sensing information without compression. Consequently, we can obtain the average number of bits to be transmitted by each node in the network as follows:

$$h_{\text{avg}} = \frac{N_{\text{NR}} h_{\text{avg,NR}} + (N - N_{\text{NR}}) h_X}{N}. \qquad (12)$$

Now, we calculate the total energy consumed by sensor nodes to transport all the information possessed by one sensing event to the sink node, E_{total}. Given that $E(L)$ in (2) is the minimum energy required to forward a bit over the distance L, E_{total} can be represented as

$$E_{\text{total}} = N\alpha_3 h_X + \frac{N}{D} \iint_D E(d) h_{\text{avg}}\, dS + N_{\text{NR}}\alpha_{12} h_X \qquad (13)$$

if we use an ideal data gathering structure. The first term on the right-hand side covers the energy consumed by sensor nodes to sense an event and the third term covers the energy consumed by non-RDT nodes to receive data from their RDT nodes. For simplicity, we assume that the deployment area is disk-shaped and that the sink node is located in the center of the deployment area. If we let r_D be the radius of D, we have

$$
\begin{aligned}
E_{\text{total}} = \left(N\alpha_3 + N_{\text{NR}}\alpha_{12}\right) h_X \\
+ N h_{\text{avg}} \left(\alpha_1 \frac{n}{(n-1)\, d_m} \frac{2}{3} r_D - \alpha_{12}\right).
\end{aligned}
\qquad (14)
$$

4. Performance Evaluation

We now verify the numerical analysis by simulation and compare the performance of the proposed scheme with that of MCCDGT. The authors in [9] used a simplified model for the data correlation, in which the compression ratio is constant. Herein, we use the spatial correlation model described in (4), in which the compression ratio varies depending on the distance between nodes; thus, more practical environments are considered. In addition, the weight of each path between

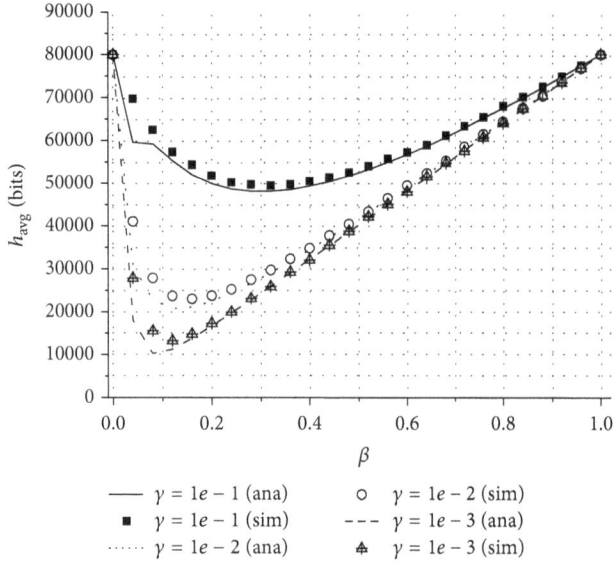

FIGURE 2: The average number of bits to be transmitted by each node in the network (h_{avg}) for different β and γ values.

sensor nodes is calculated using (1). We set the simulation parameters as follows: $N = 100$, $r_D = 30$ m, $h_X = 10$ kbytes, $\alpha_{11} = 20$ nJ/bit, $\alpha_{12} = 10$ nJ/bit, $\alpha_2 = 500$ pJ/bit/m^2, and $n = 2$. α_3 is assumed to be negligibly small. The maximum transmission range of sensor nodes is 20 m and R_{th} is 10 m.

In Figure 2, we show the average number of bits to be transmitted by each node (h_{avg}) in the numerical analysis and the simulation results of the proposed scheme for different values of β and γ. For any given β, h_{avg} decreases when γ is low, that is, when the sensing information of sensor nodes is highly correlated, because the compression ratio of data also becomes lower. On the other hand, for any given γ, there exists β that minimizes h_{avg}. If β is too small, the number of non-RDT nodes that are covered by their RDT nodes decreases. This means that most of the sensing information cannot be compressed. As a result, h_{avg} increases. If β exceeds a certain threshold, the number of RDT nodes increases. Consequently, h_{avg} also increases because the amount of energy consumed by RDT nodes to transmit uncompressed data becomes larger than the amount of energy saved by compressing data of non-RDT nodes. The numerical results approximate well to those of the simulation.

In Figure 3, we show the total energy consumed by sensor nodes to transport all the information derived from one sensing event to the sink node (E_{total}) in the proposed scheme, when the ideal data gathering structure and the shortest path tree (SPT) data gathering structure are used. Because the ideal data gathering structure is not available in practice, E_{total} in the ideal structure is calculated using (14). In the simulation, we use the SPT structure as the data gathering structure of the proposed scheme. For any given β and γ, E_{total} in the SPT structure is always greater than the ideal case. E_{total} in the ideal structure can be considered as the upper bound of E_{total} for the proposed scheme. The effects of β and γ can be explained as in the case of h_{avg}.

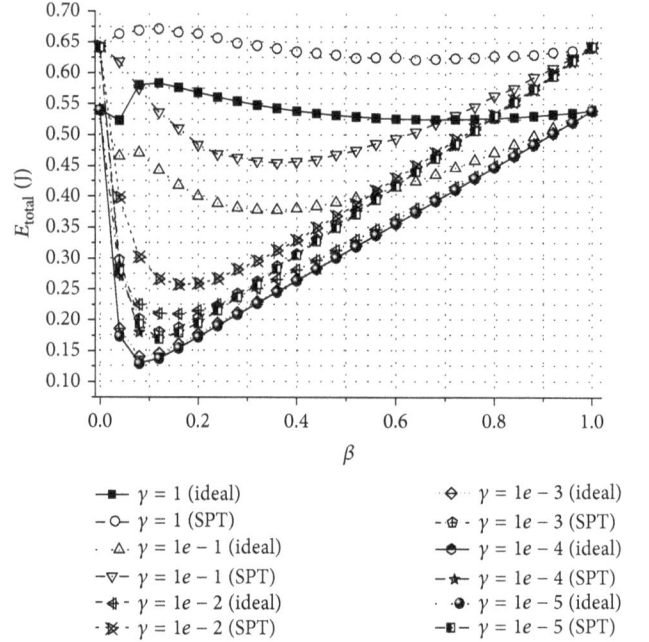

FIGURE 3: The total energy consumed by sensor nodes to transport all the information derived from one sensing event to the sink node (E_{total}) for different β and γ values when an ideal routing and a SPT routing are used.

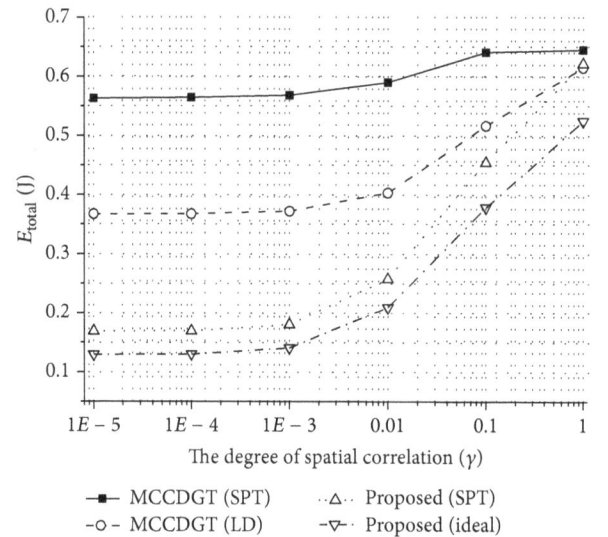

FIGURE 4: The total energy consumed by sensor nodes to transport all the information derived from one sensing event to the sink node (E_{total}) for different γ values.

In Figure 4, we compare (a) E_{total} of the proposed scheme using the ideal and SPT data gathering structures when β that minimizes E_{total} is used to (b) E_{total} of MCCDGT using the leaves deletion (LD) algorithm, which is a fully distributed and practical method. We can find the value of β that minimizes E_{total} of the proposed scheme using (12). In addition, we show the performance of MCCDGT when using only the SPT algorithm, which is the optimal

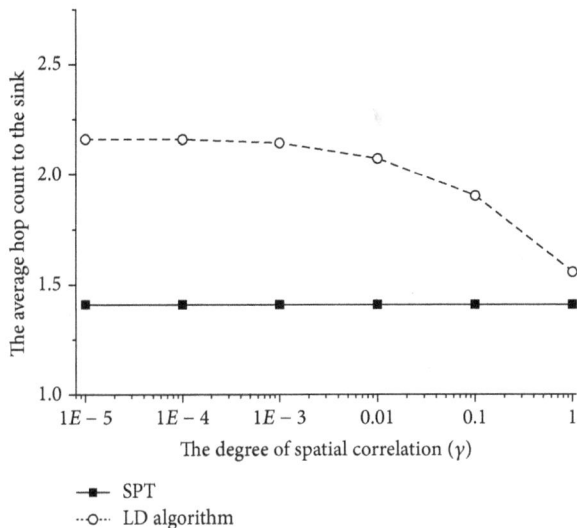

FIGURE 5: The average hop count from each sensor node to the sink node for different γ values.

solution when stores of sensed information are independent of each other. As we can see in Figure 4, the proposed scheme outperforms MCCDGT with the LD algorithm when the sensing information is highly or moderately correlated. If γ becomes larger, the performance of the proposed scheme and MCCDGT with the LD algorithm converges to those of MCCDGT with the SPT algorithm because the amount of energy that can be saved by compressing data becomes negligible.

Due to the fact that we separate the data gathering structure and the data compression scheme, we obtain an additional advantage from the proposed scheme: the average hop count from each sensor node to the sink node can be reduced. In Figure 5, we compare the average hop count to the sink node of the SPT data gathering structure and the structure that is constructed by the LD algorithm. If sensing information is more correlated, more leaf nodes become child nodes of other leaf nodes when the LD algorithm is used because deleting leaf nodes is more energy efficient in that case. Thus, data gathering paths from each sensor node to the sink node are detoured. Consequently, the average hop count to the sink node increases. Meanwhile, we can use an optimal data gathering structure, such as SPT, in the proposed scheme without considering data compression. When the network increases in size, the difference between the average hop count and the sink node in the two structures may be significant. Therefore, the proposed method has an important strength in multimedia sensor network applications that are usually delay sensitive because an increase in the hop count usually causes more delay.

5. Consideration for MAC

Many TDMA-based protocols for providing unicast or broadcast transmissions in sensor networks have been proposed. However, TDMA-based protocols have some disadvantages, especially in large-scale multihop networks. They require time synchronization for the entire network and incur overhead for slot allocation. In addition, the schedule needs to be changed when the network topology changes. TDMA-based protocols are suitable for use only in small, specific networks.

In CSMA-based protocols without specific broadcast transmission support, such as IEEE 802.11 or S-MAC [12], the hidden node problem may be significant, depending on the carrier sense threshold [17]. If a collision occurs due to hidden nodes when a non-RDT node is receiving data from its RDT node, the non-RDT node cannot compress its information because the received data is corrupted. As a result, the energy consumed by the non-RDT node to receive the data from its RDT node becomes a loss, and the performance of the proposed scheme decreases, though it still runs normally.

To avoid the degradation due to the hidden node problem, we can tune the sensing threshold of the carrier [17]. However, such tuning may not be possible due to hardware limits; besides, it may delay the transmission of data by nodes. Meanwhile, we can use MAC protocols that support reliable broadcast transmissions, some of which are addressed in [18, 19]. In particular, Robcast [19] guarantees high reliability of broadcast transmissions while keeping low energy consumption regardless of the additional control signaling. We can adapt these protocols to the proposed framework with a few modifications.

6. Conclusion

In this paper, we propose a data compression scheme for the wireless sensor networks that exploits the broadcasting characteristic of the wireless medium. We analyze the performance of the proposed scheme numerically and verify the results using simulations. Further, we compare the performance of the proposed scheme with that of MCCDGT. The simulation results show that our scheme outperforms the other schemes when sensing information is correlated. Finally, we discuss broadcast transmission support in the MAC layer for the proposed scheme.

Acknowledgment

This research was supported by the MSIP (Ministry of Science, ICT & Future Planning), Korea, under the R&D program supervised by the KCA (Korea Communications Agency) (KCA-2013-11-912-03-001).

References

[1] M. C. Vuran, Ö. B. Akan, and I. F. Akyildiz, "Spatio-temporal correlation: theory and applications for wireless sensor networks," *Computer Networks*, vol. 45, no. 3, pp. 245–259, 2004.

[2] R. Dai and I. F. Akyildiz, "A spatial correlation model for visual information in wireless multimedia sensor networks," *IEEE Transactions on Multimedia*, vol. 11, no. 6, pp. 1148–1159, 2009.

[3] M. Alaei and J. M. Barcelo-Ordinas, "Node clustering based on overlapping FoVs for wireless multimedia sensor networks," in

Proceedings of the IEEE Wireless Communications and Network-ing Conference (WCNC '10), pp. 1–6, Sydney, Australia, April 2010.

[4] M. C. Vuran and I. F. Akyildiz, "Spatial correlation-based col-laborative medium access control in wireless sensor networks," *IEEE/ACM Transactions on Networking*, vol. 14, no. 2, pp. 316–329, 2006.

[5] H. Le, H. Guyennet, and V. Felea, "OBMAC: an overhearing based MAC protocol for wireless sensor networks," in *Pro-ceedings of the International Conference on Sensor Technologies and Applications (SENSORCOMM '07)*, pp. 547–553, Valencia, Spain, October 2007.

[6] Y. Iima, A. Kanzaki, T. Hara, and S. Nishio, "Overhearing-based data transmission reduction for periodical data gathering in wireless sensor networks," in *Proceedings of the International Conference on Complex, Intelligent and Software Intensive Sys-tems (CISIS '09)*, pp. 1048–1053, Fukuoka, Japan, March 2009.

[7] D. Slepian and J. K. Wolf, "Noiseless coding of correlated infor-mation sources," *IEEE Transactions on Information Theory*, vol. 19, no. 4, pp. 471–480, 1973.

[8] A. Scaglione and S. Servetto, "On the interdependence of rout-ing and data compression in multi-hop sensor networks," *Wire-less Networks*, vol. 11, no. 1-2, pp. 149–160, 2005.

[9] R. Cristescu, B. Beferull-Lozano, M. Vetterli, and R. Wat-tenhofer, "Network correlated data gathering with explicit communication: NP-completeness and algorithms," *IEEE/ACM Transactions on Networking*, vol. 14, no. 1, pp. 41–54, 2006.

[10] M. Bhardwaj, T. Garnett, and A. P. Chandrakasan, "Upper bounds on the lifetime of sensor networks," in *Proceedings of the International Conference on Communications (ICC '01)*, pp. 785–790, Helsinki, Finland, June 2000.

[11] S. Pattem, B. Krishnamachari, and R. Govindan, "The impact of spatial correlation on routing with compression in wireless sensor networks," *ACM Transactions on Sensor Networks*, vol. 4, no. 4, article 24, 2008.

[12] W. Ye, J. Heidemann, and D. Estrin, "Medium access control with coordinated adaptive sleeping for wireless sensor net-works," *IEEE/ACM Transactions on Networking*, vol. 12, no. 3, pp. 493–506, 2004.

[13] Y. Chen and Q. Zhao, "On the lifetime of wireless sensor net-works," *IEEE Communications Letters*, vol. 9, no. 11, pp. 976–978, 2005.

[14] Z. Cheng, M. Perillo, and W. B. Heinzelman, "General network lifetime and cost models for evaluating sensor network deploy-ment strategies," *IEEE Transactions on Mobile Computing*, vol. 7, no. 4, pp. 484–497, 2008.

[15] B. Liu and D. Towsley, "A study of the coverage of large-scale sensor networks," in *Proceedings of the IEEE International Conference on Mobile Ad-Hoc and Sensor Systems*, pp. 475–483, October 2004.

[16] C. Sevgi and A. Kocyigit, "On determining cluster size of ran-domly deployed heterogeneous WSNs," *IEEE Communications Letters*, vol. 12, no. 4, pp. 232–234, 2008.

[17] A. Bachir, D. Barthel, M. Heusse, and A. Duda, "Hidden nodes avoidance in wireless sensor networks," in *Proceedings of the International Conference on Wireless Networks, Communica-tions and Mobile Computing*, pp. 612–617, June 2005.

[18] K. Tang and M. Gerla, "Random access MAC for efficient broad-cast support in ad hoc networks," in *Proceedings of the IEEE Wireless Communications and Networking Conference (WCNC '00)*, pp. 454–459, September 2000.

[19] M. Demirbas and S. Balachandran, "RoBcast: a singlehop reli-able broadcast protocol for wireless sensor networks," in *Proceedings of the 27th International Conference on Distri-buted Computing Systems Workshops (ICDCSW '07)*, Toronto, Canada, June 2007.

3

Presence Service in IMS

David Petras, Ivan Baronak, and Erik Chromy

Institute of Telecommunications, Faculty of Electrical Engineering and Information Technology, Slovak University of Technology in Bratislava, Ilkovičova 3, 812 19 Bratislava, Slovakia

Correspondence should be addressed to Erik Chromy; chromy@ut.fei.stuba.sk

Academic Editors: C. L. Hsu and A. Manikas

This paper describes the presence service, which is located in the IP multimedia subsystem. This service allows making many applications for different groups of people. The paper describes differences between a network without the service and with the service. The biggest change is an increased number of transmitted messages. The presence uses some part of the IP multimedia subsystem control layer, which is shown in communication between the user and the server. The paper deals with the number of generated messages depending on the behaviour of the users. This is described by a mathematical model using discrete Markov chains.

1. Introduction

Originally, each service or group of services had its own network for its own use. The idea of the next generation networks (NGN) comes with the development of technology. This network would be shared by all services. IP multimedia subsystem (IMS) is an implementation of NGN [1–7]. IMS allows interworking between circuit switching and packet switching networks. IMS has many advantages. New services do not need to change the structure of the network. Quality of service is guaranteed through QoS parameters.

The presence service has two roles: to inform the user about the status of others and to inform others about the user's status [8]. It is transmitted through session initiation protocol for instant messaging and presence leveraging extension (SIMPLE) [9]. This protocol allows the transmission of messages over the network without changes, when the service is deployed. Messages are created as XML documents. These look like PIDF [10] and their extension like RPID [11].

Within the scope of the project "Support of Center of Excellence for SMART Technologies, Systems, and Services II" funded by structural funds from the European union, we have built the most modern IP multimedia subsystem lab at the Institute of Telecommunications. In this lab we can also conduct research aimed for services.

IMS consists of three layers: transport, control, and application (Figure 1). The transport layer is the lowest. This layer

has two roles. First role is to secure an access of devices from different types of networks (GPRS, UMTS, IP, PSTN, etc.) through gateways like Media Gateway (MGW), Signaling Gateway (SGW). Second role is to transfer messages from the user to the control layer or from one user to another user through IP data network. The control layer is the core of the system. It controls the communication and creates connections between users. It directs messages through three call season control function (CSCF) servers. P-CSCF is an entry point to IMS. I-CSCF provides registration and interworking between two IMS. S-CSCF is a central point of direction. It communicates with the application servers. Home subscriber server (HSS) is a database where user profiles and data for the service are stored. The Subscriber location function (SLF) selects the database from several HSSs. The application layer provides services. There are three types of servers. SIP application server is used for applications using the SIP protocol. OSA application server is independent of the protocol through API between the server and the control layer. CAMEL application server is used for applications from legacy network [12].

2. Presence Architecture

Presence architecture is shown in Figure 2. It has three levels: agents, entities, and a server. The agents collect information from various sources. The agents are various programmes.

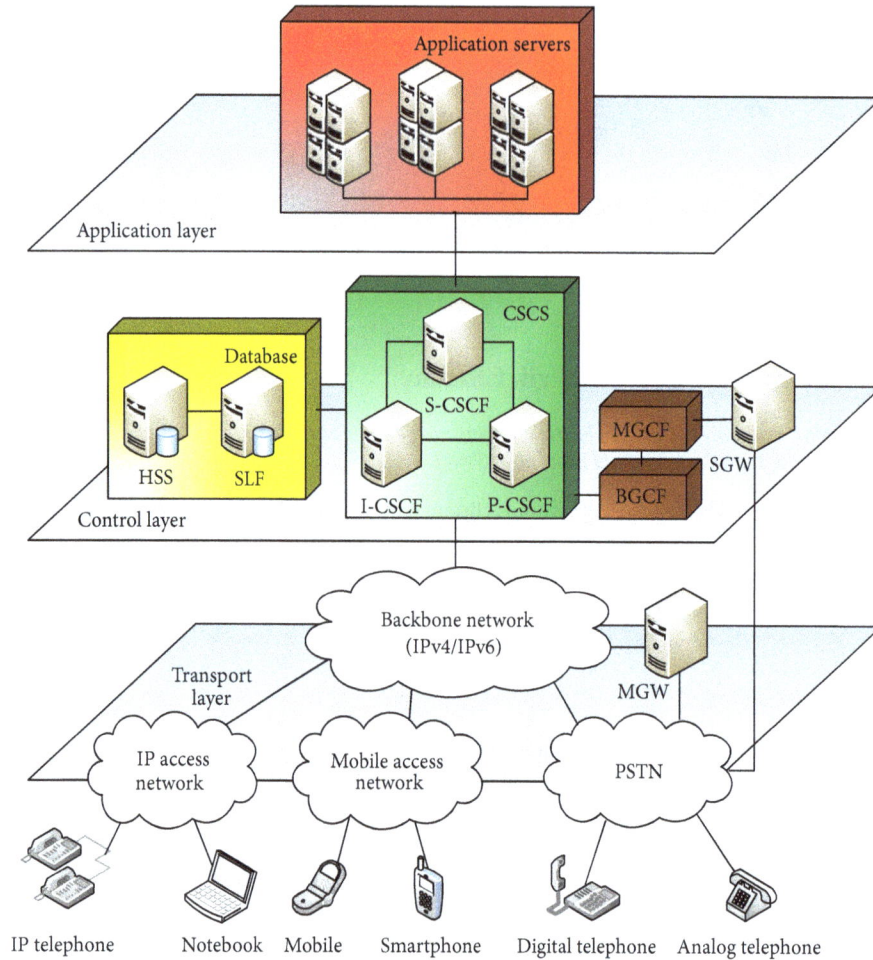

FIGURE 1: Architecture of IMS.

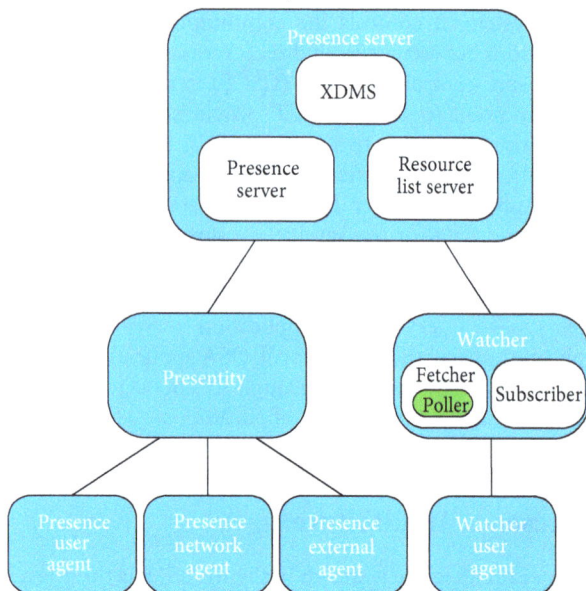

FIGURE 2: Presence architecture.

The presence user agent collects information from user devices. Presence network agent collects information from network elements. Presence external agent collects information from other networks. Watcher presence agent provides information to the watcher.

Entities are characterized by the fact that they can process the SIP messages (UE, S-CSCF, and AS). Entities are divided into two types. Presentity (presence entity) provides information about itself and the watcher observes the status of the others. Watchers are divided into three groups. The fetcher is only interested in the current status. Poller is a special kind of Fetcher, which observes status in certain time intervals. The subscriber also observes the changes in the presence of entities [13]. The server collects and sends information about users, which is stored in XML documents. The presence server receives messages and assigns it to the correct user. The resource list server creates lists of users for watchers and sends their status together. The XML document manager server (XDMS) supports other parts of the presence server. For example, XDMS knows that the watcher is authorized to observe the presence entity. Application server is designed

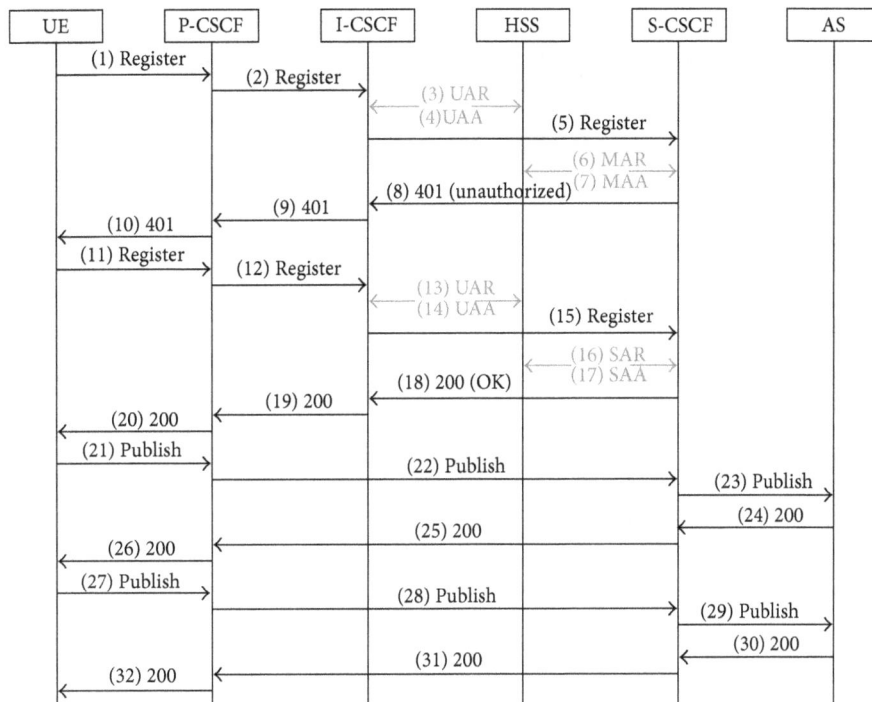

FIGURE 3: Publication status.

so that it could control the number of messages. One of the possibilities is periodic sending of messages. If there are more messages than the server can send, it puts them into a waiting queue. If the waiting queue is full, then the server deletes the messages [14].

2.1. Communication. There are two processes of exchanging messages in the presence service. Process of publishing is shown in Figure 3. This exchange of messages has two parts. The first is registration (messages 1–20) and the second is publication of status (messages 21–32). S-CSCF is assigned to the user during registration. User equipment (UE) is a telephony device, which enters IMS through P-CSCF. P-CSCF through I-CSCF determines where to send the Register request. The information about where to send the message and about the user profile for S-CSCF is stored in the HSS. First, S-CSCF sends a response 401(unauthorized). After receiving answers, UE creates another Register request, after which the user will have successfully registered. A detailed description of the registration is in [15]. In messages 21–23 UE (presentity) sends its whole status in request Publish to the application server (presence server). Messages pass only through P-CSCF and S-CSCF after the registration. S-CSCF knows where the server is according to the initial filter a criteria (iFC). The filter is obtained from the HSS during the registration. Presence server sends confirmation message 200 (OK) as soon as possible, to prevent resending messages. When changing status, UE sends another request Publish, which will go the same way as the first one. The form of the messages is described in [16]. The message itself contains only the change of the status.

The process of subscribing is shown in Figure 4. The figure describes a situation, when the watcher is in another IMS network like presence server. Process of registration is the same as in the previous figure and therefore is not listed. Entry Point is I-CSCF to another IMS network. UE (watcher) creates request Subscribe. The filter is in the request [17]. In the filter, there is information about what the watcher wants to know. UE enters into its IMS network through P-CSCF. It continues to S-CSCF. S-CSCF sends subscribe from the watcher presence network to the presentity presence network. I-CSCF finds S-CSCF and S-CSCF sends request to AS, where there is a list of contacts with status. Upon receiving the request, the application server verifies the user's authority. If it is correct, the application server sends response 200 (OK). AS sends request Notify with the body where it contains the information about the presentity status. Type of watcher is Subscribe in Figure 4. If one of the presentity, which the watcher observes, changes its state, server sends another Notify message without request.

3. Deployment of the Service

Deployment of the presence service means three issues. Application server must be in the application layer. This server receives the information from agents, it stores the information to XML documents, and it sends the information to watcher according to the filter in the request. Agents must be added to the network. User agents are applications on the user's devices. Network agents are applications on network elements in the control layer (S-CSCF and HSS) and on servers in application layer (position server and

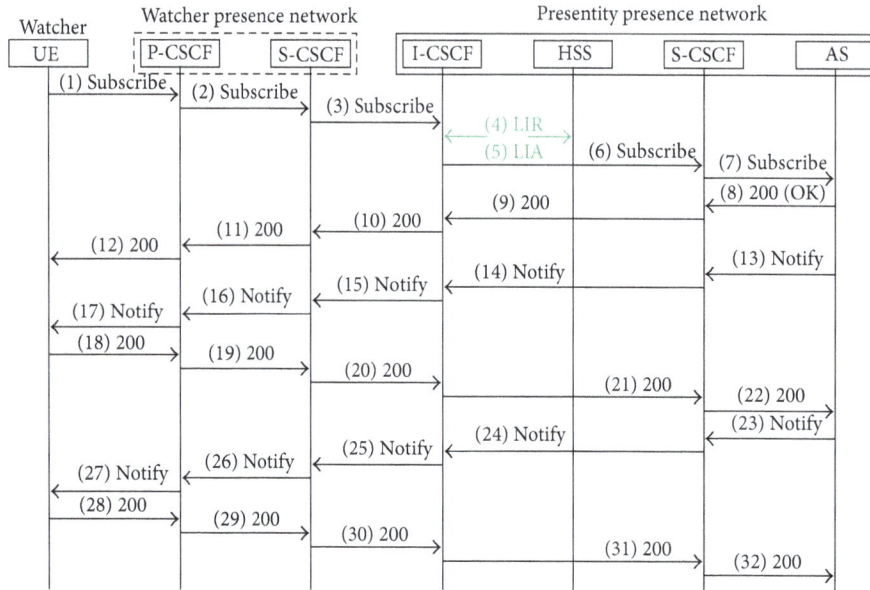

FIGURE 4: Subscribe status.

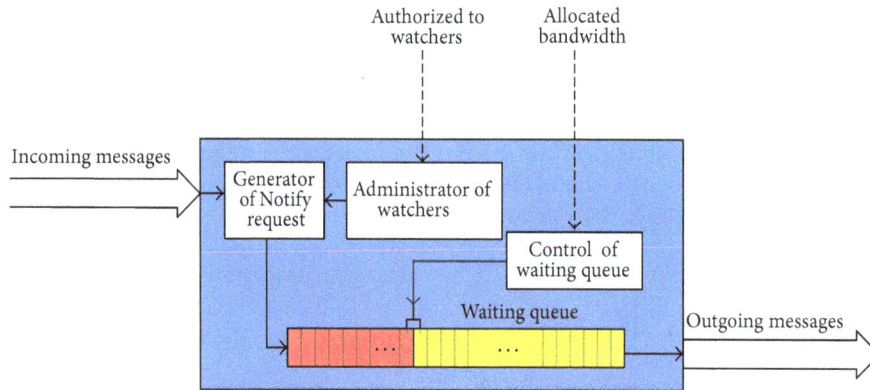

FIGURE 5: Logical scheme of the presence server.

application server of any other service). External agents collect information from other networks. Another important issue is the increased number of transmitted SIP messages. Fifty percent of transmitted messages are related to this service [18]. That is why it is most important to focus on the number of transmitted messages.

3.1. Messages. Presence service uses three types of SIP requests. publish is sent when presentity logs on (pub_login), logs off (pub_logout), modifies status (pub_modify), and refreshes status (pub_refresh) if it does not change the state. subscribe is sent when presentity starts (sub_initial), ends (sub_terminal), and refreshes (sub_refresh) to subscribe information from presence server. notify is sent by a server; that server notifies status of presentity to watchers (notify). These are eight situations, when someone sends a request [19]. The largest representation has request notify. Number of

notify is given by (1). It is important to create a mechanism to control the number of sent Notify messages [20]. These messages occur after the server receives a message publish; hence it is important to focus on Publish messages:

$$
\begin{aligned}
r\_notify \\
= \text{Watchers} \cdot (r\_pub\_login + r\_pub\_logout \\
+ r\_pub\_modify + r\_pub\_refresh).
\end{aligned}
\tag{1}
$$

3.2. Presence Server. In this paper, server creates Notify messages in Figure 5. Incoming messages are Publish requests. Generator of Notify request creates messages to send. Number of messages depends on incoming messages and the number of authorized watchers. Requests Notify are placed in the waiting queue. Messages are sent from the server periodically to avoid network congestion. It is assigned bandwidth for service. The bandwidth divided the waiting

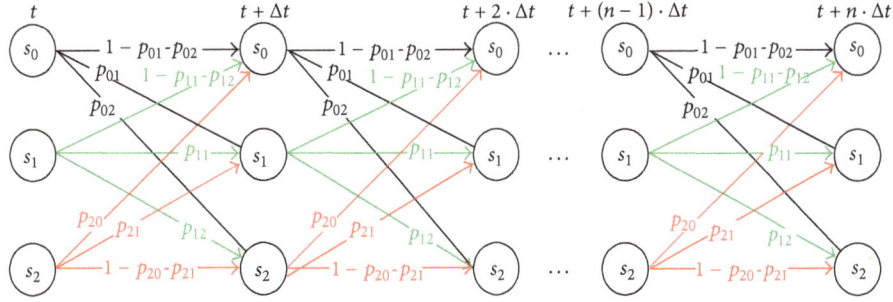

FIGURE 6: Model of making messages.

queue into two parts. Messages in yellow part of waiting queue are sent over time Δt, and messages in the red part of waiting queue must wait for send in next period. If the waiting queue is full and other messages arrive, these massages will be deleted.

4. Model for Creation of Messages

Creation of messages can be described by Markov chain shown in Figure 6. This model describes how many messages are created for time Δt in dependency on the average time the users spend in individual states and their numbers. Users can be in three states. s_0 represents online presentity that has unchanged status from time of its login or last change. s_1 represents online presentity that changed its previous status. s_2 represents offline presentity. Probability of transition from one state to another is given by exponential distribution [21] p_{ij} ($i = 0, 1, 2; j = 0, 1, 2$) as follows:

$$p_{ij} = \int_0^{\Delta t} \lambda_{ij} \cdot e^{-\lambda_{ij} x} dx, \qquad (2)$$

where λ_{ij} is given by the average time of creation message t_{ij},

$$\lambda_{ij} = \frac{1}{t_{ij}}. \qquad (3)$$

The probability, in which the user changes your state is given by the matrix:

	s_0	s_1	s_2
s_0	$1 - p_{01} - p_{02}$	p_{01}	p_{02}
s_1	$1 - p_{11} - p_{12}$	p_{11}	p_{12}
s_2	p_{20}	p_{21}	$1 - p_{20} - p_{21}$

$$\qquad (4)$$

States means the following:

(i)

$$P\left(s_0, t \mid s_0, \Delta t\right) = 1 - p_{01} - p_{02} \qquad (5)$$

is the probability in which online presentity does not change your presence status over time Δt.

(ii)

$$P\left(s_0, t \mid s_1, \Delta t\right) = p_{01} \qquad (6)$$

is the probability in which online presentity changes your presence status over time Δt.

(iii)

$$P\left(s_0, t \mid s_2, \Delta t\right) = p_{02} \qquad (7)$$

is the probability in which online presentity goes offline over time Δt.

(iv)

$$P\left(s_1, t \mid s_0, \Delta t\right) = 1 - p_{11} - p_{12} \qquad (8)$$

is the probability in which online presentity does not change your presence status over time Δt.

(v)

$$P\left(s_1, t \mid s_1, \Delta t\right) = p_{11} \qquad (9)$$

is the probability in which online presentity changes your presence status over time Δt.

(vi)

$$P\left(s_1, t \mid s_2, \Delta t\right) = p_{12} \qquad (10)$$

is the probability in which online presentity goes offline over time Δt.

(vii)

$$P\left(s_2, t \mid s_0, \Delta t\right) = p_{20} \qquad (11)$$

is the probability in which offline presentity goes online over time Δt.

(viii)

$$P\left(s_2, t \mid s_1, \Delta t\right) = p_{21} \qquad (12)$$

is the probability in which offline presentity changes your presence status over time Δt.

(ix)

$$P\left(s_2, t \mid s_2, \Delta t\right) = 1 - p_{20} - p_{21} \qquad (13)$$

is the Probability in which offline presentity stays offline over time Δt.

States s_0 and s_1 have same probability, because it is a state when presentity is online. Dividing is only for the purpose of illustrating a different message. Offline presentity cannot go to state s_1, because this state is created by the change of the online state. If we change parameters of publish modify and publish refresh, the number of other messages stays the same.

4.1. Meaning of Transition between States. Messages are created when someone goes from one state to another. The number of messages pub_modify is given by (14), the number of pub_login is given by (15), and the number of pub_logout is given by (16). Number of messages pub_refresh is counted differently. Probability of creation of messages is given by (17), where R is the time after which the user sends a message, if not, it changes its state, t_m is the average time of change presence status, and t_{of} is the average time of user log off. The number of messages pub_refresh is given by (18). The number of messages s_pub_x over interval $\langle T_1, T_2 \rangle$ is given by (19), where x is the one has type of messages.

$$\text{pub\_modify}(t)$$
$$= s_0(t) \cdot P(s_0, t \mid s_1, t - \Delta t) \tag{14}$$
$$+ s_1(t) \cdot P(s_1, t \mid s_1, t - \Delta t),$$

$$\text{pub\_login}(t) = s_2(t) \cdot P(s_2, t \mid s_0, t - \Delta t), \tag{15}$$

$$\text{pub\_logout}(t) = s_0(t) \cdot P(s_0, t \mid s_2, t - \Delta t) \tag{16}$$
$$+ s_1(t) \cdot P(s_1, t \mid s_2, t - \Delta t),$$

$$P_{\text{ref}} = \left(1 - \int_0^R \frac{1}{t_m} \cdot e^{(-1/t_m)x} dx\right)$$
$$+ \left(1 - \int_0^R \frac{1}{t_{of}} \cdot e^{(-1/t_{of})x} dx\right), \tag{17}$$

$$\text{pub\_refresh}(t) = s_0(t) \cdot P_{\text{ref}} + s_1(t) \cdot P_{\text{ref}}, \tag{18}$$

$$\text{s\_pub\_x} = \int_{T_1}^{T_2} \text{pub\_x}(t)\, dt. \tag{19}$$

The number of notify messages is given by the number of online watchers and the number of Publish messages as follows:

$$\text{notify}(t)$$
$$= \text{watchers}(t) \cdot \big(\text{pub\_refresh}(t) + \text{pub\_modify}(t)$$
$$+ \text{pub\_login}(t) + \text{pub\_logout}(t)\big). \tag{20}$$

5. Using the Model

A network with 550 000 users is given. They are assigned into online users in states s_0 and s_1 and offline users in the state s_2. t_{on} is the average time of the user being in an online state. t_{off} is the average time of the user being in the offline state.

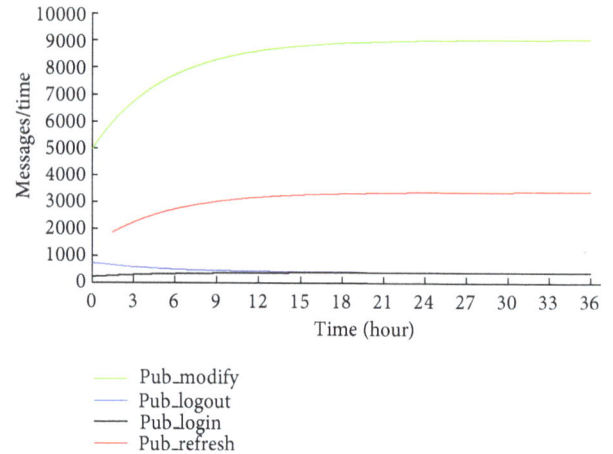

FIGURE 7: Messages in system A.

If t_{on} is more than t_{off}, then the number of online users increases. The number of online users decreases in another case. t_m is the average time of change of user state. Time Δt represents time, when server collects information and subsequently sends them periodically.

System A describes a situation, where the number of online users increases with time. The number of messages created over Δt is shown in Figure 7. The system is characterized by the following:

$s_0 = 150\,000,$

$s_1 = 50\,000,$

$s_2 = 350\,000,$

$t_{\text{on}} = 480$ minutes,

$t_{\text{off}} = 240$ minutes,

$t_m = 20$ minutes,

$\Delta t = 0.5$ minutes,

$R = 45$ minutes.

System B describes a situation, where the number of online users decreases over time. The number of messages is shown in Figure 8. The system is characterized by the following:

$s_0 = 150\,000,$

$s_1 = 50\,000,$

$s_2 = 350\,000,$

$t_{\text{on}} = 240$ minutes,

$t_{\text{off}} = 480$ minutes,

$t_m = 20$ minutes,

$\Delta t = 5$ minutes,

$R = 45$ minutes.

Figure 9 shows the number of messages pub_modify at different average times of change status t_m. Decrease in t_m would mean adding more applications to the network.

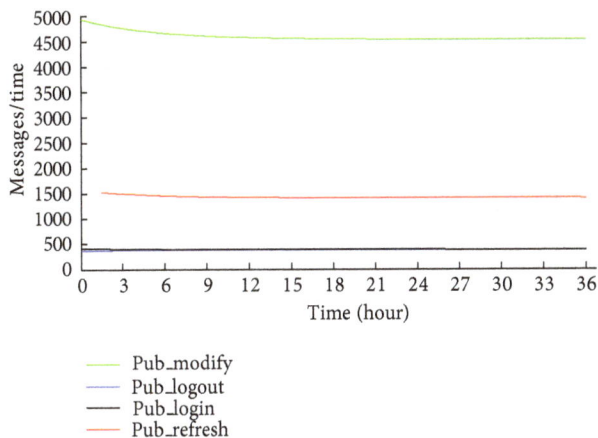

FIGURE 8: Messages in system B.

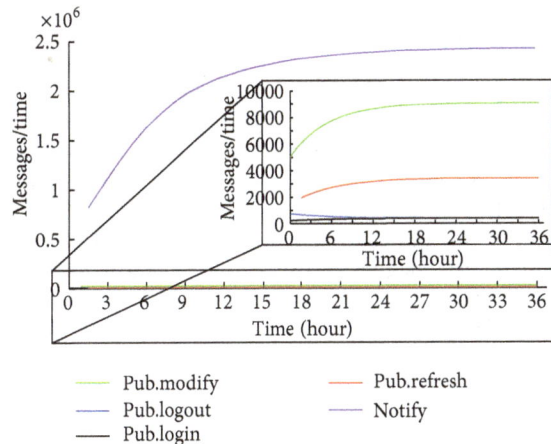

FIGURE 10: System A with Notify messages.

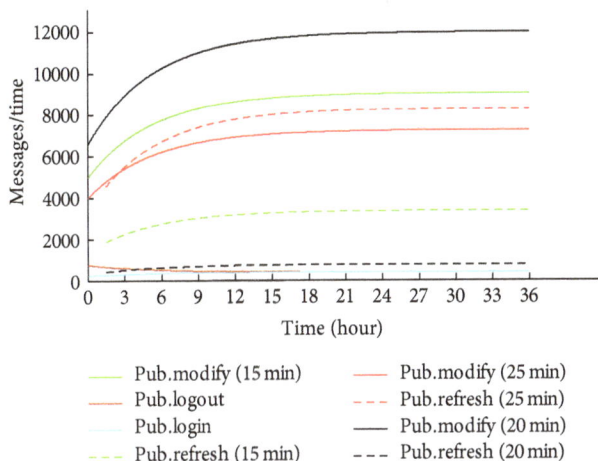

FIGURE 9: System A with increasing amount of messages.

If changes are more frequent, the amount of messages is increased.

Figure 10 shows the number of Notify messages, where the number of watchers of one's presentity is increasing with the number of online users. The number of messages is calculated by (19), where the number of watchers is given by (21). k represents the ratio of the amount of users in the telephone list and the amount of all users.

One has

$$watchers\,(t) = \left(s_0 + s_1\right) * k. \tag{21}$$

For Figure 10, $k = 0.005$.

6. Conclusion

Presence service is one of the key services in IMS. It allows the creation of a huge amount of applications, which can share information. Protection of this information is important. We can find some information about protection in [22, 23]. We have to consider some aspects before its deployment. The service requires creation of agents on the user's devices and

some blocks of IMS. Agents collect information and send it to the presence server. The presence server must be at the application layer in which the server processes incoming messages and sends information about the user's state. It is necessary to create a mechanism that controls the number of transmitted messages, because a huge amount of messages can overload the network. We need to determine the number of messages transmitted by the network before designing this mechanism. The model described in this paper can be used for this. This model displays the number of incoming and outgoing messages from the network. Users are in various states at the beginning, and gradually they log out, log in, and change their presence status. The change of the number of transmitted messages is related to this. If we observe this long enough without changing probability, the model will be in a stable state. That means that the same number of users changes their state in every step. We can determine the expected number of messages from the model, and according to this, we can design the size of the waiting queue on the presence server. We can see the ratio of messages if we want to assign different priorities too.

Acknowledgments

This paper is a part of the research activities conducted at the Slovak University of Technology in Bratislava, Faculty of Electrical Engineering and Information Technology, Institute of Telecommunications, within the scope of the project VEGA no. 1/0106/11 "Analysis and proposal for advanced optical access networks in the NGN converged infrastructure utilizing fixed transmission media for supporting multimedia services" and "Support of Center of Excellence for SMART Technologies, Systems, and Services II, ITMS 26240120029, cofunded by the ERDF."

References

[1] G. Camarillo and M. Garcia-Martin, *The 3G IP Multimedia Subsystem (IMS): Merging the Internet and the Cellular Worlds*, John Wiley & Sons, New York, NY, USA, 2nd edition, 2006.

[2] A. Al-Hezmi, T. Magedanz, J. J. Pallares, and C. Riede, "Evolving the convergence of telecommunication and TV services over NGN," *International Journal of Digital Multimedia Broadcasting*, vol. 2008, Article ID 843270, 11 pages, 2008.

[3] M. J. Sharma and V. C. M. Leung, "IP multimedia subsystem authentication protocol in LTE-heterogeneous networks," *Human-Centric Computing and Information Sciences*, vol. 2, p. 16, 2012.

[4] Z. S. Khan, M. Sher, and K. Rashid, "Presence based secure instant messaging mechanism for IP multimedia subsystem," in *Computational Science and Its Applications (ICCSA '11)*, vol. 6786 of *Lecture Notes in Computer Science*, pp. 447–457, Springer, 2011.

[5] M. Voznak and J. Rozhon, "Approach to stress tests in SIP environment based on marginal analysis," *Telecommunication Systems*, vol. 52, no. 3, pp. 1583–1593, 2013.

[6] Z. Bečvář, P. MacH, and B. Šimák, "Improvement of handover prediction in mobile WiMAX by using two thresholds," *Computer Networks*, vol. 55, no. 16, pp. 3759–3773, 2011.

[7] P. Zahradník, B. Šimák, M. Vlček, and M. Kopp, "Band-pass filters for direct sampling receivers," in *Proceedings of the 11th International Conference on Networks (ICN '12)*, pp. 44–49, IARIA, Saint Gilles, Reunion Island, 2012.

[8] M. Poikselka, G. Mayer, H. Khartabil, and A. Niemi, *The IMS: IP Multimedia Concepts and Services in the Mobile Domain*, John Wiley & Sons, New York, NY, USA, 2004.

[9] S. Leggio, "SIP for instant messaging and presence leveraging extensions," 2005, http://antoine.fressancourt.free.fr/exjobb/BB_SIP.pdf.

[10] H. Sugano, S. Fujimoto, G. Klyne, A. Bateman, W. Carr, and J. Peterson, "RFC 3863: presence information data format (PIDF)," 2004.

[11] H. Schulzrinne, U. Columbia, V. Gurbani, P. Kyzivat, and J. Rosenberg, "RFC 4480: RPID: rich presence extensions to the presence information data format (PIDF)," 2006.

[12] M. Poikselka, G. Mayer, H. Khartabil, and A. Niemi, *The IMS: IP Multimedia Concepts and Services in the Mobile Domain*, John Wiley & Sons, New York, NY, USA, 2004.

[13] M. Day, J. Rosenberg, and H. Sugano, "RFC 2778: a model for presence and instant messaging," 2000.

[14] M. Wuthnow, M. Stafford, and J. Shih, *IMS: A New Model for Blending Applications*, Taylor & Francis, Boca Raton, Fla, USA, 2010.

[15] G. Camarillo and M. Garcia-Martin, *The 3G IP Multimedia Subsystem (IMS): Merging the Internet and the Cellular Worlds*, John Wiley & Sons, New York, NY, USA, 2nd edition, 2006.

[16] A. Niemi, M. Lonnfors, and E. Leppanen, "RFC, 5264: publication of partial presence information," 2008.

[17] S. Kumar Singh and H. Schulzrinne, "Presence," 2006, /presence_simplified.pdf.

[18] C. Urrutia-Valdés, A. Mukhopadhyay, and M. El-Sayed, "Presence and availability with IMS: applications architecture, traffic analysis, and capacity impacts," *Bell Labs Technical Journal*, vol. 10, no. 4, pp. 101–107, 2006.

[19] C. Chi, R. Hao, D. Wang, and Z. Cao, "IMS presence server: traffic analysis & performance modelling," in *Proceedings of the 16th IEEE International Conference on Network Protocols (ICNP '08)*, pp. 63–72, Orlando, Fla, USA, October 2008.

[20] J. Liao, J. Wang, T. Li, J. Wang, and X. Zhu, "A token-bucket based notification traffic control mechanism for IMS presence service," *Computer Communications*, vol. 34, no. 10, pp. 1243–1257, 2011.

[21] Z. Cao, C. Chi, R. Hao, and Y. Xiao, "User behavior modeling and traffic analysis of IMS presence servers," in *Proceedings of IEEE Global Telecommunications Conference (GLOBECOM '08)*, pp. 2469–2473, New Orleans, La, USA, December 2008.

[22] F. Rezac and M. Voznak, "Penetration tests in next generation networks," in *Mobile Multimedia/Image Processing, Security, and Applications*, vol. 8406 of *Proceedings of SPIE*, Baltimore, Md, USA, 2012.

[23] J. Safarik, M. Voznak, F. Rezac, and L. Macura, "Malicious traffic monitoring and its evaluation in VoIP infrastructure," in *Proceedings of the 35th International Conference on Telecommunications and Signal Processing (TSP '12)*, pp. 259–262, Prague, Czech Republic, 2012.

4

Many-to-Many Multicast Routing Schemes under a Fixed Topology

Wei Ding,[1] Hongfa Wang,[1] and Xuerui Wei[2]

[1] *Zhejiang Water Conservancy and Hydropower College, Hangzhou, Zhejiang 310018, China*
[2] *Department of Mathematics, Shaoxing University, Shaoxing, Zhejiang 312000, China*

Correspondence should be addressed to Wei Ding; dingweicumt@163.com

Academic Editors: A. Bogliolo and J. Zheng

Many-to-many multicast routing can be extensively applied in computer or communication networks supporting various continuous multimedia applications. The paper focuses on the case where all users share a common communication channel while each user is both a sender and a receiver of messages in multicasting as well as an end user. In this case, the multicast tree appears as a terminal Steiner tree (TeST). The problem of finding a TeST with a quality-of-service (QoS) optimization is frequently NP-hard. However, we discover that it is a good idea to find a many-to-many multicast tree with QoS optimization under a fixed topology. In this paper, we are concerned with three kinds of QoS optimization objectives of multicast tree, that is, the minimum cost, minimum diameter, and maximum reliability. All of three optimization problems are distributed into two types, the centralized and decentralized version. This paper uses the dynamic programming method to devise an exact algorithm, respectively, for the centralized and decentralized versions of each optimization problem.

1. Introduction

Multicast routing has been increasingly used in computer or communication networks supporting various multimedia applications, such as real-time audio and video conferences, entertainment, and distance learning, [1, 2]. Multicast routing is known as a *multicast tree* [3], which can reduce the usage of network resource, such as network cost and bandwidth. Multicast tree can be reduced to be a *Steiner tree* in mathematics [4].

In a real world, a large number of continuous multimedia applications drive the consumers to advance their *quality of service* (QoS) requirements [2, 5, 6] (e.g., cost, delay, and bandwidth). As we all know, the minimum cost multicast tree (Steiner tree) problem is NP-hard [7] and the minimum diameter multicast tree (Steiner tree) problem is polynomial solvable [8]. Furthermore, the multicast tree problem with additional QoS requirements is frequently harder to solve. For example, in past decade, a number of heuristics [9, 10] and distributed algorithms [11, 12] have been devised for finding a minimum cost delay-constrained multicast tree. Surprisingly, the *fixed topology* version of this problem, in which the

configuration of multicast tree is given in advance with a tree topology, is easier than the classic version. Wang and Jia [13] designed a pseudo-polynomial-time algorithm, and Xue and Xiao [14] devised a full polynomial time approximation scheme. Not only that, we believe the idea of under a fixed topology will play an important role in exploring a desired multicast routing with a variety of QoS requirements.

In many practical settings, each destination is a *terminal*. A terminal means an end user, who takes charge of receiving and sending data but not branching them, that is, it can not serve as a *relay node* in charge of copying and branching data to other terminals. In a word, a terminal is a source, a receiver, or both. For instance, a member in video conference not only receives all the others' real-time images but also sends its real-time images to all the others. In fact, this is a type of general paradigm of *many-to-many* multicast routing. Provided that all terminals share a common communication channel, one many-to-many multicast tree can be reduced to a *terminal Steiner tree* (TeST) [15]. The minimum cost TeST problem is known to be NP-hard [15] and the minimum diameter TeST problem is polynomial solvable [8]. Many-to- many multicast tree includes two types, the *centralized* and *decentralized*. In the former, a network node with a bootstrap serves as a server in charge of receiving a terminal's data and then copying and branching them to all the other terminals by using multicast, resulting in a centralized multicast tree, essentially a rooted TeST (with the root at the server node). The centralized multicast tree problem is in fact *one-to-many* multicast tree problem [13, 14, 16]. In the latter, a terminal sends its data directly to all the others using a common channel, resulting in a decentralized one, an unrooted TeST in essence.

To the best of our knowledge, there have been a number of studies on the unrestricted many-to-many multicast tree mentioned above (terminal Steiner tree) [8, 15, 17–19]; however few studies on the QoS restricted version [16] due to its vast difficulty. Fortunately, we discover that the idea of under a fixed topology could provide us with a new way of studying the restricted version. In this paper, we will study the unrestricted many-to-many multicast tree problem with the objective of cost minimized or delay minimized and/or reliability maximized under a fixed topology, respectively. The approaches and results presented in the rest of the paper will contribute to study the many-to-many multicast tree with QoS restrictions under a fixed topology in the recent future.

The rest of this paper is organized as follows. In Section 2, we introduce the architecture of many-to-many multicast tree under a fixed topology. In Section 3, we make preliminaries, including defining three kinds of metrics of QoS of multicast tree, three QoS optimization problems, and two Euclidean graphs. We present an exact algorithm, respectively, for the centralized and decentralized version of the minimum cost problem in Section 4, the minimum delay problem in Section 5, and maximum reliability problem in Section 6. In Section 7, we give an example to illustrate all the algorithms. In Section 8, we conclude the paper with some research topics.

2. Architecture of Many-to-Many Multicast Tree under a Fixed Topology

A computer or communication network is frequently modeled as an undirected graph [20]. Let $G = (V, E, \omega)$ be an undirected edge-weighted graph with a subset $S \subset V$ of terminals and $T = (U, F)$ be a sample TeST. Here, V and E denote the node set and edge set of G, U and F denote the node set and edge set of T, and $\omega(e)$ denotes the weight on e for every $e \in E$. For any edge $f = \{u_i, u_j\} \in F$, we say that f is realized in G once its two endpoints u_i and u_j are mapped to two different nodes v_i and v_j in G. Note that u_i and u_j are not allowed to be mapped to a same node of G. In essence, f is mapped to a simple path in G connecting v_i and v_j, often a so-called optimal path with respect to some optimization objective, for example, a shortest path if $\omega(e)$ represents the length of e. Also, we use $R(f)$ to denote a realization of f in G, and use $\pi^*[v_i, v_j]$ to denote an optimal path in G between v_i and v_j for any a pair of nodes v_i and v_j.

Given any two different nodes x and y of T, there is a unique path in T between them. Specially, if x and y are both leaves, we call the path a *leaf-to-leaf path* of T, denoted as

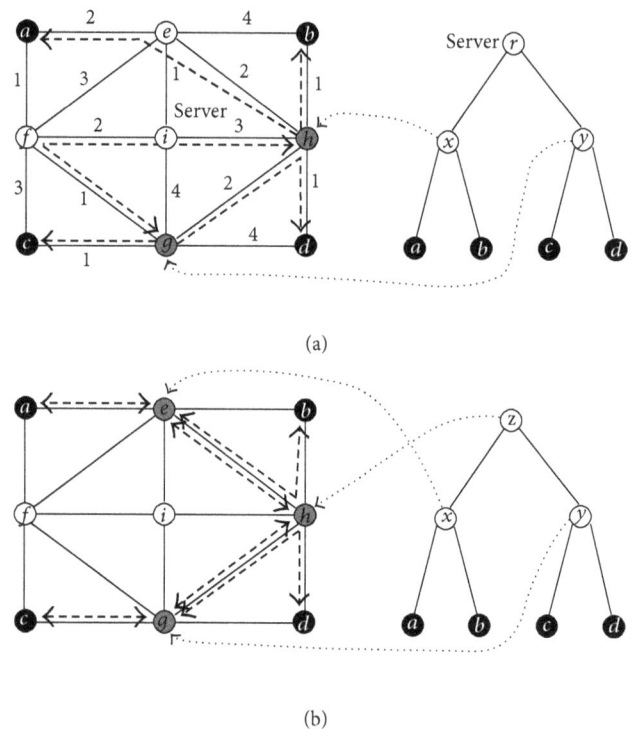

(a)

(b)

FIGURE 1: Ignore the dashed lines on the left top subfigure to obtain a sample graph where the number on each edge represents its length. The right top subfigure is a sample rooted TeST and the right bottom is a sample unrooted TeST. An example centralized many-to-many multicast tree under a fixed topology is distinguished by dashed lines on the left top subfigure and an example decentralized one is distinguished on the left bottom subfigure.

$L[x, y]$. In addition, the path in T between the root node r and a leaf x is called a *root-to-leaf path* of T, denoted as $C[r, x]$. Let $R(x, y)$ denote a realization of $L[x, y]$ in G and $R(r, x)$ a realization of $C[r, x]$. It is clear that $R(x, y)$ (resp. $R(r, x)$) is composed of all the realizations of edges on $L[x, y]$ (resp. $C[r, x]$) respectively. Furthermore, all the realizations of the edges on T form a realization of T in G, denoted as $R(T)$. A realization of T in G forms a many-to-may multicast tree under a fixed topology T in G essentially.

In this paper, we are concerned with the centralized and decentralized multicast trees under a fixed topology, namely, the realization of a given rooted and unrooted TeST topology; see Figure 1. On the top subfigure, the left graph shows a network with each edge having a weight denoting its length and the right graph shows a given rooted TeST topology. A realization of the rooted TeST in the network is distinguished by dashed edges. Considering that every leaf of the TeST is required to be mapped to a fixed node (terminal) and its root is required to be mapped to the server node, the essential work of realizing the TeST in the network is to map all the nonleaves except the root of the TeST to some nonterminals. If we arrange all the nonleaves in the order of from the 2nd level to top and from left to right on a level and label them by numbers $1, 2, \ldots, |U \setminus S|$ in sequence, we can denote all the nonleaves by an ordering (x, y, r), and then use another ordering (h, g, i) to record all the selected nonterminals in the graph. Likewise, on the bottom subfigure, the left shows a same network as above and the right shows an unrooted TeST topology. A realization of the TeST is distinguished by dashed edges. Since each leaf of the TeST is required to be mapped to a fixed node (terminal), the essence of realizing the TeST in the network is to map all the nonleaves of the TeST to some nonterminals. We can transform an unrooted tree into a rooted tree by assigning any a nonleaf as the root, and similarly use (x, y, z) to denote all the nonleaves and (e, g, h) to record all the selected nonterminals in the graph. In general, we can use an ordering $(u_1, \ldots, u_{|U \setminus S|})$, $u_i \in U \setminus S$ to denote all the nonleaves of a given TeST and use another ordering $(v_1, \ldots, v_{|U \setminus S|})$, $v_i \in V \setminus S$ to record all the selected nonterminal nodes in sequence, that is, a multicast tree under a fixed topology. In this paper, we always use $|\cdot|$ to denote the cardinality of a set.

Let T be a rooted tree, T_u the subtree of T rooted at u, and $R(T_u)$ a realization of T_u. When u is a nonleaf node, we let $\Delta(u)$ be the set of all the children of u, and then immediately obtain that

$$\sum_{u \in U \setminus S} |\Delta(u)| = |F|. \tag{1}$$

3. Fundamental Preliminaries

In the section, we make some fundamental preliminaries, which will help us to analyze the problems and understand the algorithms proposed in the following.

3.1. Metrics. First of all, we define three kinds of metrics of QoS of tree, including the *cost, delay,* and *reliability* of tree.

3.1.1. Cost. We are given an undirected graph $G = (V, E, c)$ where $c(e)$ represents the cost on e for every $c(\cdot)$. Let π be a path in G. The cost of π is equal to the sum of all the costs of edges on π, that is, $c(\pi) = \sum_{e \in \pi} c(e)$. Hence, the cost of edge realization $R(f)$ of $f = \{u_i, u_j\}$ in which u_i and u_j are mapped to two different nodes v_i and v_j in G, respectively, is denoted as $c(\pi^*[v_i, v_j])$ where $\pi^*[v_i, v_j]$ is a shortest path in G between v_i and v_j.

The *cost* of $R(T)$ is defined as the sum of all the costs of edge realizations of T, that is,

$$c(R(T)) = \sum_{\{u_i, u_j\} \in F} c\left(\pi^*\left[v_i, v_j\right]\right). \tag{2}$$

Let $R^c(T)$ denote a minimum cost realization of T. Thus,

$$c\left(R^c(T)\right) = \min_{R(T)} \sum_{\{u_i, u_j\} \in F} c\left(\pi^*\left[v_i, v_j\right]\right). \tag{3}$$

3.1.2. Delay. We are given an undirected graph $G = (V, E, d)$ where every edge $e \in E$ has a weight $d(e)$ representing the delay on e. The delay of π is equal to the sum of all the delays of edges on π, that is, $d(\pi) = \sum_{e \in \pi} d(e)$. Then, the delay of edge realization $R(f)$ is denoted as $d(\pi^*[v_1, v_2])$ in which $\pi^*[v_i, v_j]$ is a minimum delay path in G between v_i and v_j.

The delay of $R(x, y)$ is equal to the sum of all the delays of edge realizations of $L[x, y]$, that is,

$$d(R(x, y)) = \sum_{\{u_i, u_j\} \in L[x, y]} d\left(\pi^*\left[v_i, v_j\right]\right). \tag{4}$$

Similarly, we have

$$d(R(r, x)) = \sum_{\{u_i, u_j\} \in C[r, x]} d\left(\pi^*\left[v_i, v_j\right]\right). \tag{5}$$

The maximum delay of leaf-to-leaf path realization of T is called the *diameter* of $R(T)$, that is,

$$\text{diam}(R(T)) = \max_{\forall x, y \in S, x \neq y} d(R(x, y)). \tag{6}$$

Let $R^d(T)$ denote a minimum delay realization of T. So,

$$\text{diam}\left(R^d(T)\right) = \min_{R(T)} \max_{\forall x, y \in S, x \neq y} d(R(x, y)). \tag{7}$$

The maximum delay of root-to-leaf path realization of T is called the *radius* of $R(T)$, that is,

$$\text{radi}(R(T)) = \max_{\forall x \in S} d(R(r, x)). \tag{8}$$

Likewise,

$$\text{radi}\left(R^d(T)\right) = \min_{R(T)} \max_{\forall x \in S} d(R(r, x)). \tag{9}$$

Note that realizations with a minimum diameter or radius are both denoted by $R^d(T)$. We can differentiate according to the context where it appears.

3.1.3. Reliability. We are given an undirected graph $G = (V, E, p)$ where every edge $e \in E$ has an independent working probability $p(e)$ while all the nodes are immune to failures. The working probability of π is equal to the product of all the working probabilities of edges on π, that is, $p(\pi) = \prod_{e \in \pi} p(e)$. Then, the working probability of edge realization $R(f)$ is denoted as $p(\pi^*[v_i, v_j])$ where $\pi^*[v_i, v_j]$ is a maximum reliability path in G between v_i and v_j.

The working probability of $R(x, y)$, named its *reliability*, is equal to the product of all the working probabilities of edge realizations of $L[x, y]$, that is,

$$p(R(x, y)) = \prod_{\{u_i, u_j\} \in L[x, y]} p\left(\pi^*\left[v_i, v_j\right]\right). \tag{10}$$

Likewise,

$$p(R(r, x)) = \prod_{\{u_i, u_j\} \in C[r, x]} p\left(\pi^*\left[v_i, v_j\right]\right). \tag{11}$$

The minimum reliability of leaf-to-leaf path realization of T is called the *diameter reliability* of $R(T)$, that is,

$$\mathrm{dia}p\left(R\left(T\right)\right) = \min_{\forall x,y \in S, x \neq y} p\left(R\left(x,y\right)\right). \quad (12)$$

Let $R^r(T)$ denote a maximum reliability realization of T. Thus,

$$\mathrm{dia}p\left(R^r\left(T\right)\right) = \max_{R(T)} \min_{\forall x,y \in S, x \neq y} p\left(R\left(x,y\right)\right). \quad (13)$$

The minimum reliability of root-to-leaf path realization of T is called the *radius reliability* of $R(T)$, that is,

$$\mathrm{rad}p\left(R\left(T\right)\right) = \min_{\forall x \in S} p\left(R\left(r,x\right)\right). \quad (14)$$

Similarly,

$$\mathrm{rad}p\left(R^r\left(T\right)\right) = \max_{R(T)} \min_{\forall x \in S} p\left(R\left(r,x\right)\right). \quad (15)$$

Note that realizations with a maximum diameter or radius reliability are both denoted by $R^r(T)$. We can differentiate according to the context where it appears.

3.2. Problem Definitions. The focus of this paper is three optimization problems of many-to-many multicast tree under a fixed topology from the perspective of three metrics above, which are formally defined as follows.

First of all, we use an INPUT (abbreviated to I) to simplify the statements of problems: an undirected graph $G = (V, E)$, a subset $S \subset V$ of terminals, and a sample TeST topology as $T = (U, F)$ (see Figure 1). Let

$$|V| = n, \quad |E| = m, \quad |U| = \alpha, \quad |F| = \beta, \quad |S| = \lambda. \quad (16)$$

Moreover, we define $\overline{S} = S \cup \{r\}$.

Problem 1. Given an INPUT with every edge $e \in E$ having a nonnegative weight $c(e) \geq 0$ representing the cost on e, the *minimum cost many-to-many multicast tree under a fixed TeST topology problem* (MCMP) aims to find an ordering $(v_1, \ldots, v_{m-k})^*$, $v_i \in V \setminus S$ with a minimum cost.

Problem 2. Given an INPUT with every edge $e \in E$ having a positive weight $d(e) > 0$ denoting the delay on e, the *minimum delay many-to-many multicast tree under a fixed TeST topology problem* (MDMP) aims to find an ordering $(v_1, \ldots, v_{m-k})^*$, $v_i \in V \setminus S$ with a minimum delay.

Problem 3. Given an INPUT with every edge $e \in E$ having a weight $0 < p(e) < 1$ representing the working probability on e, the *maximum reliability many-to-many multicast tree under a fixed TeST topology problem* (MRMP) aims to find an ordering $(v_1, \ldots, v_{m-k})^*$, $v_i \in V \setminus S$ with a maximum reliability.

3.3. Euclidean Graphs. Given an undirected graph $G = (V, E, \omega)$ with every edge $e \in E$ having a weight $\omega(e)$, we construct two types of *Euclidean graphs* based on G in the following.

3.3.1. Shortest-Path Graph. When $\omega(e)$, $\forall e \in E$ in $G = (V, E, \omega)$ represents the cost $c(e)$ or delay $d(e)$ on e, the path between v_i and v_j with a minimum weight (e.g., cost or delay) is called a *shortest path* (SP), denoted as $\pi^*[v_i, v_j]$. So,

$$\omega\left(\pi^*\left[v_i, v_j\right]\right) = \min_{\pi[v_i, v_j]} \sum_{e \in \pi[v_i, v_j]} \omega(e). \quad (17)$$

All-pairs SPs in G form a set, denoted as Σ, that is,

$$\Sigma = \left\{\pi^*\left[v_i, v_j\right] : \forall v_i, v_j \in V, v_i \neq v_j\right\}. \quad (18)$$

Next we construct a new graph $\Sigma(G) = (V, \Sigma, \omega)$ from G such that the edge between v_i and v_j just represents an SP $\pi^*[v_i, v_j]$ in G. We call $\Sigma(G)$ as a *Shortest-Path Graph* (SPG). Evidently, $\Sigma(G)$ is a complete graph.

3.3.2. Maximum-Reliability-Path Graph. When $\omega(e)$, $\forall e \in E$ in $G = (V, E, \omega)$ represents the working probability $p(e)$ on e, the path between v_i and v_j with a maximum working probability is called a *maximum reliability path* (MRP), denoted as $\pi^*[v_i, v_j]$. So,

$$p\left(\pi^*\left[v_i, v_j\right]\right) = \max_{\pi[v_i, v_j]} \prod_{e \in \pi[v_i, v_j]} p(e). \quad (19)$$

All-pairs MRPs in G form a set, denoted as Π, that is,

$$\Pi = \left\{\pi^*\left[v_i, v_j\right] : \forall v_i, v_j \in V, v_i \neq v_j\right\}. \quad (20)$$

Next, we construct a complete graph $\Pi(G) = (V, \Pi, p)$ from G such that the edge between v_i and v_j just represents an MRP $\pi^*[v_i, v_j]$. We call $\Pi(G)$ a *Maximum- Reliability-Path Graph* (MRPG).

4. Minimum Cost Many-to-Many Multicast Tree under a Fixed Topology

In this section, we study the centralized and decentralized MCMP, respectively.

4.1. Centralized MCMP. According to discussions above in Section 2, the essence of finding a minimum cost multicast tree under a given TeST in the centralized MCMP is to find a minimum cost realization of the rooted TeST topology. In this subsection, we devise a polynomial-time exact algorithm for the centralized MCMP.

Let $R(T_u)$ be a realization of T_u with u mapped to v in the centralized MCMP. Let $C[u][v]$ denote the cost of $R(T_u)$.

Step_1: Use the Floyd's algorithm to find all-pairs shortest
         paths and then obtain $\Sigma_c(G) = (V, \Sigma_c, c)$;
Step_2: **for** {all $u \in U$, $v \in V$} **do**
            $C[u][v] \leftarrow 0$;
         end for
         for all $u \in U \setminus S$ **do**
            if $u = r$ **then**
               Compute $C[r][r]$ by (21);
            end for all $v \in V \setminus \overline{S}$ **do**
                  Compute $C[u][v]$ by (21);
               end for
            end if
         end for
Step_3: Trace out $(v_1, \ldots, v_{\alpha-\lambda})_C^*$ top-down from r of T;

ALGORITHM 1: $(v_1, \ldots, v_{\alpha-\lambda})_C^* = \text{MCCT}[I]$.

Step_1: Use the Floyd's algorithm to find all-pairs shortest paths
         and then obtain $\Sigma_c(G) = (V, \Sigma_c, c)$;
Step_2: **for** {all $u \in U$, $v \in V$} **do**
            $C[u][v] \leftarrow 0$;
         end for
         for {all $u \in U \setminus S$, $v \in V \setminus S$} **do**
            Compute $C[u][v]$ by (23);
         end for
Step_3: Trace out $(v_1, \ldots, v_{\alpha-\lambda})_D^*$ top-down from r of T;

ALGORITHM 2: $(v_1, \ldots, v_{\alpha-\lambda})_D^* = \text{MCDT}[I]$.

When $u \in U$ is a leaf of T, considering that T_u contains a single node, we set $C[u][u] = 1$. Otherwise, for every $u \in U \setminus \overline{S}$, we can use (21) to compute $C[u][v]$ for all $v \in V \setminus \overline{S}$,

$$C[u][v] = \sum_{u_k \in \Delta(u)} \min_{v_k \in V \setminus \overline{S}} \{C[u_k][v_k] + c(v_k, v)\}, \quad (21)$$

where $c(v_k, v)$ denotes the cost of $\pi^*[v_k, v]$. We use (21) to compute all of $C[u][v]$ by the dynamic programming method until $C[r][r]$ is obtained, and then we can trace out an exact solution of the centralized MRMP easily if some bookkeeping information is saved during the computation.

Above analysis leads to algorithm MCCT that can find a minimum cost centralized multicast tree under a TeST, which is denoted as $(v_1, \ldots, v_{\alpha-\lambda})_C^*$, in a polynomial time, see Theorem 1. We can use the approach in [21] to implement algorithm MCCT. So as to compute all $C[u][v]$, we need to get all-pairs shortest paths in G beforehand, namely, $\Sigma_c(G) = (V, \Sigma_c, c)$ where Σ_c denotes (18) with a cost on each edge of G. For the purpose, we can use the Floyd's algorithm, n times Dijkstra's algorithm, and so forth.

Theorem 1. *Given an INPUT as I, algorithm MCCT can find an optimal solution of the centralized MCMP correctly in $O(n^3 + \beta(n-\lambda)^2)$ time.*

Proof. Step_1 takes $O(n^3)$ time to find all-pairs shortest paths by using the Floyd's algorithm. Step_2 first spends $O(\alpha n)$ time

to initialize $C[u][v]$ for all $u \in U$ and $v \in V$, and then uses (21) to compute $C[u][v]$ for all $\in U \setminus S$, whose time complexity is at most

$$O\left(|\Delta(r)| \cdot |V \setminus \overline{S}|\right) + \sum_{u \in U \setminus \overline{S}} \sum_{v \in V \setminus \overline{S}} O\left(|\Delta(u)| \cdot |V \setminus \overline{S}|\right)$$

$$= O\left(|\Delta(r)| \cdot |V \setminus \overline{S}|\right) + O\left(\sum_{u \in U \setminus \overline{S}} |\Delta(u)| \cdot \sum_{v \in V \setminus \overline{S}} |V \setminus \overline{S}|\right)$$

$$= O\left(|\Delta(r)| \cdot |V \setminus \overline{S}|\right) + O\left(|V \setminus \overline{S}|^2 \cdot \sum_{u \in U \setminus \overline{S}} |\Delta(u)|\right)$$

$$\leq O\left(|V \setminus \overline{S}|^2 \cdot \sum_{u \in U \setminus S} |\Delta(u)|\right)$$

$$\overset{(1)}{=} O\left(|F|(n-\lambda-1)^2\right)$$

$$= O\left(\beta(n-\lambda)^2\right).$$

(22)

Step_3 only takes $O(\alpha - \lambda)$ time if we save some book-keepings information during the computation in Step_2. □

4.2. Decentralized MCMP. By the discussions above in Section 2, to find a minimum cost multicast tree under a given

---

Step_1: Use the Floyd's algorithm to find all-pairs shortest paths
and then obtain $\Sigma_d(G) = (V, \Sigma_d, d)$;
Step_2: **for** {all $u \in U$, $v \in V$} **do**
$\qquad Y[u][v] \leftarrow 0$;
$\qquad$ **end for**
$\qquad$ **for** all $u \in U \setminus S$ **do**
$\qquad\quad$ **if** $u = r$ **then**
$\qquad\qquad$ Compute $Y[r][r]$ by (24);
$\qquad\quad$ **else for** all $v \in V \setminus \bar{S}$ **do**
$\qquad\qquad\quad$ Compute $Y[u][v]$ by (24);
$\qquad\qquad$ **end for**
$\qquad\quad$ **end if**
$\qquad$ **end for**
Step_3: Trace out $(v_1, \ldots, v_{\alpha-\lambda})_C^\#$ top-down from r of T;

ALGORITHM 3: $(v_1, \ldots, v_{\alpha-\lambda})_C^\# = \text{MDCT}[I]$.

TeST in the decentralized MCMP is essentially to find a minimum cost realization of the unrooted TeST topology. Since an unrooted TeST can be transformed into a rooted TeST by assigning it any nonleaf as its root, we can adapt the algorithm MCCT for finding a minimum cost decentralized multicast tree under an unrooted TeST, which is denoted as $(v_1, \ldots, v_{\alpha-\lambda})_D^*$. The resultant algorithm is named as MCDT. The main differences between MCDT and MCCT are that $v_k \in V \setminus \bar{S}$ in (21) is changed to $v_k \in V \setminus S$ in

$$C[u][v] = \sum_{u_k \in \Delta(u)} \min_{v_k \in V \setminus S} \{C[u_k][v_k] + c(v_k, v)\}, \quad (23)$$

and MCDT removes the restriction of r mapped to r from MCCT. Based on Theorem 1., we calculate the time complexity of algorithm MCDT, shown in Theorem 2.

Theorem 2. *Given an INPUT as I, algorithm MCDT can find an optimal solution of the decentralized MCMP correctly in $O(n^3 + \beta(n - \lambda)^2)$ time.*

Proof. It is similar to the proof of Theorem 1. $\qquad\square$

5. Minimum Delay Many-to-Many Multicast Tree under a Fixed Topology

In this section, we study the centralized and decentralized MDMP, respectively.

5.1. Centralized MDMP. In the centralized MDMP, to find a minimum delay multicast tree under a given TeST topology is to find a minimum delay realization of the rooted TeST. In this subsection, we design a polynomial-time exact algorithm for the centralized MDMP.

Let $R(T_u)$ be a realization of T_u with u mapped to v in the centralized MDMP. The delay of $R(T_u)$ is equal to the radius of $R(T_u)$. We let $Y[u][v]$ denote the radius of $R(T_u)$. When

$u \in U$ is a leaf of T, we set $Y[u][u] = 0$. Otherwise, for each $u \in U \setminus \bar{S}$, we can use (24) to compute $Y[u][v]$ for all $v \in V \setminus \bar{S}$,

$$Y[u][v] = \max_{u_k \in \Delta(u)} \min_{v_k \in V \setminus \bar{S}} \{Y[u_k][v_k] + d(v_k, v)\}, \quad (24)$$

where $d(v_k, v)$ denotes the delay of $\pi^*[v_k, v]$. We use (24) to compute $Y[u][v]$ recursively until $Y[r][r]$ is achieved, and then we can trace out an exact solution of the centralized MDMP easily if some bookkeeping information is saved during the whole computation.

Based on above analysis, we devise algorithm MDCT to find a minimum delay centralized multicast tree under a TeST, which is denoted as $(v_1, \ldots, v_{\alpha-\lambda})_C^\#$, in a polynomial time; see Theorem 3. We also can use the way in [21] to execute algorithm MDCT. In order to compute all $Y[u][v]$, we need to get all-pairs shortest paths in G beforehand, namely, $\Sigma_d(G) = (V, \Sigma_d, d)$ in which Σ_d denotes (18) with a delay on every edge of G. Here, we can use the Floyd's algorithm to get $\Sigma_d(G)$.

Theorem 3. *Given an INPUT as I, algorithm MDCT can find an optimal solution of the centralized MDMP correctly in $O(n^3 + \beta(n - \lambda)^2)$ time.*

Proof. It is similar to the proof of Theorem 1. $\qquad\square$

5.2. Decentralized MDMP. The essence of finding a minimum delay multicast tree under a given TeST in the decentralized MDMP is to find a minimum delay realization of the unrooted TeST topology.

First of all, we can always use the method in [22] to transform an unrooted tree into a rooted tree and further into a binary tree, denoted as $T^B = (U^B, F^B)$. Clearly, $|U^B \setminus S| = |S| - 1 = \lambda - 1$. For any nonleaf $u \in U^B \setminus S$, let u_l and u_r denote its left and right child, respectively. Let $R(T_u^B)$ be a realization of T_u^B with u mapped to v in the decentralized MDMP. The delay of $R(T_u^B)$ is equal to the diameter of $R(T_u^B)$ and denoted by $X[u][v]$. When $u \in U^B$ is a leaf of T_u^B, we set $X[u][u] = 0$

Step_1: Use the Floyd's algorithm to find all-pairs shortest paths
and then obtain $\Sigma_d(G) = (V, \Sigma_d, d)$;

Step_2: **for** {all $u \in U^B$ and $v \in V$} **do**

$\qquad X[u][v] \leftarrow 0; Y[u][v] \leftarrow 0;$

end for

for {all $u \in U^B \setminus S$ and $v \in V \setminus S$} **do**

$\qquad$ Compute $Y[u][v]$ by (26) and $X[u][v]$ by (25);

end for

Step_3: Trace out $(v_1, \ldots, v_{\alpha-\lambda})_D^{\#}$ top-down from r of T;

ALGORITHM 4: $(v_1, \ldots, v_{\alpha-\lambda})_D^{\#} = \text{MDDT}[I]$.

---

since T_u^B has a single node. Otherwise, for each $u \in U^B \setminus S$, we can use (25) to compute $X[u][v]$ for all $v \in V \setminus S$,

$$
\begin{aligned}
&X[u][v] \\
&= \min_{v_l, v_r \in V \setminus S} \max \left\{ X[u_l][v_l], X[u_r][v_r], \right. \\
&\qquad\qquad \left. Y[u_l][v_l] + Y[u_r][v_r] + d(v_l, v) + d(v_r, v) \right\},
\end{aligned}
\tag{25}
$$

where u_l is mapped to v_l and u_r is mapped to v_r, $d(v_l, v)$ and $d(v_r, v)$ denote the delay of $\pi^*[v_l, v]$ and $\pi^*[v_r, v]$, respectively, and both of $Y[u_l][v_l]$, $Y[u_r][v_r]$ can be derived from

$$
Y[u][v] = \max \left\{ \begin{array}{l} \min_{v_l \in V \setminus S} \{Y[u_l][v_l] + d(v_l, v)\} \\ \min_{v_r \in V \setminus S} \{Y[u_r][v_r] + d(v_r, v)\} \end{array} \right\}.
\tag{26}
$$

We can use (25) to compute $X[u][v]$ recursively, and then we can trace out an exact solution of the decentralized MDMP, denoted as $(v_1, \ldots, v_{\alpha-\lambda})_D^{\#}$, if some bookkeeping information are is during the whole computation. The resulting algorithm is called MDDT, whose time complexity is presented in Theorem 4.

Theorem 4. *Given an INPUT as I, algorithm MDDT can find an optimal solution of the decentralized MDMP correctly in $O(n^3 + \lambda(n - \lambda)^3)$ time.*

Proof. It is similar to the proofs of Theorems 1 and 2. Step_2 takes $O((2\lambda - 1)n)$ time to initialize $X[u][v]$ and $Y[u][v]$ for all $u \in U^B$ and $v \in V$ and then computes $X[u][v]$ and $Y[u][v]$ for all $u \in U^B \setminus S$ and $v \in V \setminus S$, whose time complexity is at most

$$
\begin{aligned}
&\sum_{u \in U^B \setminus S} \sum_{v \in V \setminus S} \left(O(|V \setminus S|) + O(|V \setminus S|^2) \right) \\
&= O\left(|U^B \setminus S| \cdot |V \setminus S|^3 \right) \\
&= O\left(\lambda(n - \lambda)^3 \right).
\end{aligned}
\tag{27}
$$

Therefore the time complexity of algorithm MDDT is no more than $O(n^3 + \lambda(n - \lambda)^3)$. $\qquad\square$

Input: an undirected edge-weighted graph $G = (V, E, p)$;

Output: all-pairs MRP's of G (namely $\Pi(G) = (V, \Pi, p)$);

Step_1: **for** {$i = 1, 2, \ldots, |V|$ and $j = 1, 2, \ldots, |V|$} **do**

$\qquad p^{(0)}(i, j) \leftarrow P_{i,j};$

end for

Step_2: **for** $k = 1, 2, \ldots, |V|$ **do**

$\qquad$ **for** {$i = 1, 2, \ldots, |V|$ and $j = 1, 2, \ldots, |V|$} **do**

$\qquad\qquad$ **if** $i = j$ **then**

$\qquad\qquad\qquad p^{(k)}(i, j) \leftarrow \infty;$

$\qquad\qquad$ **else**

$\qquad\qquad\qquad$ Compute $p^{(k)}(i, j)$ by (29);

$\qquad\qquad$ **end if**

$\qquad$ **end for**

end for

PROCEDURE 1: Procedure CMRP.

6. Maximum Reliability Many-to-Many Multicast Tree under a Fixed Topology

In this section, we study the centralized and decentralized MRMP, respectively.

6.1. Constructing an MRPG. MRPG based on G will play an important role in the design of algorithm for MRMP in G. According to the discussions in Section 3, MRPG based on G is an Euclidean graph comprising all-pairs MRPs in G. So the key work of constructing MRPG is to devise an efficient algorithm for finding all-pairs MRPs. In this section, we present such an algorithm with a cubic time.

Firstly, we introduce a fundamental Lemma 5.

Lemma 5. *Given an undirected graph $G = (V, E, p)$ with every edge $e \in E$ having an independent working probability, for any path $\pi[v_i, v_j]$ composed of two subpaths $\pi[v_i, v_k]$ and $\pi[v_k, v_j]$, $\pi[v_i, v_j]$ is an MRP in G if and only if both $\pi[v_i, v_k]$ and $\pi[v_k, v_j]$ are MRPs in G.*

Proof. On one hand, if $\pi[v_i, v_j]$ is an MRP, we can verify that the combination of $\pi'[v_i, v_k]$ and $\pi[v_k, v_j]$ forms a more reliable path than $\pi[v_i, v_j]$ provided that $\pi'[v_i, v_k]$ is a more reliable path than $\pi[v_i, v_k]$ or the combination of $\pi[v_i, v_k]$ and $\pi'[v_k, v_j]$ forms a more reliable path than $\pi[v_i, v_j]$ provided

---

Step_1: Use procedure CMRP to find all-pairs maximum
     reliability paths and then obtain $\Pi(G) = (V, \Pi, p)$;
Step_2: for {all $u \in U$, $v \in V$} do
     $\varphi[u][v] \leftarrow 0$;
    end for
    for all $u \in U \setminus S$ **do**
      if $u = r$ **then**
        Compute $\varphi[r][r]$ by (30);
      else for all $v \in V \setminus \bar{S}$ **do**
        Compute $\varphi[u][v]$ by (30);
      end for
      end if
    end for
Step_3: Trace out $(v_1, \ldots, v_{\alpha-\lambda})^{\triangle}_C$ top-down from r of T;

ALGORITHM 5: $(v_1, \ldots, v_{\alpha-\lambda})^{\triangle}_C = \mathrm{MRCT}[I]$.

---

that $\pi'[v_k, v_j]$ is a more reliable path than $\pi[v_k, v_j]$. This causes a contradiction. On the other hand, if $\pi[v_i, v_k]$ and $\pi[v_k, v_j]$ are both MRP's, we can verify that either $\pi'[v_i, v_k]$ is more reliable than $\pi[v_i, v_k]$ or $\pi'[v_k, v_j]$ is more reliable than $\pi[v_k, v_j]$ provided that $\pi'[v_i, v_j]$ consisting of $\pi'[v_i, v_k]$ and $\pi'[v_k, v_j]$ is a more reliable path than $\pi[v_i, v_j]$. This causes a contradiction. □

From Lemma 5, we claim that the most reliable paths in G satisfy the *triangle inequality*, based on which we can design a dynamic programming algorithm for finding all-pairs MRP's.

For any edge $e = \{v_i, v_j\} \in E$, we rewrite $p(e)$ to be $p(v_i, v_j)$. We can use (28) to construct a probability matrix $P = (P_{i,j})_{n \times n}$,

$$P_{i,j} = \begin{cases} 1 & \text{if } i = j, \\ p(v_i, v_j) & \text{if } i \neq j, \{v_i, v_j\} \in E, \\ 0 & \text{if } i \neq j, \{v_i, v_j\} \notin E. \end{cases} \quad (28)$$

We use $p^{(k)}(i, j)$ to denote the working probability of a current MRP between v_i and v_j after v_k is introduced. We set $p^{(0)}(i, j) = P_{i,j}$ initially and then use (29) to compute $p^{(k)}(i, j)$ for k from 1 to n.

$$p^{(k)}(i, j) = \max \left\{ p^{(k-1)}(i, j), \right.$$
$$\left. p^{(k-1)}(i, k) \times p^{(k-1)}(k, j) \right\}. \quad (29)$$

Finally, $p^{(n)}(i, j)$ is the working probability of the MRP in G between v_i and v_j.

In essence, above idea of using (29) recursively forms our dynamic programming algorithm for finding all-pairs MRP's, namely, constructing $\Pi(G) = (V, \Pi, p)$, which is described as procedure CMRP (Procedure 1). The time complexity of CMRP is shown in Lemma 6.

Lemma 6. *Given an undirected edge-weighted graph $G = (V, E, p)$ with n nodes and m edges in which every edge has*

an independent working probability, procedure CMRP can find all-pairs maximum reliability paths in $O(n^3)$ time.

Proof. Step_1 spends $O(n^2)$ time to initialize $p^{(0)}(i, j)$ for $i = 1, 2, \ldots, n$ and $j = 1, 2, \ldots, n$. For each $k = 1, 2, \ldots, n$, Step_2 spends $O(n^2)$ time to compute $p^{(k)}(i, j)$ for $i = 1, \ldots, n$ and $j = 1, \ldots, n$. Therefore, the time complexity of CMRP is $O(n^3)$. □

6.2. Algorithms. In this section, we present an exact algorithm for the centralized and decentralized MRMP, respectively.

6.2.1. Centralized MRMP. The essence of finding a maximum reliability multicast tree under a TeST topology in the centralized MRMP is to find a maximum reliability realization of the rooted TeST.

Let $R(T_u)$ be a realization of T_u with u mapped to v in the centralized MRMP. The reliability of $R(T_u)$ refers to its radius reliability, which is denoted as $\varphi[u][v]$. When $u \in U$ is a leaf of T, we set $\varphi[u][u] = 0$. Otherwise, for each $u \in U \setminus \bar{S}$, we can use (30) to compute $\varphi[u][v]$ for all $v \in V \setminus \bar{S}$,

$$\varphi[u][v] = \min_{u_k \in \Delta(u)} \max_{v_k \in V \setminus \bar{S}} \left\{ \varphi[u_k][v_k] \times p(v_k, v) \right\}, \quad (30)$$

where $p(v_k, v)$ denotes the reliability of $\pi^*[v_k, v]$. We use (30) to compute $\varphi[u][v]$ recursively until $\varphi[r][r]$ is got, and then we can trace out an exact solution of the centralized MRMP if some bookkeeping information is saved during the whole computation.

Above analysis can be described as algorithm MRCT. It can find a maximum reliability centralized multicast tree under a TeST, denoted as $(v_1, \ldots, v_{\alpha-\lambda})^{\triangle}_C$, in a polynomial time; see Theorem 7. We use the method in [21] to perform algorithm MRCT. In order to compute $\varphi[u][v]$, we need to get all-pairs maximum reliability paths in G beforehand, namely, $\Pi(G) = (V, \Pi, p)$. This work can be accomplished by procedure CMRP.

Step_1: Use procedure CMRPto find all-pairs maximum
         reliability paths and then obtain $\Pi(G) = (V, \Pi, p)$;
Step_2: for {all $u \in U^B$ and $v \in V$} do
             $\varphi[u][v] \leftarrow 0; \psi[u][v] \leftarrow 0$;
         end for
         for {all $u \in U^B \setminus S$ and $v \in V \setminus S$} do
             Compute $\varphi[u][v]$ by (32) and $\psi[u][v]$ by (31);
         end for
Step_3: Trace out $(v_1, \ldots, v_{\alpha-\lambda})_D^\triangle$ top-down from r of T;

ALGORITHM 6: $(v_1, \ldots, v_{\alpha-\lambda})_D^\triangle = \text{MRDT}[I]$.

Theorem 7. *Given an INPUT as I, algorithm MRCT can find an optimal solution of the centralized MRMP correctly in* $O(n^3 + \beta(n-\lambda)^2)$ *time.*

Proof. It is similar to the proof of Theorem 1. □

6.2.2. Decentralized MRMP. To find a maximum reliability multicast tree under a TeST in the decentralized MDMP is essentially to find a maximum reliability realization of the unrooted TeST. We can use the way in [22] to transform an unrooted tree into a rooted tree and further into a binary tree $T^B = (U^B, F^B)$. And some definitions and notations therein are still used here. Let $R(T_u^B)$ be a realization of T_u^B with u mapped to v in the decentralized MRMP. The reliability of $R(T_u^B)$ refers to its diameter reliability, which is denoted as $\psi[u][v]$. When $u \in U^B$ is a leaf of T_u^B, we set $\psi[u][u] = 0$. Otherwise, for each $u \in U^B \setminus S$, we can use (31) to compute $\psi[u][v]$ for all $v \in V \setminus S$,

$$\psi[u][v]$$

$$= \max_{v_l, v_r \in V \setminus S} \min \{\psi[u_l][v_l], \psi[u_r][v_r],$$

$$\varphi[u_l][v_l] \times \varphi[u_r][v_r] \times p(v_l, v) \times p(v_r, v)\}, \quad (31)$$

where u_l is mapped to v_l and u_r is mapped to v_r, $p(v_l, v)$ and $p(v_r, v)$ denote the reliability of $\pi^*[v_l, v]$ and $\pi^*[v_r, v]$, respectively, and both of $\varphi[u_l][v_l], \varphi[u_r][v_r]$ can be obtained by using

$$\varphi[u][v] = \min \left\{ \begin{array}{l} \max_{v_l \in V \setminus S} \{\varphi[u_l][v_l] \times p(v_l, v)\} \\ \max_{v_r \in V \setminus S} \{\varphi[u_r][v_r] \times p(v_r, v)\} \end{array} \right\}. \quad (32)$$

We can use (31) to compute $\psi[u][v]$ recursively, and then we can trace out an exact solution of the decentralized MRMP, denoted as $(v_1, \ldots, v_{\alpha-\lambda})_D^\triangle$, if some bookkeeping information is saved during the whole computation. This leads to algorithm MRDT, whose time complexity is shown in Theorem 8.

Theorem 8. *Given an INPUT as I, algorithm MRDT can find an optimal solution of the decentralized MRMP correctly in* $O(n^3 + \lambda(n-\lambda)^3)$ *time.*

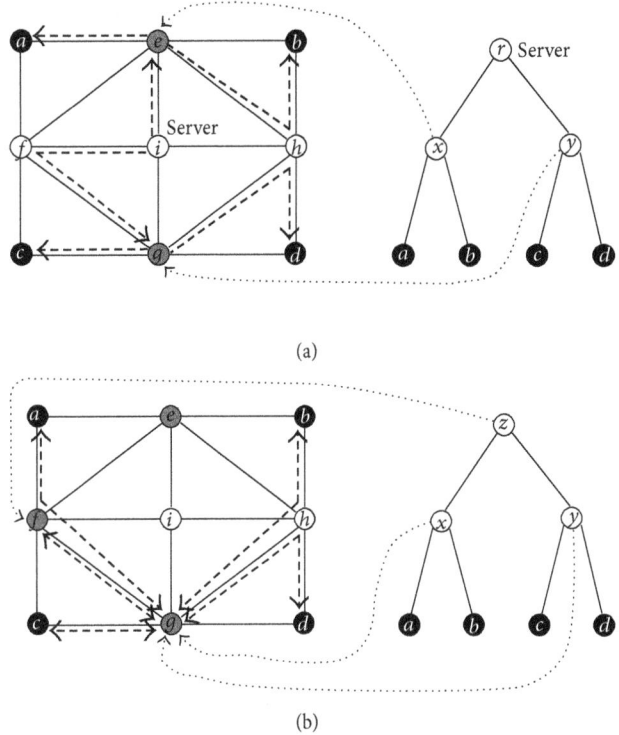

(a)

(b)

FIGURE 2: For ease of view, we neglect the numbers on edges. The example network for MCMP and the binary tree topology are both the same as those shown in Figure 1 and the integer on every edge represents its quantity of cost. The minimum cost multicast tree for the centralized MCMP is drawn with the bold dashed edges on the top subfigure and the one for the decentralized MCMP is drawn on the bottom subfigure.

Proof. It is similar to the proof of Theorem 4. □

7. Illustrative Examples

In this section, we take the network and the binary tree topology shown in Figure 1 as an example to illustrate the six algorithms proposed above (Algorithms 1–6).

Suppose the integer on every link of the network in Figure 1 represents the quantity of cost. For instance, each unit of cost is 0.8 dollar; then the cost on edge $\{b, e\}$ is 3.2 dollars. We apply algorithm MCCT to solve the centralized MCMP. MCCT first uses the Floyd's algorithm to find all-pairs minimum cost paths in the given network, that is, constructing the SPG $\Sigma_c(G) = (V, \Sigma_c, c)$ with respect to the link costs of network then computes the values of (21) recursively, and finally terminates with an optimal ordering (e, g, i) of the centralized MCMP. So we know that x, y, r are mapped to e, g, i, respectively, and then derive an optimal solution from $\Sigma_c(G)$ distinguished by bold dashed edges on the top subfigure of Figure 2; for example, the communication between x and b is established by the minimum cost path e-h-b and the communication between r and y is established by the minimum cost path i-f-g. Similarly we can apply algorithm MCDT to solve the decentralized MCMP. MCDT first constructs $\Sigma_c(G)$, then computes the values of (23)

(a)

(b)

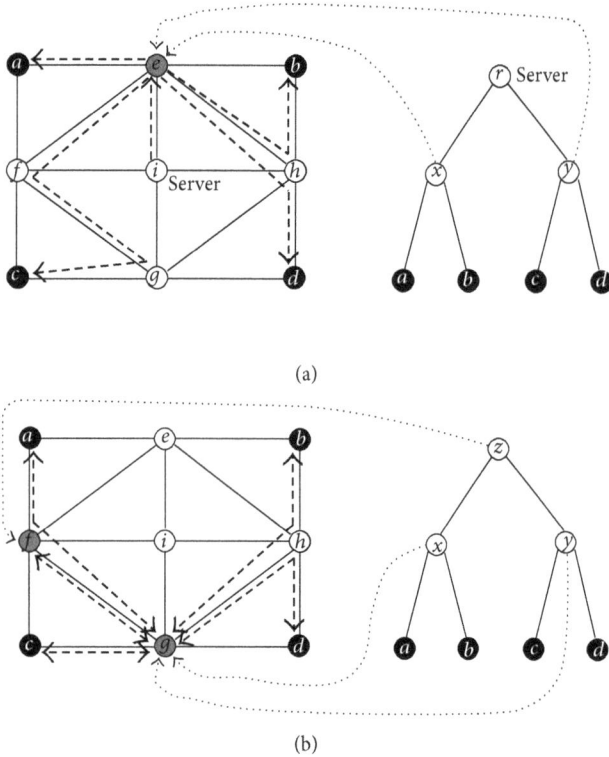

FIGURE 3: The example network for MDMP and the binary tree topology are both the same as those shown in Figure 1 and the integer on every edge represents its amount of delay. The minimum delay multicast tree for the centralized MDMP is drawn with the bold dashed edges on the top subfigure and the one for the decentralized MDMP is drawn on the bottom subfigure.

(a)

(b)

FIGURE 4: The example network for MRMP and the binary tree topology are both the same as those shown in Figure 1 and the integer on every edge represents the parameter of the linear probability function. The maximum reliability multicast tree for the centralized MRMP is drawn with the bold dashed edges on the top subfigure and the one for the decentralized MRMP is drawn on the bottom subfigure.

recursively, and finally ends with (g, g, f). Hence we can derive an optimal solution to the decentralized MCMP from $\Sigma_c(G)$, distinguished by bold dashed edges on the bottom subfigure of Figure 2.

Suppose the integer on every link of the network in Figure 1 represents its amount of delay. For example, every unit of delay is 1 ms, then the delay on link $\{e, f\}$ is 3 ms. We apply algorithm MDCT to solve the centralized MDMP. MDCT first uses the Floyd's algorithm to find all-pairs minimum delay paths in the given network, that is, constructing the SPG $\Sigma_d(G) = (V, \Sigma_d, d)$ with respect to the link delays of network, then computes the values of (24) recursively, finally ends with an optimal ordering (e, e, i) of the centralized MDMP. So we know that x, y, r are mapped to e, e, i, respectively, then derive an optimal solution from $\Sigma_d(G)$ distinguished by dashed edges on the top subfigure of Figure 3. Similarly, we apply MDDT to the decentralized MDMP. MDDT first constructs $\Sigma_d(G)$, and then computes the values of (25) and (26) recursively, finally terminates with (g, g, f). We can derive an optimal solution to the decentralized MDMP from $\Sigma_d(G)$, which is distinguished by bold dashed edges on the bottom subfigure of Figure 3 and equal to the optimal solution to the decentralized MCMP.

Suppose that every link $e \in E$ of the network given in Figure 1 has an independent working probability whose value

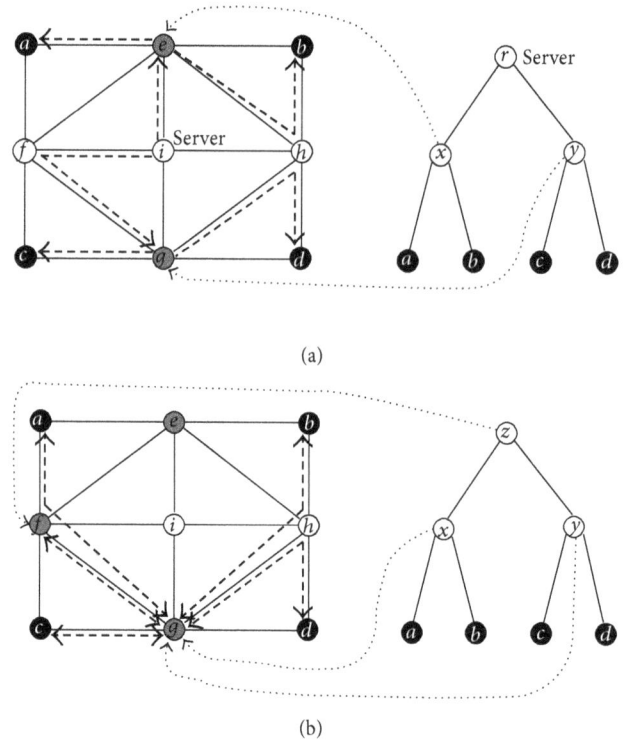

is a linear function in the integer $\delta(e)$ on e, formulated as $p(e) = 1 - 0.01 \times \delta(e)$. For instance, the working probability of $\{a, e\}$ is 0.98. We apply algorithm MRCT to solve the centralized MRMP. MRCT first uses procedure CMRP to find all-pairs maximum reliability paths in the network, that is, constructing the MRPG $\Pi(G) = (V, \Pi, p)$ with respect to the link probabilities of the given network, then computes the values of (30) recursively, finally terminates with an optimal ordering (e, g, i) of the centralized MRMP. We can get an optimal solution from $\Pi(G)$ distinguished by bold dashed edges on the top subfigure of Figure 4. Similarly, we can apply algorithm MRDT to solve the decentralized MRMP. MRDT first constructs $\Pi(G)$, then computes the values of (31) and (32) recursively, and finally ends with (g, g, f). Hence we can derive an optimal solution to the decentralized MRMP from $\Pi(G)$, distinguished by bold dashed edges on the bottom subfigure of Figure 4. Evidently, the optimal solution to the centralized MRMP is same as that to the centralized MCMP as well as the optimal solution to the decentralized MRMP is same as that to the decentralized MCMP.

8. Conclusions

This paper introduces the architecture of a many-to-many multicast tree with fixed topology, reduces it to a realization of the given TeST topology, and applies the idea of under a fixed topology to deal with three optimization problems, that is, the minimum cost, minimum delay, and maximum reliability multicast tree under a fixed topology problem. Each problem includes the centralized and decentralized versions. For the both versions of each problem, an exact algorithm is devised using the dynamic programming approach, respectively. On the condition that we are given a collection of alternative tree topologies, it is of interests to explore a best topology from all the alternative tree topologies.

Moreover, if we consider two or more weights on every link of a network, it is an interesting and important research topic how to devise an efficient algorithm for the multicast tree problem under a fixed topology with multiple objectives or with a single objective and at least one constraint.

Acknowledgments

This research was supported in part by the Key Research Grant xkyzd201207 (Ding) and the 2010 Excellent Young Teacher Grant 19093-88 (Ding) of Zhejiang Water Conservancy and Hydropower College.

References

[1] F. Kuo, W. Effelsberg, and J. J. Garcia-Luna-Aceves, *Multimedia Communications: Protocols and Applications*, Prentice Hall, Englewood Cliffs, NJ, USA, 1998.

[2] Z. Wang and J. Crowcroft, "Quality-of-service routing for supporting multimedia applications," *IEEE Journal on Selected Areas in Communications*, vol. 14, pp. 1228–1234, 1996.

[3] K. Bharath-Kumar and J. M. Jaffe, "Routing to multiple destinations in computer networks," *IEEE Transactions on Communications*, vol. 31, pp. 343–351, 1983.

[4] D. Z. Du, J. M. Smith, and J. H. Rubinstein, *Advances in Steiner Trees*, Kluwer Academic Publishers, Dordrecht, Netherlands, 2000.

[5] G. Apostolopoulos, R. Guerin, S. Kamat, and S. Tripathi, "Quality of service based routing: a performance perspective," in *Proceedings of the Applications, Technologies, Architectures, and Protocols for Computer Communication (ACM SIGCOMM '98)*, vol. 98, pp. 17–28, 1998.

[6] B. Wang and J. C. Hou, "Multicast routing and its QoS extension problems," *IEEE Network Algorithms, and Protocols*, vol. 14, pp. 22–36, 2000.

[7] M. R. Garey and D. S. Johnson, *Computers and Intractability: A Guide to the Theory of NP-Completeness.*, Freeman, San Francisco, Calif, USA, 1979.

[8] W. Ding and K. Qiu, "Algorithms for the minimum diameter terminal steiner tree problem," *Journal of Combinatorial Optimization*, 2013.

[9] V. P. Kompella, J. C. Pasquale, and G. C. Polyzo, "Multicast routing for multimedia communication," *IEEE/ACM Transactions on Networking*, vol. 1, pp. 286–292, 1993.

[10] Q. Zhu, M. Parsa, and J. Garcia-Luna-Aceves, "A source-based algorithm for delay-constrained minimal-cost multicasting," in *Proceedings of the IEEE International Conference on Computer Communications (INFOCOM '95)*, pp. 377–384, 1995.

[11] X. Jia, "A distributed algorithm of delay-bounded multicast routing for multimedia applications in wide area networks," *IEEE/ACM Transactions on Networking*, vol. 6, no. 6, pp. 828–837, 1998.

[12] V. P. Kompella, J. C. Pasquale, and G. C. Polyzos, "Two distributed algorithms for multicasting multimedia information," in *Proceedings of the IEEE Computer Communications and Networks (ICCCN '93)*, pp. 343–349, 1993.

[13] L. Wang and X. Jia, "Note fixed topology steiner trees and spanning forests," *Theoretical Computer Science*, vol. 215, pp. 359–370, 1999.

[14] G. Xue and W. Xiao, "A polynomial time approximation scheme for minimum cost delay-constrained multicast tree under a steiner topology," *Algorithmica*, vol. 41, no. 1, pp. 53–72, 2004.

[15] G. Lin and G. Xue, "On the terminal steiner problem," *Information Processing Letters*, vol. 84, pp. 103–107, 2002.

[16] M. Moh and B. Nguyen, "QoS-guaranteed one-to-many and many-to-many multicast routing," *Computer Communications*, vol. 26, pp. 652–669, 2003.

[17] Y. H. Chen, "An improved approximation algorithm for the terminal steiner tree problem," in *Computational Science and Its Applications (ICCSA '11)*, B. Murgante et al., Ed., vol. 6784 of *Lecture Notes in Computer Science*, pp. 141–151, 2011.

[18] D. E. Drake and S. Hougrady, "On approximation algorithms for the terminal steiner tree problem," *Information Processing Letters*, vol. 89, pp. 15–18, 2004.

[19] F. V. Martineza, J. C. D. Pinab, and J. Soares, "Algorithm for terminal steiner trees," *Theoretical Computer Science*, vol. 389, pp. 133–142, 2007.

[20] J. A. Bondy and U. S. R. Murty, *Graph Theory with Application.*, Macmillan, London, UK, 1976.

[21] W. Ding and G. Xue, "A linear time algorithm for computing a most reliable source on a tree network with faulty nodes," *Theoretical Computer Science*, vol. 412, no. 3, pp. 225–232, 2011.

[22] A. Tamir, "An $O(pn^2)$ algorithm for the p-median and related problems on tree graphs," *Operations Research Letters*, vol. 19, pp. 59–64, 1996.

5

An Efficient and Secure Certificateless Authentication Protocol for Healthcare System on Wireless Medical Sensor Networks

Rui Guo, Qiaoyan Wen, Zhengping Jin, and Hua Zhang

State Key Laboratory of Networking and Switching Technology, Beijing University of Posts and Telecommunications, Beijing 100876, China

Correspondence should be addressed to Rui Guo; grbupt@gmail.com

Academic Editors: Z. Cao, R. Lu, Q. Shi, and Q. Wu

Sensor networks have opened up new opportunities in healthcare systems, which can transmit patient's condition to health professional's hand-held devices in time. The patient's physiological signals are very sensitive and the networks are extremely vulnerable to many attacks. It must be ensured that patient's privacy is not exposed to unauthorized entities. Therefore, the control of access to healthcare systems has become a crucial challenge. An efficient and secure authentication protocol will thus be needed in wireless medical sensor networks. In this paper, we propose a certificateless authentication scheme without bilinear pairing while providing patient anonymity. Compared with other related protocols, the proposed scheme needs less computation and communication cost and preserves stronger security. Our performance evaluations show that this protocol is more practical for healthcare system in wireless medical sensor networks.

1. Introduction

Wireless medical sensor networks (WMSNs) have a capability of connecting patient with doctor by using of lightweight devices with limited memory, small and low power [1]. All these medical sensors collaborate together to collecting patient's physiological signals (e.g., blood pressure, blood sugar, and pulse oximeter) and send the collected data to health professional's hand-held devices (i.e., PDA, iPhone, iPad, etc.) via a wireless channel. The doctor uses these hand-held devices to observe the patient's real-time health condition.

However, the healthcare system on WMSN has many challenges, such as reliable data transmission, timely delivery of data, and power management [2]. Patient's privacy, a big concern for healthcare system, must be ensured at all sections on WMSN. The Health Insurance Portability and Accountability Act (HIPAA) of 1996 established rules for healthcare provider that it is necessary to control who is accessing to medical server's (MS's) resources and whether they are authorized to do so. Therefore, a secure authentication scheme among patient, MS, and doctor is needed to

protect the patient's privacy. So far many schemes that use cryptography have been proposed for this goal.

Most recently, Pu et al. [3] proposed a generic construction of smart card-based password authentication protocol for Telecare Medicine Information Systems (TMIS) and proved its security. Wu et al. [4] proposed a concrete efficient authentication scheme for TMIS. In their scheme, Wu et al. introduced a precomputing phase to compute costly and time-consuming exponential operations that are stored in a smart card. He et al. [5] pointed out that Wu et al.'s scheme could not resist impersonation attack and insider attack. Then, they proposed a more secure authentication scheme for TMIS. However, Wei et al. [6] demonstrated that both of Wu et al.'s scheme and He et al.'s scheme could not achieve a two-factor authentication. To overcome the weakness, Wei et al. proposed an improved authentication scheme for TMIS. Zhu [7] showed that Wei et al.'s scheme is vulnerable to an offline password guessing attack and also proposed a new authentication scheme for TMIS.

A common property of the above schemes is that the patient's identity ID is transmitted in plaintext on the public channel, which leads to impersonating attack and divulging the patient's privacy. To avoid these risks, based on the identity-based public key cryptography (ID-PKC) [8], Das et al. [9] proposed a dynamic ID-based remote client authentication scheme without any verifier table. However, Chien and Chen [10] pointed out that it fails to protect the anonymity of a user, and Ku and Chang [11] demonstrated that it is vulnerable to impersonation attack.

To address the key escrow problem [8] in ID-based authentication scheme, Xiong et al. [12] and Zhang et al. [13] proposed two certificateless authentication schemes, respectively. Unfortunately, their schemes are based on the bilinear pairing. Chen et al. [14] pointed out that the relative computation cost of the bilinear pairing is approximately twenty times higher than that of the scalar multiplication over a cyclic additive group, which is unsuitable for healthcare system on WMSN with lower computation power. Therefore, it is vitally important to present a certificateless authentication without bilinear pairing in the healthcare system.

In this paper, based on certificateless public key cryptography (CL-PKC) [15], we propose a certificateless authentication scheme without bilinear pairing in healthcare system on WMSN. Our protocol can establish a secure channel in Patient-to-MS and Doctor-to-MS with high efficiency. The proposed scheme has the following advantages: (1) it limits the power of MS to resist the malicious MS attack. (2) It ensures that the serial numbers of patient's wearable medical sensor and doctor's hand-held device can be updated in time. (3) It avoids the management of digital certificate and releases the key escrow problem by MS. (4) It achieves the Girault trust level 3 [16] as in traditional public key infrastructure (PKI). (5) It provides patient anonymity. (6) It preserves the perfect forward secrecy. (7) It can resist replay attack and impersonation attack. (8) It does not need to operate the bilinear pairing.

The remainder of this paper is organized as follows. Section 2 addresses some preliminaries such as the computational assumptions, security model, Girault's trust level, and the model of certificateless authentication. Section 3 proposes a certificateless authentication scheme and analyzes its security. Section 4 compares the proposed scheme with some other related schemes. Finally, we conclude the paper in Section 5.

2. Preliminaries

In this section, we review some fundamental backgrounds required in this paper, namely, computational assumptions, security model, Girault's trust level, and the model of certificateless authentication.

2.1. Computational Assumptions. The security of our protocol is based on the following computational assumptions:

Discrete Logarithm (DL) problem: let G be a cyclic additive group of prime order p; P is a generator of G. Given $Q \in G$, find an integer $x \in Z_p^*$ such that $Q = xP$.

The DL assumption is that there is no polynomial time algorithm that can solve the DL problem with nonnegligible probability.

Computational Diffie-Hellman (CDH) problem: let G be a cyclic additive group of prime order p; P is a generator of G. Given $Q, R \in G$ and $Q = xP$, $R = yP$ for any $x, y \in Z_p^*$, compute xyP.

The CDH assumption is that there is no polynomial time algorithm that can solve CDH problem with nonnegligible probability.

2.2. Security Model. In WMSN, we assume that attackers are "internal adversary" and "external adversary." Internal adversary is a legitimate member of WMSN, such as the malicious MS who has the ability of obtaining the private key and eavesdropping the privacy information of patient. We also assume that the external adversary is divided into four kinds. Type I adversary may capture the transmitted information between patient and doctor. By this information, Type I adversary can get the specific identity of patient. Type II adversary has a capability of extracting the secret key from the transmitted information; it may derivate the secret key in previous session by using this extracted key. Type III adversary may eavesdrop the transmitted information in public channel. Then, it transmits this information again to deceive patient (or doctor) that is provided from the legitimate doctor (or patient). Type IV adversary may capture the transmitted information and extract some important data from it. After that, it may impersonate the patient (or doctor) to communicate with the legitimate doctor (or patient).

2.3. Girault's Trust Level. Girault's trust level provides the trust hierarchy for public key cryptography, which can be used to judge the creditability of the authority (e.g., the MS in the healthcare system on WMSN).

> Level 1: the authority knows (or can easily compute) users' secret keys. Therefore, the authority can impersonate any user at any time without being detected.
>
> Level 2: the authority does not knows (or cannot easily compute) users' secret keys. Nevertheless, it can still impersonate user by generating false guarantees (e.g., false public keys).
>
> Level 3: the authority cannot compute users' secret keys, and it can be proven that it generates false guarantees of users' if it does so.

According to these definitions, we can easily find that the conventional certificateless cryptography can reach Level 2, and a traditional PKI can achieve Level 3 while the ID-PKC falls into Level 1.

2.4. Model of Certificateless Authentication. A certificateless authentication scheme consists of six probabilistic, polynomial time algorithms: *Setup, User-Key-Generation, Partial-Key-Extract, Set-Private-Key, Set-Public-Key,* and *Authentication.* These algorithms are defined as follows.

Setup. Taking security parameter k as input, the authority returns a list of public parameters param and a randomly chosen master secret key msk.

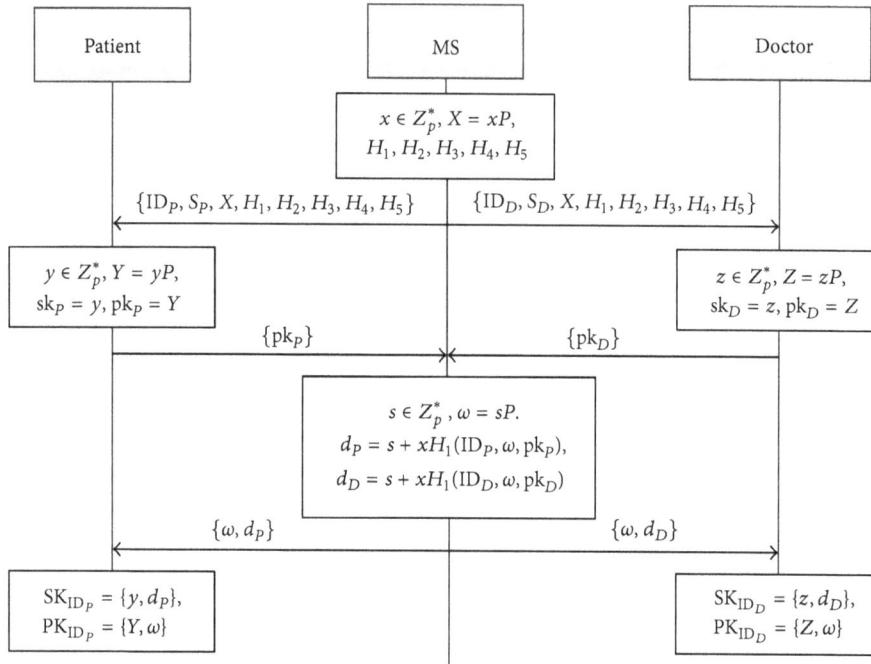

FIGURE 1: Initialization phase.

User-Key-Generation. Taking a list of public parameters param as input, the user returns a secret key sk and a public key pk.

Partial-Key-Extract. Taking param, msk, user's identity ID, and pk received from the user as inputs, the authority returns a partial private key D_{ID} and a partial public key P_{ID}.

Set-Private-Key. Taking param, D_{ID}, and sk as inputs, the user returns a private key SK_{ID}.

Set-Public-Key. Taking param, P_{ID}, and pk as inputs, the user returns a public key PK_{ID}.

Authentication. Taking identity, private key of the sender, and a list of parameters param as inputs, the receiver verifies the legality of the sender by its public key.

This model is similar to that of [15] but with a crucial difference that *User-Key-Generation* algorithm must be run prior to the *Partial-Key-Extract* algorithm, which makes the scheme achieve Girault's trust level 3.

3. Our Protocol

In this section, we propose a certificateless authentication scheme without bilinear pairing to ensure the legality of Patient and Doctor by the MS.

3.1. Construction. The proposed scheme involves three entities: Patient, Doctor, and MS. Before Patient obtains the

wearable medical sensor at the first time, MS presets the $\{\text{ID}_P, S_P\} \in \{0, 1\}^m$ and $\{\text{ID}_D, S_D\} \in \{0, 1\}^m$ into Patient's sensor and his/her doctor's health professional hand-held device through the secure channel as their identities and the serial numbers of equipments, respectively. Besides, these two serial numbers will be preserved secretly by themselves. The details of our certificateless authentication scheme are as follows.

We show the initialization phase of this protocol in Figure 1.

Setup. The MS generates a large prime p, which makes the DL and CDH problems in the cyclic additive group G with generator P of order p be intractable. Then, the MS picks $x \in Z_p^*$ uniformly at random, computes $X = xP$, and chooses hash functions

$$
\begin{aligned}
H_1 &: \{0, 1\}^m \times G^* \times G^* \longrightarrow Z_p^*, \\
H_2 &: \{0, 1\}^m \times \{0, 1\}^m \times \{0, 1\}^m \longrightarrow Z_p^*, \\
H_3 &: G^* \longrightarrow \{0, 1\}^m, \qquad H_4 : \{0, 1\}^m \longrightarrow \{0, 1\}^m, \\
H_5 &: \{0, 1\}^m \longrightarrow \{0, 1\}^*,
\end{aligned}
\tag{1}
$$

which can be achieved easily by collision-resistant hash function. Return $\{p, P, G, X, H_1, H_2, H_3, H_4, H_5\}$ as scheme parameters and the master secret key msk = $\{x\}$.

Patient/Doctor-Key-Generation. The Patient and the Doctor pick $y, z \in Z_p^*$ at random, compute $Y = yP$, $Z = zP$, and return $(\text{sk}_P, \text{pk}_P) = (y, Y)$ and $(\text{sk}_D, \text{pk}_D) = (z, Z)$, respectively.

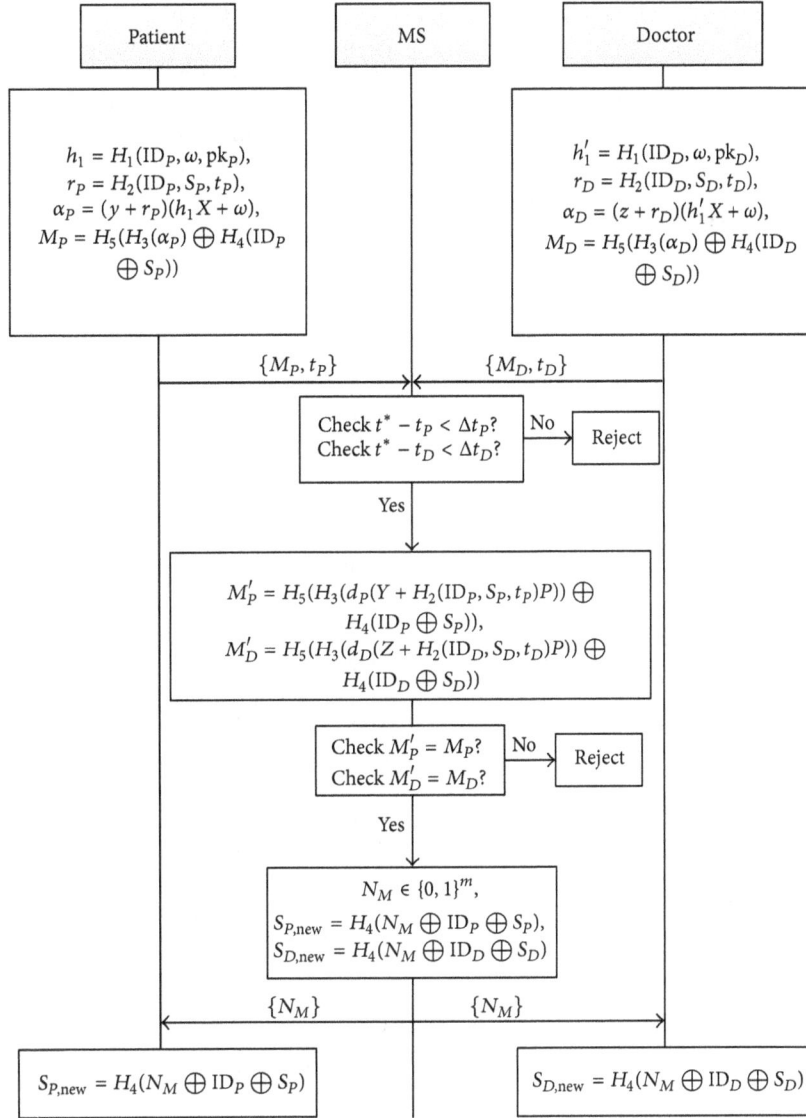

FIGURE 2: Authentication phase.

Partial-Key-Extract. The MS picks $s \in Z_p^*$ at random and computes

$$\omega = sP,$$

$$d_P = s + x H_1 (\mathrm{ID}_P, \omega, \mathrm{pk}_P), \qquad (2)$$

$$d_D = s + x H_1 (\mathrm{ID}_D, \omega, \mathrm{pk}_D).$$

Return $(P, D_{\mathrm{ID}_P}) = (\omega, d_P)$, $(P, D_{\mathrm{ID}_D}) = (\omega, d_D)$ as partial keys to be placed into Patient's sensor and the Doctor's hand-held device, respectively.

Set-Private-Key. The Patient sets $\mathrm{SK}_{\mathrm{ID}_P} = (\mathrm{sk}_P, D_{\mathrm{ID}_P}) = (y, d_P)$ as his/her private key, and the Doctor sets $\mathrm{SK}_{\mathrm{ID}_D} = (\mathrm{sk}_D, D_{\mathrm{ID}_D}) = (z, d_D)$ as his/her private key as well.

Set-Public-Key. Set $\mathrm{PK}_{\mathrm{ID}_P} = (\mathrm{pk}_P, \omega)$ and $\mathrm{PK}_{\mathrm{ID}_D} = (\mathrm{pk}_D, \omega)$ as the public keys of Patient and Doctor, respectively.

Now, we show the authentication phase in Figure 2.

Authentication

Step 1. The Patient picks the current time stamp t_P and computes

$$h_1 = H_1 (\mathrm{ID}_P, \omega, \mathrm{pk}_P), \qquad r_P = H_2 (\mathrm{ID}_P, S_P, t_P),$$

$$\alpha_P = (y + r_P) \cdot (h_1 X + \omega), \qquad (3)$$

$$M_P = H_5 (H_3 (\alpha_P) \oplus H_4 (\mathrm{ID}_P \oplus S_P)).$$

Send $\{M_P, t_P\}$ to the MS.

Step 2. The Doctor picks the current time stamp t_D and computes

$$h_1' = H_1\left(ID_D, \omega, pk_D\right), \qquad r_D = H_2\left(ID_D, S_D, t_D\right),$$

$$\alpha_D = (z + r_D) \cdot \left(h_1' X + \omega\right), \qquad (4)$$

$$M_D = H_5\left(H_3\left(\alpha_D\right) \oplus H_4\left(ID_D \oplus S_D\right)\right).$$

Send $\{M_D, t_D\}$ to the MS.

Step 3. If $(t^* - t_P) < \Delta t_P$ and $(t^* - t_D) < \Delta t_D$, where Δt_P and Δt_D denote the expected valid time interval for time delay of Patient and Doctor, the MS proceeds to the next step. Otherwise, return "*Reject.*"

Step 4. The MS computes

$$M_P' = H_5\left(H_3\left(d_P \cdot \left(Y + H_2\left(ID_P, S_P, t_P\right) \cdot P\right)\right)\right.$$

$$\left. \oplus H_4\left(ID_P \oplus S_P\right)\right),$$

$$\qquad (5)$$

$$M_D' = H_5\left(H_3\left(d_D \cdot \left(Z + H_2\left(ID_D, S_D, t_D\right) \cdot P\right)\right)\right.$$

$$\left. \oplus H_4\left(ID_D \oplus S_D\right)\right).$$

If M_P' is equal to M_P, Patient is a legal one. Otherwise, return "*Reject.*" In addition, if M_D' is equal to M_D, Doctor is a legal one. Otherwise, return "*Reject.*"

Step 5. The MS picks $N_M \in \{0, 1\}^m$ uniformly at random and updates the serial numbers of Patient and Doctor as follows:

$$S_{P,\text{new}} = H_4\left(S_P \oplus N_M \oplus ID_P\right),$$

$$S_{D,\text{new}} = H_4\left(S_D \oplus N_M \oplus ID_D\right). \qquad (6)$$

Send $\{N_M\}$ to Patient and Doctor.

Step 6. By using of $\{N_M\}$, Patient computes

$$S_{P,\text{new}} = H_4\left(S_P \oplus N_M \oplus ID_P\right) \qquad (7)$$

for updating the serial number of his/her wearable medical sensor.

Step 7. After obtaining $\{N_M\}$, Doctor computes

$$S_{D,\text{new}} = H_4\left(S_D \oplus N_M \oplus ID_D\right) \qquad (8)$$

for updating the serial number of his/her hand-held device.

3.2. Security Analysis

Theorem 1. *This certificateless authentication scheme is secure in the following possible attacks, provided that H_1 is a collision-resistance hash function and DL and CDH problems are intractable.*

Proof

Anonymity. In the proposed scheme, the partial key $d_P = s + xH_1(ID_P, \omega, pk_P)$ is used instead of ID_P to ensure the Patient's anonymity. Since ID_P is never transmitted as plaintext form in the public channel, Type I adversary cannot find the real identity ID_P of Patient. That is, when Patient transmits his/her health information, their real identity ID_P can only be computed as $d_P = s + xH_1(ID_P, \omega, pk_P)$ to be transmitted, where s is a random value, H_1 is a collision-resistant hash function, and x is the master secret key which is preserved by MS. Therefore, Type I adversary cannot trace Patient.

Perfect Forward Secrecy. To extract $\{M_P, M_D\}$ without the knowledge of the values $\{r_P, y, d_P, r_D, z, d_D\}$, Type II adversary should solve the DL problem and the CDH problem from public parameters. Moreover, $r_P = H_2(ID_P, S_P, t_P)$ and $r_D = H_2(ID_D, S_D, t_D)$ will be different in every session for the reason of time stamps $\{t_P, t_D\}$ and the updated serial numbers $\{S_P, S_D\}$. Therefore, Type II adversary cannot receive the previous value $\{r_P, y, d_P, r_D, z, d_D\}$ and the protocol enjoys the perfect forward security.

Replay Attack. During the data transmission, Type III adversary may eavesdrop $\{M_P, M_D\}$ and impersonate the legitimate Patient and Doctor to transmit $\{M_P, M_D\}$ to MS. After each session is over, the serial numbers of the Patient's sensor and Doctor's hand-held device have been updated to be the new serial numbers $\{S_{P,\text{new}}, S_{D,\text{new}}\}$, which can be used to generate the new messages $\{M_{P,\text{new}}, M_{D,\text{new}}\}$. Hence, Type III adversary cannot pass the verification by retransmitting $\{M_P, M_D\}$ in the new session. Moreover, there are time stamps $\{t_P, t_D\}$ in this scheme, which ensures the freshness of $\{M_P, M_D\}$.

Impersonation Attack. The impersonation attack fails due to the secret serial number. Provided that Type IV adversary wants to impersonate the legitimate Patient and Doctor, it must produce the relative $\{M_P, M_D\}$ for passing the verification of MS. However, in order to generate the exactly $\{M_P, M_D\}$, Type IV adversary needs to obtain the current serial numbers $\{S_P, S_D\}$ first of all, which are preserved secretly by Patient and Doctor and updated in time in the end of *Authentication* phase. Therefore, Type IV adversary has no capability to impersonate the legitimate Patient and Doctor to generate the correct $\{M_P, M_D\}$.

Malicious MS Attack. The malicious MS cannot obtain the private keys to eavesdrop the privacy information of patient. This authentication scheme is proposed on the base of CL-PKC, and the private keys (SK_{ID_P}, SK_{ID_D}) generated by Patient and Doctor consist of partial private keys (d_P, d_D) and the secret values (y, z). The malicious MS cannot obtain (y, z) from public parameters for the intractable of DL and CDH problems. Therefore, our scheme can resist the malicious MS attack.

Achieve Girault's Trust Level 3. The *Patient/Doctor-Key-Generation* must be run prior to *Partial-Key-Extract*. In this way, the *Partial-Key-Extract* algorithm includes (pk_P, pk_D) generated by Patient and Doctor as input. Therefore, provided

TABLE 1: Functionality comparisons.

Properties	[7]	[12]	[13]	Ours
User anonymity	No	No	No	Yes
Perfect forward secrecy	No	Yes	Yes	Yes
Replay attack resistance	Yes	No	No	Yes
Impersonation attack resistance	Yes	Yes	Yes	Yes
Malicious server attack resistance	Yes	Yes	Yes	Yes
No certificate management	No	Yes	Yes	Yes
Trust level	1	2	3	3

TABLE 2: Cryptographic operation time.

Fast-Tate-Pairing	Exponential	Scalar multiplication
2.66 ms	3.75 ms	0.94 ms

TABLE 3: Efficiency comparisons.

Scheme	Authen	Bandwidth	Time
[7]	4E	48 bytes	15 ms
[12]	6P + 6E + 21S	96 bytes	58.2 ms
[13]	2P + 10S	72 bytes	14.72 ms
Ours	8S	28 bytes	7.52 ms

that the MS replaces $(\mathrm{pk}_P, \mathrm{pk}_D)$, there will exist two working keys $(\mathrm{pk}_P, \mathrm{pk}'_P)$ and $(\mathrm{pk}_D, \mathrm{pk}'_D)$ for Patient and Doctor, respectively. Furthermore, two working public keys $(\mathrm{PK}_{\mathrm{ID}_P}, \mathrm{PK}'_{\mathrm{ID}_P})$ binding only one identity ID_P can result from two partial private keys (the same to Doctor), and only the MS could generate these two working partial private keys. Hence, it can be proven that MS generates false guarantees of Patient and Doctor, which means that our scheme achieves Girault's trust level 3 (the same level as is enjoyed in a traditional PKI).

Thus, to sum up the analysis above, we complete the proof of Theorem 1.

4. Comparisons

In this section, we evaluate some performance issues of our protocol with related works in functionality and efficiency.

4.1. Functionality Comparisons. Table 1 demonstrates the functionality comparisons between the proposed scheme and others [7, 12, 13]. Zhu's, Xiong et al.'s, and Zhang et al.'s protocols do not provide user anonymity. Moreover, the schemes in [12, 13] are insecure against the replay attack. However, as shown in Table 1, our scheme not only provides user anonymity but also achieves all security requirements. Furthermore, our scheme does not need an additional certificate to bind the user to its public key.

4.2. Efficiency Comparisons. In this subsection, we compare the proposed scheme with others on the computation complexity of authentication (Authen), bandwidth of the largest message (Bandwidth), and operation time in authentication (Time). Without considering the addition of two points, hash function and exclusive-OR operations, each scheme has three types of operations, that is, pairing (P), exponentiation (E), and scalar multiplication (S).

We evaluate the cryptographic operations by using of MIRACL (version 5.6.1, [17]), a standard cryptographic library, on a laptop using the Intel Core i5-2400 at a frequency of 3.10 GHz with 3 GB memory, and then obtain the average running time in Table 2. For pairing-based schemes, we use the Fast-Tate-Pairing in MIRACL, which is defined over the MNT curve E/F_q [18] with embedding degree 4, and q is a 160-bit prime. For ECC-based scheme, we employed the parameter secp192r1 [19], where $p = 2^{192} - 2^{64} - 1$. Moreover, the length of an element in multiplication group is set to be 1024 bits.

We compare the computation cost of different protocols with the method in [20]. For example, to finish the authentication in [12], six pairing operations, six exponentiations in Z_p^*, and twenty-one scalar multiplications are needed; thus, the operation time is $2.66 \times 6 + 3.75 \times 6 + 0.94 \times 21 = 58.2$ ms. Assuming the bit size of the identity, the point in additional group and the output of one-way hash function are all 192 bits. We also assume that the size of timestamp is 32 bits. In [12], the largest message contains three points in additional group and one identification; thus, the bandwidth of it is $(192 \times 3 + 192)/8 = 96$ bytes. The detailed comparison results are demonstrated in Table 3.

From Table 3, we know that the largest bandwidth of our scheme is only 28 bytes and the whole operation time in authentication is only 7.52 ms, which shows that our protocol is suitable for the lightweight devices (with limited memory, small and low power) in the healthcare system on WMSN.

5. Conclusions

In this paper, we propose a secure certificateless authentication scheme to ensure the legality of Patient and Doctor in healthcare system on WMSN. Meanwhile, this protocol also provides patient anonymity and resists the malicious MS attack to meet the privacy requirements in HIPAA. Our certificateless authentication protocol achieves a lower communication and computational overhead and stronger security than others. By the performance evaluation, the results show that our protocol is suitable for healthcare system on WMSN.

Acknowledgments

This work is supported by NSFC (Grants nos. 61272057, 61202434, 61170270, 61100203, 61003286, and 61121061), the Fundamental Research Funds for the Central Universities (Grants nos. 2012RC0612, 2011YB01).

References

[1] R. S. H. Istepanian, E. Jovanov, and Y. T. Zhang, "Introduction to the special section on m-Health: beyond seamless mobility and global wireless health-care connectivity," *IEEE Transactions on Information Technology in Biomedicine*, vol. 8, no. 4, pp. 405–414, 2004.

[2] F. Bellifemine, G. Fortino, R. Giannantonio, R. Gravina, A. Guerrieri, and M. Sgroi, "SPINE: a domain-specific framework for rapid prototyping of WBSN applications," *Software, Practice and Experience*, vol. 41, no. 3, pp. 237–265, 2011.

[3] Q. Pu, J. Wang, and R. Y. Zhao, "Strong authentication scheme for telecare medicine information systems," *Journal of Medical Systems*, vol. 36, no. 4, pp. 2609–2619, 2012.

[4] Z.-Y. Wu, Y.-C. Lee, F. Lai, H.-C. Lee, and Y. Chung, "A secure authentication scheme for telecare medicine information systems," *Journal of Medical Systems*, vol. 36, no. 3, pp. 1529–1535, 2012.

[5] D. B. He, J. H. Chen, and R. Zhang, "A more secure authentication scheme for telecaremedicine information systems," *Journal of Medical Systems*, vol. 36, no. 3, pp. 1989–1995, 2012.

[6] J. H. Wei, X. X. Hu, and W. F. Liu, "An improved authentication scheme for telecare medicine information systems," *Journal of Medical Systems*, vol. 36, no. 6, pp. 3597–3604, 2012.

[7] Z. A. Zhu, "An efficient authentication scheme for telecare medicine information systems," *Journal of Medical Systems*, vol. 36, no. 6, pp. 3833–3838, 2012.

[8] A. Shamir, "Identity-based cryptosystems and signature schemes," in *Proceedings of the Advances in Cryptology (CRYPTO '85)*, pp. 47–53, 1985.

[9] M. L. Das, A. Saxena, and V. P. Gulati, "A dynamic ID-based remote user authentication scheme," *IEEE Transactions on Consumer Electronics*, vol. 50, no. 2, pp. 629–631, 2004.

[10] H. Y. Chien and C. H. Chen, "A remote authentication scheme preserving user anonymity," in *Proceedings of the International Conference on AINA*, vol. 2, 2005.

[11] W. C. Ku and S. T. Chang, "Impersonation attack on a dynamic ID-based remote user authentication scheme using smart cards," *IEICE Transactions on Communications*, vol. E88-B, no. 5, pp. 2165–2167, 2005.

[12] H. Xiong, Z. Chen, and F. G. Li, "Provably secure and efficient certificateless authenticated tripartite key agreement protocol," *Mathematical and Computer Modelling*, vol. 55, no. 3-4, pp. 1213–1221, 2012.

[13] L. Zhang, F. Zhang, Q. Wu, and J. Domingo-Ferrer, "Simulatable certificateless two-party authenticated key agreement protocol," *Information Sciences*, vol. 180, no. 6, pp. 1020–1030, 2010.

[14] L. Chen, Z. Cheng, and N. P. Smart, "Identity-based key agreement protocols from pairings," *International Journal of Information Security*, vol. 6, no. 4, pp. 213–241, 2007.

[15] S. S. Al-Riyami and K. G. Paterson, "Certificateless public key cryptography," in *Proceedings of the Advances in Cryptology (ASIACRYRT '03)*, pp. 452–473, 2003.

[16] M. Girualt, "Self-certified public keys," in *Proceedings of the Advances in Cryptology (EUROCRYPTO '91)*, pp. 490–497.

[17] M. Scott, "Miracl library," http://certivox.com/.

[18] A. Miyaji, M. Nakabayashi, and S. Takano, "New explicit conditions of elliptic curve traces for FR-reduction," *IEICE Transactions on Fundamentals of Electronics, Communications and Computer Sciences*, vol. 84, no. 5, pp. 1234–1243, 2001.

[19] The Certicom Corporation, SEC2: Recommended elliptic curve domain parameters, 2000.

[20] K. Ren, W. Lou, K. Zeng, and P. J. Moran, "On broadcast authentication in wireless sensor networks," *IEEE Transactions on Wireless Communications*, vol. 6, no. 11, pp. 4136–4144, 2007.

6

Deadline-Aware Energy-Efficient Query Scheduling in Wireless Sensor Networks with Mobile Sink

Murat Karakaya

Department of Computer Engineering, Atilim University, Incek, 06836 Ankara, Turkey

Correspondence should be addressed to Murat Karakaya; kmkarakaya@gmail.com

Academic Editors: A. Bogliolo, C.-M. Chen, W. Ding, C.-H. Ko, A. Manikas, and D. Zheng

Mobile sinks are proposed to save sensor energy spent for multihop communication in transferring data to a base station (sink) in Wireless Sensor Networks. Due to relative low speed of mobile sinks, these approaches are mostly suitable for delay-tolerant applications. In this paper, we study the design of a query scheduling algorithm for query-based data gathering applications using mobile sinks. However, these kinds of applications are sensitive to delays due to specified query deadlines. Thus, the proposed scheduling algorithm aims to minimize the number of missed deadlines while keeping the level of energy consumption at the minimum.

1. Introduction

A Wireless Sensor Network (WSN) can be defined as a network of sensor nodes deployed to monitor a field with wireless communication capability and base stations (sinks) to gather information from sensors for uploading to a remote central. Usually sensor nodes are powered by unrechargeable batteries. When a sensor depletes its battery, it becomes nonfunctional which can affect the connectivity and correctness of WSN. Therefore, energy consumption is a crucial factor affecting the life time of a WSN.

Various energy conservation approaches have been proposed and implemented so far to maximize network lifetime as surveyed in [1]. One common way to decrease the energy consumption in the communications between sensors and static sink (SS) is using multihop forwarding instead of direct connection. Multihop communication has the advantage of using low power in transmitting data to a nearby sensor. However, one of the main problems of applying Multihop paradigm in WSN is that the sensors around SS deplete their energy very fast due to high forwarding data traffic. When that occurs, the sink becomes unreachable and WSN is nonfunctional. As an alternative to Multihop communication, researchers have proposed to mobilize sinks to collect data from sensors [2]. These mobile sinks (MSs) are capable of moving in the monitored field and contacting sensors using either one hop (direct) or limited number of hops communication methods. As a result, the required energy for transferring data from sensors to sink is reduced considerably and the life time of WSN is extended significantly [3]. Unfortunately, this method suffers from relatively high delay time in uploading data from sensors due to lower speed of MS compared to speed of radio communication. Therefore, this approach is suitable for delay tolerant applications and not a good option for real or near real time applications such as location-based querying or target tracking [4].

In this paper, we present a novel way of using MS for a near real time application, namely, query-based data gathering with deadlines, by trading off delay in response with energy consumption in Multihop communication. In this class of applications, location-based queries are submitted to WSN, and responses should be collected before the specified deadline expires. For this reason, we design a query scheduling algorithm to exploit MS deterministic mobility for saving energy in communication and to exploit speed of Multihop communication for minimizing delay caused by slow MS motion, whenever any of them is feasible. Thus, our algorithm balances the system throughput and energy consumption by optimizing the number of hops and duration of response time. The algorithm is very simple yet successful and applicable to the situations where either controlling MS moves is not possible (e.g., geographic conditions) or feasible (e.g., attached to a public transport vehicle).

The paper is organized as follows: related work is presented in Section 2. Section 3 provides the details of WSN model and the proposed query scheduling algorithm. Simulation model and results of simulation tests are discussed in Section 4. Finally, in Section 5 conclusion and future work direction are presented.

2. Related Work

The possible ways of initiating data transfer from sensors to a sink can be categorized into four classes: event-driven, time-driven, query-based, and hybrid [5]. In event-driven data collection approach, data needs to be collected whenever a specific event is detected and then forwarded to a sink (base station). Similarly, query-based data collection is triggered upon receiving a user's query, and response should be routed to a sink. On the other hand, in a time-driven approach, data collected by sensors is transferred to a sink periodically. As a last option, one can combine the above approaches to create hybrid approaches.

For any data collection approach, sensor readings should be transferred to some sink which can be classified as static sink (SS) and mobile sink (MS) according to mobility. Sink mobility affects the communication pattern between sensor and sink [6]. When static sink is used, sensors mostly depend on Multihop communication to forward their data to the sink. However, if sink can move, it can visit sensors and collect data from them directly.

MS mobility patterns can be listed as controlled, random, or deterministic [6]. In controlled mobility, MS next stop can be decided dynamically by an algorithm depending on some network or performance parameters. However, MS can follow some paths constructed by selecting next sensor randomly with a distribution probability in a random mobility model. In some other cases, MS moves deterministically on predefined or predictable paths (roads, railways, etc.) in regular time intervals.

In this work, we focus on query-based data collection in WSN using mobile sink with deterministic mobility pattern. Likewise, in [7, 8] the authors propose a query propagation and collection method minimizing energy and time using MS. They show that selecting the shortest path for delivering response packet is very important to minimize the total energy spending for the query. They calculate an optimum location on the path to submit a query such that query packet would arrive to target sensor, and then response would follow the shortest path to reach to the collect point just before MS arrives. Unfortunately, in the work they do not consider query deadlines, and their solution cannot be easily applied to handle the deadlines. Moreover, they assume that the target sensor location should always be ahead of MS moving direction. Therefore, if a target sensor is located on the other direction of the MS move, the proposed method cannot calculate any collection point at all. In reality, target sensor can be located at any place in the monitored area. In our solution, we propose a query scheduling algorithm to

minimize the energy consumption and meet query deadlines for more realistic and generic scenario.

In another similar work, authors model a WSN in which sensors store collected data in their finite memory [9]. MS collects data from these sensors via one-hop communication pattern and frees their memory. If MS is late and sensor memory gets full, the sensor removes all the readings from memory and restarts to collect data. The task is to schedule MS to visit each sensor before the sensor memory overflows with the collected data. Thus, there is a deadline for visiting each sensor. In their solution, assuming that the initial capacity of each sensor memory and sensor sampling rates are known, the authors attempt to create an MS route such that MS would be able to visit each sensor before an overflow occurs. In our case, we aim to create energy-optimized paths for packets, while the queries to sensors are random. For MS mobility model, they assume that MS has a controlled mobility, while we assume it to be a deterministic mobility.

There are other query based data collection approaches which mainly focus on designing a high level interface for query—response interactions between application and sensors. Unfortunately, these approaches do not work on details of underlying network topology, communication requirements, and energy consumption issues [5].

3. Deadline-Aware Energy-Efficient Query Scheduling Algorithm

The *Deadline-Aware Energy-Efficient Query Scheduling Algorithm* (DES) has twofold objective: providing responses before the given deadlines and consuming minimum possible amount of sensor energy in the Multihop communication. Below we first present the WSN model and the problem and then provide the proposed solution details.

3.1. WSN Model and Problem Definition. We assume that a WSN has been deployed for monitoring some environmental changes such as heat and mobility, as shown in Figure 1. WSN has three important components: sensor nodes, mobile sink, and Multihop routing protocol. In this model, each sensor node has its own limited memory to store the readings, and the location of the sensors is known. Underlying Multihop routing protocol can route messages from a source node to a destination node via a shortest path. Mobile sink consistently moves back and forth with a fixed velocity on a predefined route between two *Route Ends* (RE). Actually, MS could be embedded onto a regularly moving object such as public transportation vehicle (bus, train, or ship).

Any sensor node which is onehop away from the route can directly communicate with passing by MS. These sensors are called *Gateway Sensors* (GSs). Similarly, MS can forward a query to a target sensor by submitting it to any GS around itself. Via direct connection WSN central management authority can send queries to MS and demand responses to be uploaded. In these queries the *Target Sensor* (TS) and the deadline of the response are specified. Query deadline is the time before which the query should be executed and the response should be delivered to the remote central.

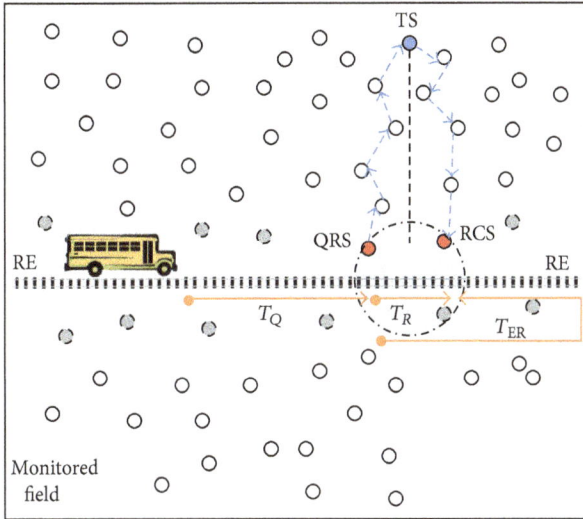

○ Sensor
◔ Gateway sensor

FIGURE 1: Assumed WSN model.

QRS	RCS	TS	Deadline	Query

RCS	Deadline	Response

FIGURE 2: Query and response message content.

The queries with different deadlines arrive dynamically as MS keeps moving on the path.

Whenever the remote central initiates a query, MS attempts to create a schedule such that the communication cost (energy consumption) for forwarding messages would be minimum and response would be uploaded before the given deadline (the algorithm can be run by remote central or MS. In this work we assume that MS runs the algorithm whenever it receives a query). A schedule is composed of *Query Release Sensor* (QRS) to inject a query into the WSN and *Response Collect Sensor* (RCS) to collect the response.

While preparing schedule, MS keeps moving and releases queries to related QRS when it is in one-hop proximity of QRS. This trip, from current MS location when it received the query to QRS location, is named *Query Release Trip* (T_Q). WSN communication protocol takes care of forwarding the query message from QRS to *Target Sensor* (TS). TS processes the query, prepares the response message, and forwards it to RCS (query and response message contents are given in Figure 2). Intermediate sensors in the routing path drop any messages whose deadline are already expired.

As MS is passing by RCS, it attempts to collect the response message from the sensor. The trip from QRS to RCS is called *Response Collect Trip* (T_R). If the response message has already arrived to RCS, MS can collect and upload it to remote central. On the other hand, if the response message has not arrived at RCS yet, MS takes *an Extended Response Collect Trip* (T_{ER}) by moving to the route end (RE) and comes

back to RCS once more. Thus, MS tries to collect the response message from RCS for the second time.

To meet the deadline in the WSN model given above, we need to ensure that MS reaches RCS before deadline and the response message reaches RCS before MS. Similarly to minimize the energy consumption in transferring query and response packages, we have to select QRS and RCS from GS such that they have the shortest path (minimum number of hops away) from the target sensor. However, these two conditions could not be always satisfied. Thus, the main scheduling problem is to find QRS and RCS such that MS can obtain the response before the deadline and routing costs the minimum energy consumption. Below we provide the details of Deadline-Aware Energy-Efficient Query Scheduling (DES).

3.2. Solution. DES first attempts to construct *Least Energy Consuming Schedule* (LECS) using the shortest paths between QRS-TS and TS-RCS. If it is not possible due to deadline or message routing, it can attempt to create either *Optimum Energy Consuming Schedule* (OECS) keeping TS-RCS path shortest but extending QRS-TS path or *Maximum Energy Consuming Schedule* (MECS) modifying both TS-RCS and QRS-TS paths to gain more time. When none of the routing paths are feasible, DES drops the request immediately.

As summarized in Algorithms 1, 2, and 3, DES considers MS current location (L_{MS}), movement direction (Dir_{MS}), and speed ($Speed_{MS}$) along with route ends location (L_{RE}), target sensor location (L_{TS}), query size ($Size_Q$), response size ($Size_R$), data transfer rate ($Rate_{Data}$), query processing time ($Proc_Q$), expected numbers of hops from QRS to destination (Hop_D), expected number of hops from destination to RCS (Hop_{RCS}), and query deadline ($Deadline_Q$) for creating feasible schedules.

The details of creating schedules are given below.

3.2.1. Constructing LECS. If deadline permits to create the shortest routing path between QRS to RCS via TS, the energy saving would be maximized since the number of hops are expected to be minimum as seen in Figure 3. Thus, LECS attempts to choose QRS and RCS among the gateway sensors which are nearest to TS. For this reason in Algorithm 1, the nearest point location (L_{NP}) on the MS route to L_{TS} is calculated, and one-hop away gateway sensors around it are candidates for being *QRS* or *RCS*. If there is only one candidate sensor it is selected as both QRS and RCS. However, if there are more than one, DES selects the sensors which MS will contact first as QRS and other sensor which MS will contact latest as RCS. The reason for this selection is to maximize the chances of receiving the response message before MS arrives at RCS.

After deciding QRS and RCS, LECS checks several parameters to see if the query deadline can be met with the current schedule. The duration of Query Release Trip Time (T_Q), Response Collect Trip Time (T_R), and Extended Response Collect Trip Time (T_{ER}) is computed as in (1), (2), and (3), respectively (MS movement direction is important factor for calculating T_Q (see Figure 3)). Moreover, *Expected Query Forwarding Time* (T_{EQF}), time required to forward

(1) Select QRS and RCS
(2) Calculate T_Q, T_R, T_{ER}, T_{EQF}, and T_{ERF}
(3) **if** $T_Q + T_R >$ Deadline$_Q$ **then**
(4) **failed**   ▷ MS is late
(5) 　run MECS algorithm
(6) **else**
(7) 　**if** $T_{EQF} + T_{ERF} <= T_R$ **then**
(8) 　　**successful** ▷ Schedule QRS and RCS
(9) 　**else**
(10) 　　**if** $(T_{EQF} + T_{ERF} <= T_{ER})$ and $(T_Q + T_{ER} <=$ Deadline$_Q)$ **then**
(11) 　　　**successful** 　▷ Schedule QRS and RCS
(12) 　　**else**
(13) 　　　**failed**   ▷ Response is late
(14) 　　　run OECS algorithm
(15) 　　**end if**
(16) 　**end if**
(17) **end if**

ALGORITHM 1: Calculate Least Energy Consuming Schedule.

Require: RCS
Ensure: LECS found: $T_Q + T_R <=$ Deadline$_Q$ and $T_{EQF} + T_{ERF} > T_R$
(1) select AQRS s.t. updated $T_{EQF} + T_{ERF} < T_R$ holds
(2) **if** AQRS is feasible **then**
(3) 　**successful** 　▷ Schedule AQRS and RCS
(4) **else**
(5) 　select AQRS s.t. updated $T_{EQF} + T_{ERF} < T_{ER}$
　　　and $T_Q + T_{ER} <=$ Deadline$_Q$ holds
(6) 　**if** AQRS is feasible **then**
(7) 　　**successful** 　▷ Schedule AQRS and RCS
(8) 　**else**
(9) 　　**failed**   ▷ Response is late
(10) 　　run MECS algorithm
(11) 　**end if**
(12) **end if**

ALGORITHM 2: Calculate Optimum Energy Consuming Schedule.

(1) Select ARCS s.t. Dis(ARCS, TS) is minimum
　　and updated $T_Q + T_R <$ Deadline$_Q$
(2) Select AQRS s.t. updated $T_{EQF} + T_{ERF} < T_R$
(3) **if** (ARCS and AQRS) is feasible **then**
(4) 　**successful** ▷ Schedule AQRS and ARCS
(5) **else**
(6) 　Select ARCS and AQRS s.t. Dis(ARCS, AQRS) is maximum,
　　updated $T_Q + T_R <$ Deadline$_Q$, and updated $T_{EQF} + T_{ERF} < T_R$
(7) 　**if** (ARCS and AQRS) is feasible **then**
(8) 　　**successful** 　▷ Schedule AQRS and ARCS
(9) 　**else**
(10) 　　**reject query** 　　▷ Deadline is short
(11) 　**end if**
(12) **end if**

ALGORITHM 3: Calculate Maximum Energy Consuming Schedule.

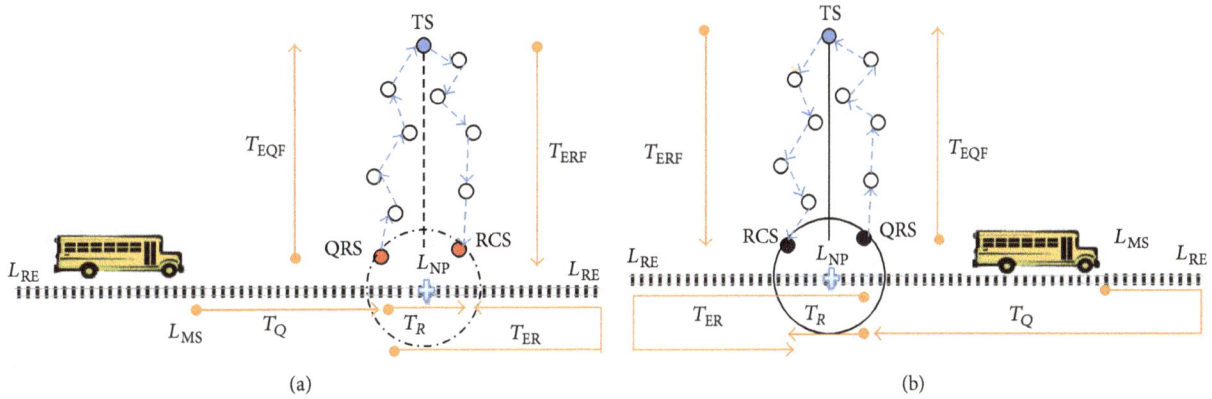

FIGURE 3: Least Energy Consuming Schedule when MS moves in the direction of TS (a) and when it moves in the other direction (b).

a query message from L_{QRS} to TS, and *Expected Query Processing and Response Forwarding Time* (T_{ERF}), time needed to process the query and forward the response message to RCS, are estimated as in (4) and (5), respectively. In these equations, we estimated the number of hops for given two locations using the sensor density, communication range, and the bandwidth.

After calculating all these parameters, LECS first checks if MS has enough time to reach RCS location before the deadline. If deadline allows, MS should ensure the time needed for query forwarding and processing, and response forwarding would be less than trip time to RCS. In some cases response message would be late to arrive at RCS, and as a second chance, we might allow MS to move up to RE and come back to RCS. When response can reach RCS before the extended response collect trip time, we should check if deadline does not still expire (one can suggest that even in an extended trip time it is not enough, and deadline does not expire; therefore, MS can execute another tour on the path back to RCS once more. Since the communication speed is much more than the MS speed, we ignore this case).

According to all these conditions we either schedule QRS and RCS successfully, or we call other scheduling algorithms to calculate alternative paths. If LECS algorithm fails to create a shortest routing path due to late arrival of response message to RCS, it calls OECS to select an alternative QRS (AQRS) such that MS can release query earlier and WSN would have more time to route the response message to RCS. On the other hand when LECS algorithm fails because of deadline expiration before MS finishes collect trip, MECS algorithm is called for choosing alternative QRS (AQRS) and alternative RCS (ARCS) such that MS and the response message would meet at ARCS before the deadline. Consider

$$
T_Q = \begin{cases} \dfrac{\text{Dis}\left(L_{MS}, L_{QRS}\right)}{\text{Speed}_{MS}} \\ \quad \text{if Dir}_{MS} \text{ is towards QRS} \\ \dfrac{\text{Dis}\left(L_{MS}, L_{RE}\right) + \text{Dis}\left(L_{RE}, L_{QRS}\right)}{\text{Speed}_{MS}} \\ \quad \text{if Dir}_{MS} \text{ is reverse direction of QRS,} \end{cases} \tag{1}
$$

$$
T_R = \frac{\text{Dis}\left(L_{QRS}, L_{RCS}\right)}{\text{Speed}_{MS}}, \tag{2}
$$

$$
T_{ER} = \frac{\text{Dis}\left(L_{QRS}, L_{RE}\right) + \text{Dis}\left(L_{RE}, L_{RCS}\right)}{\text{Speed}_{MS}}, \tag{3}
$$

$$
T_{EQF} = \frac{\text{Size}_Q}{\text{Rate}_{Data}} * \text{Hop}_D, \tag{4}
$$

$$
T_{ERF} = \frac{\text{Size}_R}{\text{Rate}_{Data}} * \text{Hop}_{RCS}. \tag{5}
$$

3.2.2. Constructing OECS. As seen in Algorithm 2, OECS algorithm searches an alternative routing path when response message is late to arrive at RCS before MS passes by even though deadline allows MS to reach RCS. Therefore, OECS algorithm relocates QRS such that query message forwarding would begin earlier than before and reply message can arrive to RCS on time with optimum energy consumption in Multihop communications (see Figure 4). Since the response data size is expected to be larger than that of query, to save more energy in routing messages, response forwarding path should be kept shortest in the first place. Given densely and uniformly deployed sensors, we may construct minimum energy consuming path for forwarding response messages by creating a shortest path between TS and RCS, as in LECS algorithm. As a result, optimum route can be constructed if we can construct a routing path such that query forwarding would take some more hops, but response forwarding takes still the least number of hops, and response would be ready for MS to collect [7, 8].

Using RCS selected by LECS algorithm, OECS algorithm calculates alternative QRS (AQRS) location between L_{MS} and L_{NP} such that total time required for T_{EQF} and T_{ERF} would be less or equal to response collect trip time (T_R). If such a location is feasible, AQRS is selected among gateway sensors one hop away from this location. Otherwise, we can select the nearest sensor from L_{MS} as AQRS and test if an extended response collect trip can be run before deadline, and result

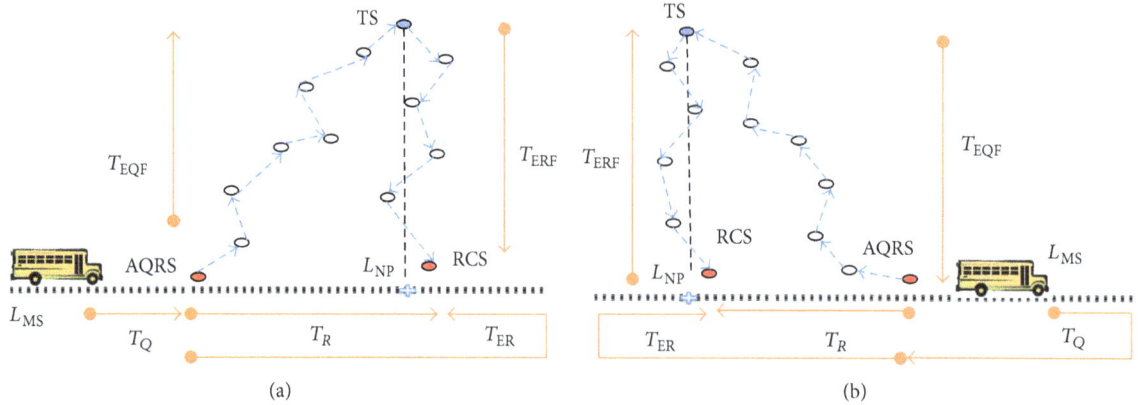

FIGURE 4: Optimum energy consuming schedule when MS moves in the direction of TS (a) and when it moves in the other direction (b).

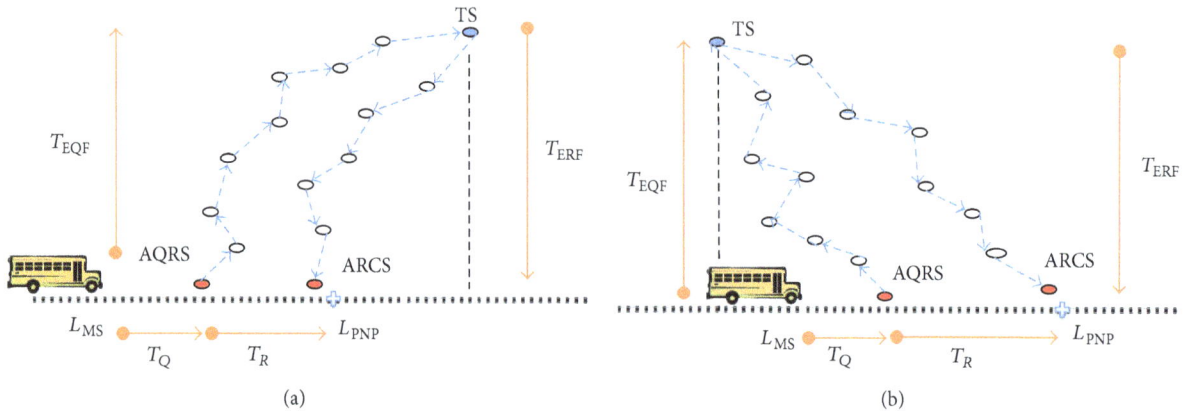

FIGURE 5: Most Energy Consuming Schedule when MS moves in the direction of TS (a) and when it moves in the other direction (b).

would be ready at RCS before MS reaches it. If this option fails as well, we call MECS to create an alternative schedule.

3.2.3. Constructing MECS. As discussed above, LECS attempts to use minimum energy by constructing shortest routing paths for query and response message delivery, whereas OECS consumes least energy for response messages but more energy for delivering query to gain time. Whenever these two algorithms fail to create a feasible solution, as a last resort, they call MECS algorithm. The routing path constructed by MECS algorithm costs more energy to gain time by attempting to select an alternative QRS (AQRS) as well as an alternative RCS (ARCS) in the hope that response message can be reachable by MS before the deadline.

As shown in Figure 5, MECS first decides the possible nearest point (L_{PNP}) on the route to TS where MS can get before deadline finishes. The sensors in the vicinity of L_{PNP} are candidates for alternative RCS (ARCS). Then alternative QRS (AQRS) location is calculated such that response message arrives to ARCS before MS.

Considering deadline if such AQRS location is not available, MECS algorithm recalculates AQRS and ARCS locations such that the distance between these two locations would be the largest. We hope that while MS moves from

AQRS to ARCS, the query and reply message can be forwarded up to ARCS before deadline. If MECS algorithm again fails to find such AQRS and ARCS locations, the query will be rejected.

4. Simulation Model and Results

This section presents the evaluation of Energy-Efficient Deadline-Aware Scheduling with respect to several performance metrics and two other scheduling methods.

4.1. Simulation Model. Table 1 summarizes important simulation parameters and their default values which generally apply to those used in similar studies such as [7, 8, 10]. Below we discuss each simulation parameter in detail.

WSN. We assume a realistic deployment area of 1000×500 meters. We consider random topology in which sensors are deployed randomly on the monitored field. MS moves on a fixed route located in the middle of monitored field from east to west.

Sensors. The number of sensors is 1500 as default. The radio range of sensors is set to be 50 meters with a data rate of

TABLE 1: Simulation parameters and default values.

Parameters	Definition	Default setting
W	Width of monitored field	1000 m
H	Height of monitored field	500 m
N	Number of sensors	1771
T	Sensor topology	Grid
ST	Simulation time	54000 s (15 h)
RT	MS route location	Horizontal center
RR	Radio range	50 m
DR	Data transfer rate	256 Kb/s
EC	Data transfer cost	50 nJ/bit
BP	Initial battery power	0.01 J
QP	Query processing time	0.1 s
PC	Query processing cost	100 nJ
MS	MS speed	40 km/h
QS	Query size	32 Byte
RS	Response size	256 Byte
QA	Query arrival rate	Exp. (mean = 30 s)
DL	Deadline	min. 17 s/max. 74 s

256 Kb/s. We measure the energy consumed at the sensor network as an absolute value in Joules. Each sensor begins with a full battery of 0.1 joule and dissipates 50 nJ per bit of battery energy for transferring data packages. When the sensor is idle the energy consumption is assumed to be 40 nJ/s. Query processing takes 0.1 seconds and costs 100 nJ.

Simulation. Each simulation runs 15 simulated hours, and to find the results of observed performance metrics, every set of experiments is executed for ten times.

MS. MS speed is fixed at 40 km/h.

Query and Response. Size of query and response data packet is assumed to be 32 byte and 256 byte, respectively. Query arrival distribution follows an exponential distribution with a mean of 30 seconds. We have defined two different deadline values for queries: *shorter* and *longer*. *Shorter deadline value* is set as expected minimum data transfer time for the distance equals to half of monitored field height. Similarly *longer deadline value* is chosen as expected minimum data transfer time from one route end to sensors located at the furthest point of the monitored field. For the current setting of simulation experiments, these values are calculated about 17 and 74 seconds for shorter and longer deadlines, respectively. Target sensors of queries are selected randomly.

4.2. Performance Metrics. Performance metrics are as follows.

(i) *Generated Query*: Number of queries created by remote central during the simulation run time.

(ii) *Submitted Query*: Number of queries submitted to WSN by MS.

(iii) *Rejected Query*: Number of queries which could not be submitted to WSN due to some reason (short

deadline, nonexisting connection to target sensor, empty battery, etc.).

(iv) *Received Query*: Number of queries whose response arrives at MS on time.

(v) *Missed Deadline*: Number of queries whose response could not get to MS before the deadline exceeds.

(vi) *Successful Query Ratio*: Ratio of Received Query to Generated Query.

(vii) *Network Life Time*: Duration between the time that simulation begins and the time that any sensor's battery power gets lower than a specified level. The levels are sliced as 10% of the starting battery capacity.

(viii) *Average Energy Consumption Per Submitted Query*: Average amount of energy consumed for forwarding query and response messages in WSN during simulation time.

For a given query generation rate distribution, any scheduling algorithm is subject to similar number of queries. However, the number of submitted query will depend on different parameters such as the current network connections, sink position, battery power of sensors, and scheduling algorithm. Ideally all generated queries should be submitted to WSN. On the other hand, to save sensor energy, scheduling algorithms can reject submitting queries whose response would not arrive on time. If the implemented scheduling algorithm is successful in selecting these kinds of queries, WSN network life would extend. Otherwise, if it fails in prediction, then system throughput will be decreased considerably. Contrary to the prediction of late response arrival, a scheduling algorithm could submit all the generated queries hoping that the responses would arrive on time.

Thus, to compare success of our algorithm, *MS with Deadline-Aware Energy-Efficient Scheduling* (MS/DES), we implemented two other data collecting approaches using *Static Sink* (SS) and *MS with Immediate Scheduling* (MS/IS). In SS approach sink is fixed and located in the middle of the monitored field to submit queries and collect responses. The other approach, MS/IS, also uses a mobile sink but with a simplified scheduling algorithm. In MS/IS, query messages are immediately forwarded to one of the gateway sensors expecting to collect responses from some gateway sensor which has the shortest path from the target sensor. Thus, MS/IS only focuses on minimizing energy consumption of response forwarding without concerning about energy consumption of query forwarding and satisfying deadline. In the simulation experiments, all parameters are set the same for these three approaches.

4.3. Base Experiment Results. For the default values of parameters given in Section 4.1 with the minimum deadline value (17 sec.), we obtained the following results presented in Figures 6, 7, and 8.

In Figure 6, the effects of differences in scheduling and sink model on the results are observed. Since MS/IS does not refuse any query and submits them immediately, the numbers of missed deadlines are very high. On the other

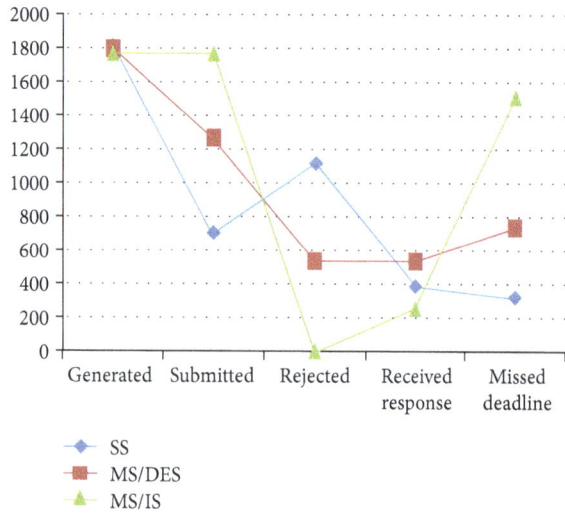

FIGURE 6: Results of various performance metrics related with query and response numbers.

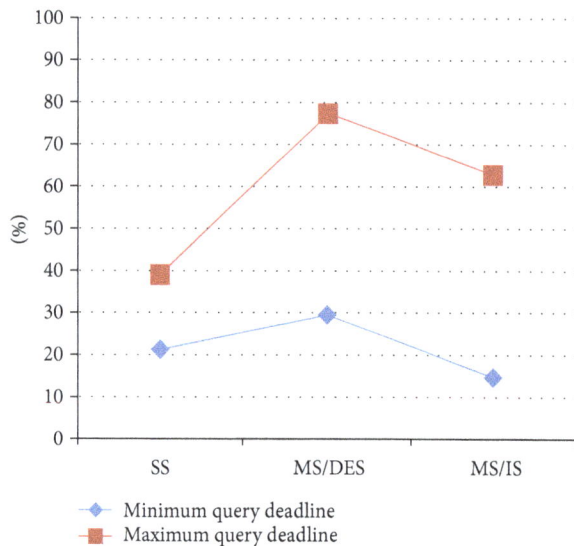

FIGURE 7: Results of Successful Query Ratio for different query deadline values.

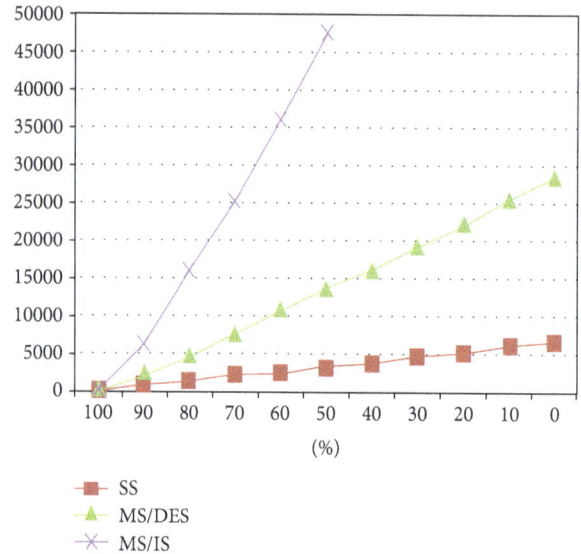

FIGURE 8: Results of Network Life Time performance metric for different battery levels.

hand, MS/DES rejects submitting some queries because it calculates that they would not arrive on time; therefore, MS/DES leads to a less number of missed deadlines compared to MS/IS. SS attempts to submit every generated query, but when the sensors around the SS deplete their batteries, SS can not submit incoming queries any longer. MS/DES can submit more queries than SS and receive more responses on time than that of the others.

Figure 7 depicts the results of Successful Query Ratio metric for different query deadline values. Thanks to its prediction and rejection of queries whose responses would be possibly late, MS/DES produces the highest Successful Query Ratio for different query deadline values as well. Interestingly, SS and MS/IS have reacted to the changes in deadline contrastingly; SS functions better with less deadline

values, whereas MS/IS produces better results with larger query deadlines. The reason behind these observation is that MS/IS does not have enough time to reach RCS with less query deadlines.

Figure 8 presents one perspective of the energy consumptions of the above mentioned methods: Network Life Time. In the figure, y-axis shows the simulation time in seconds, and x-axis denotes the ratio of remaining battery power to initial battery power. Time is recorded when the first sensor battery decreases to the specified battery power. SS depletes the sensors energy fast due to the nature of broadcasting via one-hop away neighbors. The sensors around the SS are the first sensors to deplete their battery since all the forwarded messages between WSN and SS pass over them. MS/IS uses sensor battery power efficiently since the responses with larger data size always follow the shortest path to the MS/IS path which minimizes the forwarding energy requirement. Contrary to MS/IS, MS/DES scarifies the battery energy usage for satisfying query deadline whenever it is required. Therefore, MS/DES consumes more energy to provide a better Successful Query Ratio as desired.

As a second parameter to calculating energy spending of the data collection methods, we present the Average Energy Consumption Per Submitted Query in Figure 9. As depicted in the figure, SS and MS/IS almost spend the same amount of energy per query even when the deadline differs. However, MS/DES can decrease the energy consumption when query deadline is longer by adjusting the query and response path to minimize the number of hops. Thus, MS/DES is adaptive to conditions to balance the Successful Query Ratio and energy consumption.

4.4. Experiment Results for a Larger Deadline Value. For a larger value of deadline (74 sec.), we obtained the following results presented in Figure 10. The effect of having more time

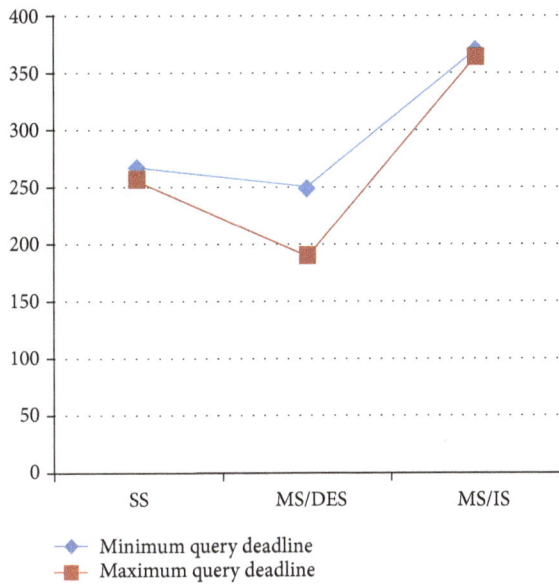

FIGURE 9: Results of Average Energy Consumption Per Submitted Query for different query deadline values.

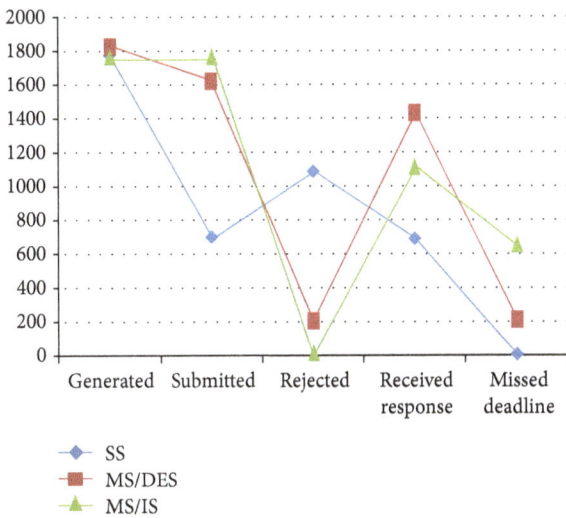

FIGURE 10: Results of various performance metrics related with query and response numbers for a larger deadline.

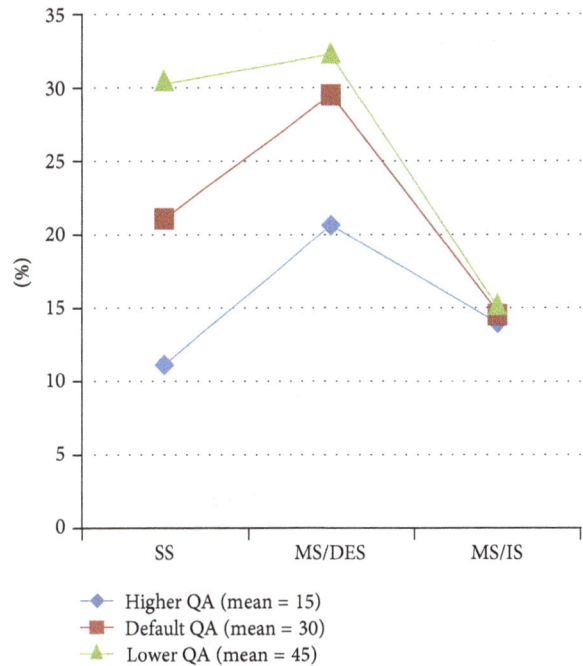

FIGURE 11: Results of Successful Query Ratio for different query loads.

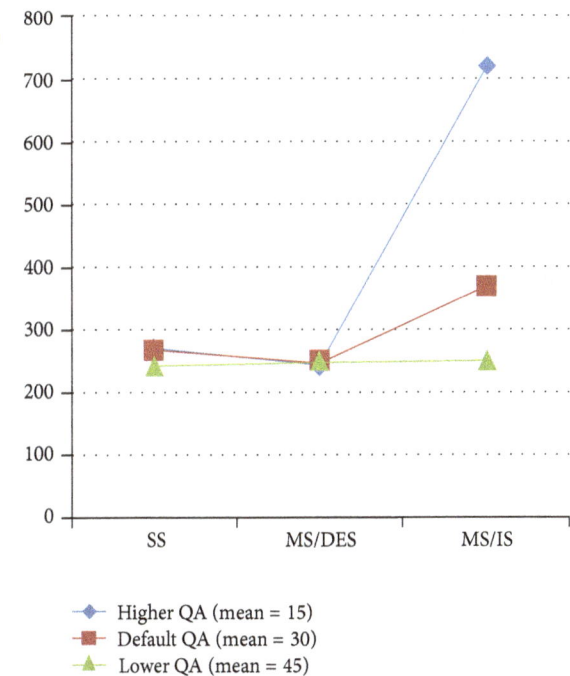

FIGURE 12: Results of Average Energy Consumption Per Submitted Query for different query loads.

to collect response messages can be viewed in the increase of number of received responses and in the decrease of missed deadlines for all three data collection methods. MS/DES provides the best result for received response parameter. Furthermore, we can notice that MS/DES now rejects less number of queries which proves that MS/DES properly decides possible response arrive time. For the Successful Query Ratio given in Figure 7, MS/DES performs twice better than that of SS and 1.2 times better than that of MS/IS.

4.5. Experiment Results for Different Query Loads. Query Arrival Rate (QA) follows exponential distribution with a mean value 30 as default. To simulate different query loads

on WSN, we change the QA mean value to 15 and 45. The results of Successful Query Ration and Average Energy Consumption Per Submitted Query performance metrics for different query load values are presented in Figures 11 and 12, respectively.

Since MS/IS just submits any generated query immediately expecting to collect the response whenever MS/IS passes by the RCS, MS/IS does not react to changes in QA when the Successful Query Ration metric is observed. However, SS produces poor success query ratio when QA is high, because, for a fixed simulation time, with a higher query arrival rate, SS needs to submit more queries which leads to early battery depletion of surrounding sensors. As a result, all forthcoming queries are to be rejected. With lower QA, SS can submit more of them via its surrounding sensors before their batteries get empty. MS/DES presents better results for Successful Query Ration metric with lower QA. However, MS/DES outperforms the other approaches for all different query arrival rates.

For Average Energy Consumption Per Submitted Query performance metric, MS/IS consumes different amount of energy, whereas MS/DES and SS consume similar amount of energy when a different level of QA is applied. Since MS/IS immediately submits incoming queries to the nearest sensor and expects responses to arrive to the route via the shortest path, the difference in Energy Consumption Per Submitted Query occurs only in the query path. When there are more queries in a higher QA, the differences are increased.

On the other hand, since SS is located at the center of WSN and query and response paths are the same mostly, Average Energy Consumption Per Submitted Query is not affected by the different query load. For MS, scheduling of queries is based on the algorithm which aims to minimize the energy consumption for each query, MS/DES can sustain the same level of energy spending successfully. Thus, even MS/DES is not fixed at a location, it could accomplish similar level of success in Average Energy Consumption Per Submitted Query performance metric as SS.

5. Conclusions

This paper introduced a scheduling algorithm for the location-based queries in WSN with a mobile sink following a deterministic mobility pattern. In WSN, Multihop communication pattern is used to disseminate the queries and the responses. The queries have associated with deadlines. The proposed scheduling algorithm aims at maximizing the number of successful queries and reducing the sensors' energy expenditure due to Multihop communication by exploiting deterministic sink mobility. For this reason, before submitting queries, the scheduling algorithm selects the release and collect sensors such that two important performance requirements can be met: the energy required to forward data packages should be minimum, and the response arrival time should not exceed the specified deadline.

We also simulated two data collection methods for the sake of comparison, namely, SS and MS/IS. We conducted extensive simulation tests, and the obtained results show that our scheduling algorithm can attain more successful queries with less amount of energy even when query load and deadline change.

As a future work, we would like to extend and adapt the algorithm to different mobility models other than linear route. We plan to apply some heuristics such as Ant Colony

Optimization techniques to decide minimum energy consuming paths while MS decides its own route.

References

[1] G. Anastasi, M. Conti, M. Di Francesco, and A. Passarella, "Energy conservation in wireless sensor networks: a survey," *Ad Hoc Networks*, vol. 7, no. 3, pp. 537–568, 2009.

[2] J. Luo, J. Panchard, M. Piórkowski, M. Grossglauser, and J. P. Hubaux, "MobiRoute: routing towards a mobile sink for improving lifetime in sensor networks," in *Sensor Networks in the 2nd IEEE/ACM DCOSS*, pp. 480–497, Springer, 2006.

[3] M. I. Khan, W. N. Gansterer, and G. Haring, "Static vs. mobile sink: the influence of basic parameters on energy efficiency in wireless sensor networks," *Computer Communications*, vol. 36, no. 9, pp. 965–978, 2013.

[4] Y. Yun and Y. Xia, "Maximizing the lifetime of wireless sensor networks with mobile sink in delay-tolerant applications," *IEEE Transactions on Mobile Computing*, vol. 9, no. 9, pp. 1308–1318, 2010.

[5] T. A. A. Alsbou, M. Hammoudeh, Z. Bandar, and A. Nisbet, "An overview and classification of approaches to information extraction in wireless sensor networks," in *Proceedings of the 5th International Conference on Sensor Technologies and Applications (SENSORCOMM '11)*, p. 255, IARIA, 2011.

[6] M. Di Francesco, S. K. Das, and G. Anastasi, "Data collection in wireless sensor networks with mobile elements: a survey," *ACM Transactions on Sensor Networks*, vol. 8, no. 1, pp. 7:1–7:31, 2011.

[7] Y. Chen, L. Cheng, C. Chen, and J. Ma, "Meeting position aware routing for query-based mobile enabled wireless sensor network," in *Proceedings of the IEEE 70th Vehicular Technology Conference Fall (VTC '09)*, pp. 1–5, September 2009.

[8] L. Cheng, Y. Chen, C. Chen, and J. Ma, "Query-based data collection in wireless sensor networks with mobile sinks," in *Proceedings of the ACM International Wireless Communications and Mobile Computing Conference (IWCMC '09)*, pp. 1157–1162, ACM Press, New York, NY , USA, June 2009.

[9] A. A. Somasundara, A. Ramamoorthy, and M. B. Srivastava, "Mobile element scheduling with dynamic deadlines," *IEEE Transactions on Mobile Computing*, vol. 6, no. 4, pp. 395–410, 2007.

[10] R. Li, C. Zheng, and Y. Zhang, "Study of power-aware routing protocal in wireless sensor networks," in *Proceedings of the International Conference on Electrical and Control Engineering (ICECE '11)*, pp. 3173–3176, 2011.

7

Cascadable Current-Mode First-Order All-Pass Filter Based on Minimal Components

Jitendra Mohan[1] and Sudhanshu Maheshwari[2]

[1] Department of Electronics & Communication Engineering, Jaypee Institute of Information Technology, Noida 201304, India
[2] Department of Electronics Engineering, Z.H. College of Engineering and Technology, Aligarh Muslim University, Aligarh 202002, India

Correspondence should be addressed to Jitendra Mohan; jitendramv2000@rediffmail.com

Academic Editors: D. S. Budimir and K. Dejhan

A novel current-mode first-order all-pass filter with low input and high output impedance feature is presented. The circuit realization employs a single dual-X-second-generation current conveyor, one grounded capacitor, and one grounded resistor, which is a minimum component realization. The theoretical results are verified using PSPICE simulation program with TSMC 0.35 μm CMOS process parameters.

1. Introduction

Current-mode circuit design using current conveyor has received a considerable attention owning to its potential advantages such as wider dynamic range, greater linearity, wide bandwidth, simple circuitry, and low power consumption [1]. Considering these advantages of current conveyor, recently several current mode first-order all-pass filters employing different types of current conveyor such as second-generation current conveyor [2–4], four terminal floating nullor [5], third-generation current conveyor [6], differential voltage current conveyor [7, 8], current differencing buffered amplifier [9], current operational amplifier [10], and dual-X second-generation current conveyor [11, 12] have been reported. These reported filters reveal some useful features depending on the individual topology as summarized in Table 1. The comparison between the proposed circuit and the previously reported circuits is based on the use of number of active elements, number of grounded passive components, and low input and high output impedance feature(s). In general, the input impedance should be lower in comparison to the output impedance to avoid loading problem while cascading such current-mode circuits to form larger system.

In this paper, a novel cascadable current-mode (CM) first-order all-pass filter is proposed. The circuit uses a

dual-X second generation multioutput current conveyor (DX-MOCCII), a grounded resistor, and a grounded capacitor, which is ideal for IC implementation. The circuit offers low-input impedance and high-output impedance feature and also free from matching constraints. Nonideal gain and parasitic effects of the DX-MOCCII on the transfer function of the proposed filter are also analysed.

2. The Proposed Circuit

Dual-X second-generation current conveyor [13] is a useful and versatile active element, which has found several applications in analog signal processing [14–18]. The DX-MOCCII symbol is shown in Figure 1 and is characterized by the following port relationships:

$$\begin{bmatrix} I_Y \\ V_{X+} \\ V_{X-} \\ I_{Z1+} \\ I_{Z2+} \\ I_{Z1-} \\ I_{Z2-} \end{bmatrix} = \begin{bmatrix} 0 & 0 & 0 \\ 1 & 0 & 0 \\ -1 & 0 & 0 \\ 0 & 1 & 0 \\ 0 & -1 & 0 \\ 0 & 0 & 1 \\ 0 & 0 & 1 \end{bmatrix} \begin{bmatrix} V_Y \\ I_{X+} \\ I_{X-} \end{bmatrix}, \qquad (1)$$

where the suffixes refer to the respective terminals. The active element is characterized by high input impedance at

TABLE 1: Comparison of various current-mode all-pass filters.

References	No. of active elements	Single active element	No. of resistors and capacitors	All-grounded passive elements	Low input impedance	High output impedance	Component matching constraint
Higashimura and Fukui [2]	1-CCII	Yes	4	No	No	Yes	Yes
Higashimura [5]	1-FTFN	Yes	3	No	No	No	Yes
Toker et al. [9]	1-CDBA	Yes	2	No	No	Yes	No
Maheshwari and Khan [6]	1-CCIII	Yes	2	No	No	No	No
Kilinç and Çam [10]	1-COA	Yes	2	No	No	Yes	No
Minaei and Ibrahim [7]	1-DVCC	Yes	3	No	No	Yes	Yes
Khan et al. [3]	2-MOCCII	No	2	Yes	No	Yes	No
Maheshwari [8]	1-DVCC	Yes	2	Yes	No	Yes	No
Minaei and Yuce [4]	2-DOCCII	No	2	Yes	Yes	Yes	No
Minaei and Yuce [11]	1-DXCCII	Yes	4	Yes	No	Yes	Yes
Beg et al. [12]	1-DX-MOCCII	Yes	4	Yes	No	Yes	Yes
Proposed Circuit	1-DX-MOCCII	Yes	2	Yes	Yes	Yes	No

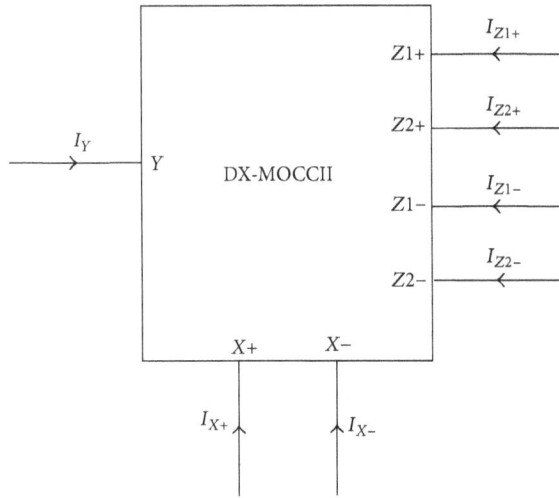

FIGURE 1: Symbol of DX-MOCCII.

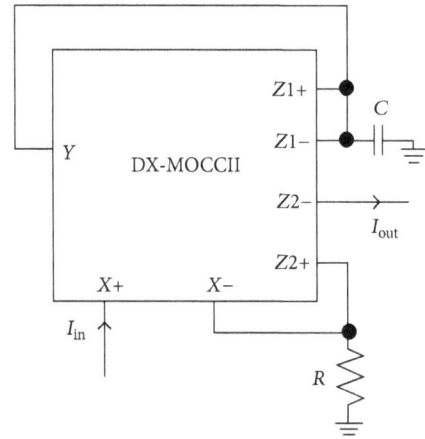

FIGURE 2: Proposed cascadable current-mode first-order all-pass filter.

the Y terminal, high output impedance at the $Z1+$, $Z2+$, $Z1-$, and $Z2-$ terminals, and low impedance at the $X+$ and $X-$ terminals.

The proposed current-mode (CM) first-order all-pass filter employing a DX-MOCCII, a grounded capacitor, and a grounded resistor is shown in Figure 2. Routine analysis of the circuit, using (1), yields the following transfer function:

$$\frac{I_{\text{out}}}{I_{\text{in}}} = -\left(\frac{s - (1/CR)}{s + (1/CR)}\right). \tag{2}$$

The frequency-dependent phase response of (2) is

$$\Phi = -2\tan^{-1}(\omega RC). \tag{3}$$

From (3), it can be seen that the proposed circuit can provide a phase shift between $0°$ and $-180°$ at output terminal (I_{out}).

The salient features of the proposed circuit are the use of single active element, two grounded passive components, and providing low input and high output impedance. As all the passive components used are in grounded form, it is suitable for integrated circuit implementation and also reduces the associated parasitic effects [19].

By interchanging the resistor (R) with a capacitor (C) in Figure 2, an additional circuit can be derived from the proposed circuit. However, the use of capacitor at the $X-$ terminal degrades the high frequency operation [20].

3. Nonideal Analysis and Parasitic Effects

3.1. Non-Ideal Analysis. Taking the nonidealities of the DX-MOCCII into account, the port relationship of the voltage and current terminals of the active element can be rewritten as

$$\begin{bmatrix} I_Y \\ V_{X+} \\ V_{X-} \\ I_{Z1+} \\ I_{Z2+} \\ I_{Z1-} \\ I_{Z2-} \end{bmatrix} = \begin{bmatrix} 0 & 0 & 0 \\ \beta_1 & 0 & 0 \\ -\beta_2 & 0 & 0 \\ 0 & \alpha_1 & 0 \\ 0 & -\alpha_2 & 0 \\ 0 & 0 & \alpha_3 \\ 0 & 0 & \alpha_4 \end{bmatrix} \begin{bmatrix} V_Y \\ I_{X+} \\ I_{X-} \end{bmatrix}. \tag{4}$$

Here, α_1 and α_2 are the current transfer gains from $X+$ terminal to $Z1+$ and $Z2+$ terminals, α_3 and α_4 are the current transfer gains from $X-$ terminal to $Z1-$ and $Z2-$ terminals,

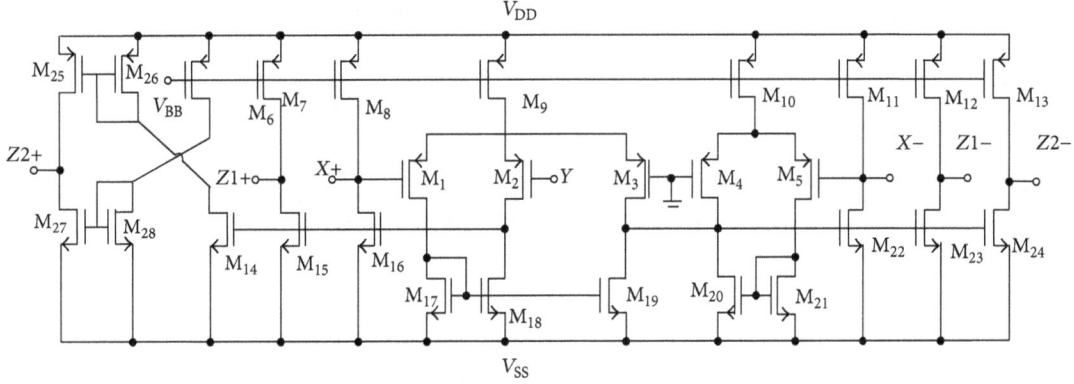

FIGURE 3: CMOS implementation of DX-MOCCII.

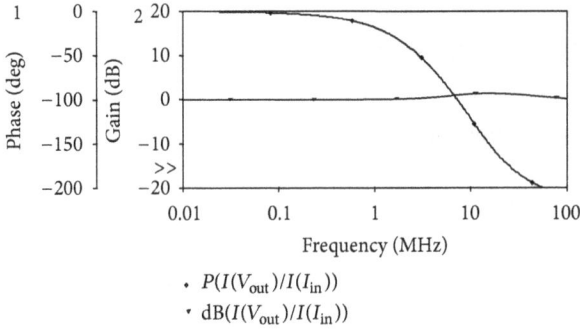

- $P(I(V_{\text{out}})/I(I_{\text{in}}))$
- $\text{dB}(I(V_{\text{out}})/I(I_{\text{in}}))$

FIGURE 4: Simulated gain and phase responses of the all-pass filter.

respectively, and β_1 and β_2 are the voltage transfer gains from input to $X+$ and $X-$ terminals, respectively. More specifically, $\alpha_1 = (1 - \varepsilon_1)$, $\alpha_2 = (1 - \varepsilon_2)$, $\alpha_3 = (1 - \varepsilon_3)$, $\alpha_4 = (1 - \varepsilon_4)$, $\beta_1 = (1 - \delta_1)$, and $\beta_2 = (1 - \delta_2)$, where ε is the current transfer error (tracking error) and δ is the voltage transfer error (tracking error) of the DX-MOCCII. However, these transfer gains differ from unity by the voltage and current tracking errors of the DX-MOCCII.

The proposed circuit is reanalyzed by taking the tracking errors of the nonideal MO-DXCCII into account, and the modified current transfer function is given as

$$\frac{I_{\text{out}}}{I_{\text{in}}} = -\alpha_2 \alpha_4 \left(\frac{s - (\beta_2 \alpha_1 / CR\alpha_2)}{s + (\beta_2 \alpha_3 / CR)} \right). \tag{5}$$

Equation (5) reveals that the nonidealities do affect the filter gain and the pole frequency as well as the zero frequency. Assuming matched current transfer gains (α_1 and α_3) the phase characteristics would not be affected. The sensitivities of pole frequency (ω_o) and gain (H) with respect to active and passive components are derived from (5). These are as follows:

$$S_{C,R}^{\omega_o} = -1, \qquad S_{\alpha_3,\beta_2}^{\omega_o} = 1, \qquad S_{\alpha_1,\alpha_2,\alpha_4,\beta_1}^{\omega_o} = 0,$$
$$S_{C,R}^{H} = 0, \qquad S_{\alpha_1,\alpha_3,\beta_1,\beta_2}^{H} = 0, \qquad S_{\alpha_2,\alpha_4}^{H} = 1. \tag{6}$$

From the results, it is evident that the sensitivities are within unity in magnitude, thus ensuring a low sensitivity performance.

TABLE 2: Aspect ratios of the transistors.

Transistors	W (μm)	L (μm)
M_1-M_2	1.4	0.7
M_3-M_5	2.8	0.7
M_{17}-M_{18}	2.4	0.7
M_{19}-M_{21}	4.8	0.7
M_6-M_{16}, M_{22}-M_{28}	9.6	0.7

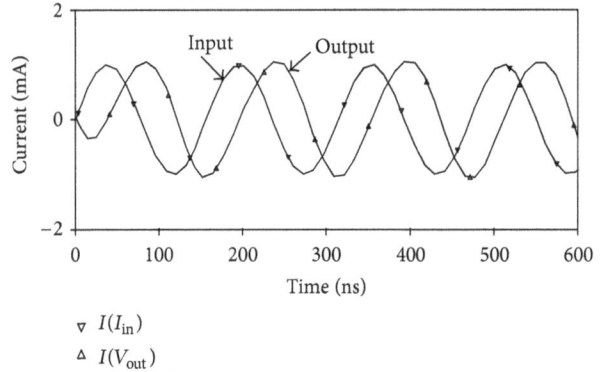

- $\triangledown\ I(I_{\text{in}})$
- $\triangle\ I(V_{\text{out}})$

FIGURE 5: Time-domain input and output responses of all-pass filter.

3.2. Parasitic Effects. Next study is carried on the effect of device parasitics on the performance of the proposed circuit. The various parasitics are a low value parasitic serial resistance R_X at X the terminal Y exhibits a high value parasitic resistance R_Y in parallel with low value capacitor C_Y, and the terminals Z exhibit a high value parasitic resistance R_Z in parallel with low value capacitance C_Z. The main among these are the Y and Z terminals parasitic capacitances and the X terminal's parasitic resistances. A reanalysis of the proposed circuit yields the modified transfer function as

$$\frac{I_{\text{out}}}{I_{\text{in}}} = -\left(\frac{sRC' - sRC_{Z2+} - 1}{s^2 RR_X C' C_{Z2+} + s\left(RC' + R_X C' + RC_{Z2+}\right) + 1} \right), \tag{7}$$

where $C' = C + C_{Z1-} + C_{Z1+}$.

From (7), the effect of capacitance C_{Z2+} becomes non-negligible at very high frequencies. Most of the parasitic

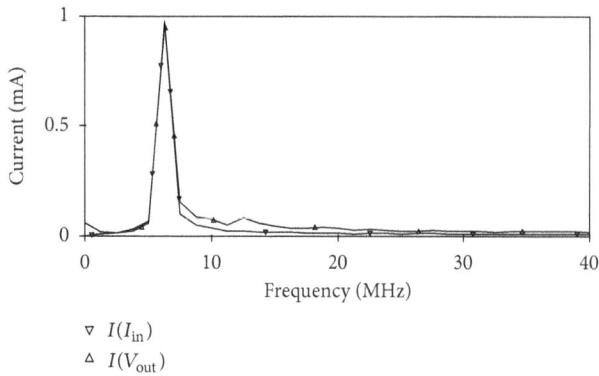

FIGURE 6: Fourier spectrum of the input and output.

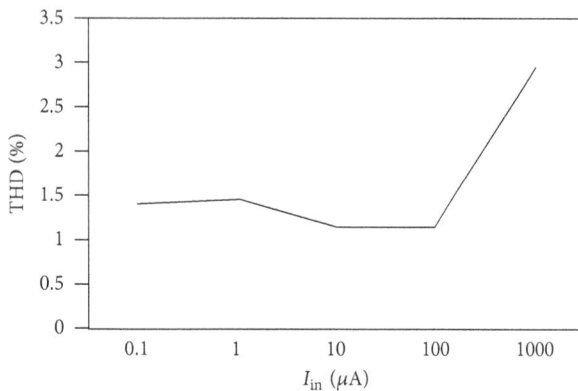

FIGURE 7: THD variation at output (I_{out}) with sinusoidal signal amplitude at 6.36 MHz.

capacitances get absorbed with the external grounded capacitor, as are in shunt with it. Also the parasitic resistance gets absorbed with the external grounded resistor, as it is in series with it. Such a merger will cause a slight deviation in circuit parameters, which can be corrected by predistorting the passive element values used in the circuit.

4. Simulation Results

To demonstrate the performance of the proposed circuit, the PSPICE simulation program is used. In the simulation, the TSMC 0.35 μm CMOS process parameters were used. The CMOS implementation of DX-MOCCII is shown in Figure 3 [13]. The aspect ratios of the CMOS transistors of the DX-MOCCII are listed in Table 2. DC supply voltages of ±1.8 V and biasing voltage of $V_{BB} = -0.7$ V were used. The proposed circuit of Figure 2 was designed with $R = 1$ kΩ and $C = 25$ pF to obtain a pole frequency of 6.36 MHz. The gain and phase responses are shown in Figure 4, where a phase shift of 90° at a pole frequency of 6.27 MHz is obtained, which is close to the theoretical designed value. The time-domain input and output responses of the circuit at the pole frequency are shown in Figure 5. Also, the Fourier spectrum of input signal and output signal is shown in Figure 6. Next, the amplitude of the input sinusoidal signal is varied from 0.1 μA to 1000 μA,

and the total harmonic distortion (THD) curve is plotted at a pole frequency of 6.36 MHz and is shown in Figure 7.

5. Conclusion

In this paper, a new current-mode cascadable all-pass filter is presented. The proposed circuit uses single DX-MOCCII, a grounded resistor, and a grounded capacitor, which is the minimum component realization for an active RC filter circuit. The circuit requires no matching constraints and low active and passive sensitivities and employs grounded passive components only, which makes it suitable for integrated circuit implementation. The circuit also exhibits the feature of low-input impedance and high-output impedance. The PSPICE simulation results of the proposed circuit are in good agreement with the theoretical results.

Acknowledgment

The authors are thankful to the academic editors for recommending this paper.

References

[1] C. Toumazou, F. J. Lidgey, and D. G. Haigh, *Analog IC Design: The Current-Mode Approach*, Peter Peregrinus, London, UK, 1990.

[2] M. Higashimura and Y. Fukui, "Realization of current mode all-pass networks using a current conveyor," *IEEE Transactions on Circuits and Systems*, vol. 37, no. 5, pp. 660–661, 1990.

[3] I. A. Khan, P. Beg, and M. T. Ahmed, "First order current mode filters and multiphase sinusoidal oscillators using CMOS MOCCIIs," *Arabian Journal for Science and Engineering*, vol. 32, no. 2, pp. 119–126, 2007.

[4] S. Minaei and E. Yuce, "All grounded passive elements current-mode all-pass filter," *Journal of Circuits, Systems and Computers*, vol. 18, no. 1, pp. 31–43, 2009.

[5] M. Higashimura, "Current-mode allpass filter using FTFN with grounded capacitor," *Electronics Letters*, vol. 27, no. 13, pp. 1182–1183, 1991.

[6] S. Maheshwari and I. A. Khan, "Novel first order all-pass sections using a single CCIII," *International Journal of Electronics*, vol. 88, no. 7, pp. 773–778, 2001.

[7] S. Minaei and M. A. Ibrahim, "General configuration for realizing current-mode first-order all-pass filter using DVCC," *International Journal of Electronics*, vol. 92, no. 6, pp. 347–356, 2005.

[8] S. Maheshwari, "Novel cascadable current-mode first order all-pass sections," *International Journal of Electronics*, vol. 94, no. 11, pp. 995–1003, 2007.

[9] A. Toker, S. Özoguz, O. Çiçekoglu, and C. Acar, "Current-mode all-pass filters using current differencing buffered amplifier and a new high-Q bandpass filter configuration," *IEEE Transactions on Circuits and Systems II*, vol. 47, no. 9, pp. 949–954, 2000.

[10] S. Kilinç and U. Çam, "Current-mode first-order allpass filter employing single current operational amplifier," *Analog Integrated Circuits and Signal Processing*, vol. 41, no. 1, pp. 47–53, 2004.

[11] S. Minaei and E. Yuce, "Unity/variable-gain voltage-mode/current-mode first-order all-pass filters using single dual-X

second-generation current conveyor," *IETE Journal of Research*, vol. 56, no. 6, pp. 305–312, 2010.

[12] P. Beg, M. A. Siddiqi, and M. S. Ansari, "Multi output filter and four phase sinusoidal oscillator using CMOS DX-MOCCII," *International Journal of Electronics*, vol. 98, no. 9, pp. 1185–1198, 2011.

[13] A. Zeki and A. Toker, "The dual-X current conveyor (DXCCII): a new active device for tunable continuous-time filters," *International Journal of Electronics*, vol. 89, no. 12, pp. 913–923, 2002.

[14] A. Zeki and A. Toker, "DXCCII-based tunable gyrator," *International Journal of Electronics and Communications*, vol. 59, no. 1, pp. 59–62, 2005.

[15] S. Minaei, "Electronically tunable current-mode universal biquad filter using dual-X current conveyors," *Journal of Circuits, Systems and Computers*, vol. 18, no. 4, pp. 665–680, 2009.

[16] S. Maheshwari and S. Maheshwari, "Multi input multi output biquadratic universal filter using Dual-X Current Conveyor (DXCCII)," in *Proceedings of the International Conference on Computer and Communication Technology (ICCCT '10)*, pp. 626–629, September 2010.

[17] S. Maheshwari and B. Chaturvedi, "High-input low-output impedance all-pass filters using one active element," *IET Circuits, Devices and Systems*, vol. 6, no. 2, pp. 103–110, 2012.

[18] S. Maheshwari and M. S. Ansari, "Catalog of realizations for DXCCII using commercially available ICs and applications," *Radioengineering*, vol. 21, pp. 281–289, 2012.

[19] M. Bhusan and R. W. Newcomb, "Grounding of capacitors in integrated circuits," *Electronic Letter*, vol. 3, pp. 148–149, 1967.

[20] A. Fabre, O. Saaid, and H. Barthelemy, "On the frequency limitations of the circuits based on second generation current conveyors," *Analog Integrated Circuits and Signal Processing*, vol. 7, no. 2, pp. 113–129, 1995.

8

The System Power Control Unit Based on the On-Chip Wireless Communication System

Tiefeng Li,[1,2] **Caiwen Ma,**[1] **and WenHua Li**[1]

[1] *Xi'an Institute of Optics and Precision Mechanics of Chinese Academy of Sciences, Xi'an 710119, China*
[2] *The Graduate University of Chinese Academy of Sciences, Beijing 100049, China*

Correspondence should be addressed to WenHua Li; 85422776@163.com

Academic Editors: D. De Vleeschauwer and B. R. Kumar

Currently, the on-chip wireless communication system (OWCS) includes 2nd-generation (2G), 3rd-generation (3G), and long-term evolution (LTE) communication subsystems. To improve the power consumption of OWCS, a typical architecture design of system power control unit (SPCU) is given in this paper, which can not only make a 2G, a 3G, and an LTE subsystems enter sleep mode, but it can also wake them up from sleep mode via the interrupt. During the sleep mode period, either the real-time sleep timer or the global system for mobile (GSM) communication sleep timer can be used individually to arouse the corresponding subsystem. Compared to previous sole voltage supplies on the OWCS, a 2G, a 3G, or an LTE subsystem can be independently configured with three different voltages and frequencies in normal work mode. In the meantime, the voltage supply monitor, which is an important part in the SPCU, can significantly guard the voltage of OWCS in real time. Finally, the SPCU may implement dynamic voltage and frequency scaling (DVFS) for a 2G, a 3G, or an LTE subsystem, which is automatically accomplished by the hardware.

1. Introduction

With the rapid increase of complexity and size of OWCS, the power consumption issue for the OWCS is increasingly becoming critical and needs to be solved quickly. As we know, the duration and stability of voltage supply are two factors that have an important impact on the power performance of the OWCS. The duration and stability of voltage supply are two factors that have an important impact on the power performance of the OWCS. In the first instance, the duration period depends on overall power consumption of the whole OWCS when the total amount of the battery's power is fixed and unchanged. An excess of power consumption can accelerate the aging of the OWCS and cut down the battery life. On the other hand, the power density of OWCS is continuously increasing with the scaling of each technology generation. It is necessary for the whole OWCS to reduce the dissipated heat because high heat dissipation increases instability, which can lead to the drastic fluctuation of voltage supply and the crash of OWCS in a short time. Early in the OWCS design stage when there is a lack of conception of saving energy, it is usually ignored by designers because the negative influence that is directly derived from power consumption is very small. However, in the development of wireless communication and the integrated circuit technology, the design of OWCS is more complicated and consumes more energy. In this case, it is often desirable to minimize power consumption to maximize OWCS lifetime. In this paper, we intend to deal with the power consumption issue of the OWCS with an efficient hardware approach and see how it is different from others' designs. In doing so, we put forward one new SPCU concept that is based on the OWCS [1, 2], which can work in both the active and sleep mode according to the actual requirement. The differences between the novel SPCU and the old design are emphasized in the introduction. Firstly, in the traditional design, software scheduler method in the operating system is still popular in reducing power consumption, which estimates CPU workload according to the frequencies of calling scheduler [3–9]. The main reason that most engineers adopt software scheduler is that they are not real CPU designers and are unable to change CPU hardware architecture at all. Thus, they have to lower power consumption in the operating system via software compensation method. However, with the quick development of CPU technology, the dominant frequency of current CPU has achieved more than 1.5G. For example, the dormant frequency of ARM Cortex A9 is about 1.6G. Obviously, this software approach is not enough in dynamic environments because it cannot accurately trace the status of CPU workload in the deadline when the frequency of CPU is higher than before. By contrast, the new SPCU that is fully presented from hardware side has a remarkable improvement in saving power. Because SPCU is integrated into the OWCS with three different voltages and frequencies, the various performance requirement of CPU can be rapidly responded by SPCU at the first moment; thus, the speed and accuracy of tracking CPU workload are largely improved in essence. The most significant is that all tasks of tracking CPU workload and predicting CPU performance can be finished by hardware automatically, which can transparently lessen the software burden of operating system in the OWCS. Secondly, many new features are firstly used in SPCU. For instance, the GSM sleep timer, the hardware DVFS, the supply voltage monitor, the reset status recorder, and the aging monitor. Thirdly, some features are optimized, such as the clock controller. Those features will be depicted in detail below.

Generally speaking, the contribution in this paper is that new SPCU aims at novelty in the realization of hardware architecture, instead of following the beaten path of conventionality. In actual application, it is proved that the reliable SPCU can keep the OWCS work with lower power consumption and increase the OWCS working cycle better. In our design, the SPCU mainly includes the transition state machine of power mode, the clock controller, the sleep timer, the voltage supply monitor, the reset and wake-up circuit, the voltage converter, the hardware DVFS, and the aging monitor.

2. The State Machine of Power Mode Transition

In conventional design [10], the state machine of power mode transition has two drawbacks: (1) only one sleep mode is deep mode, (2) the subpower domains are not switched off during deep sleep. For the previous OWCS, shallow sleep is critical for diminishing power. In addition to the shallow sleep requirement, the subpower domains should be switched off when voltage of CPU is automatically switched off during deep sleep. From the point of view of saving power consumption, it is necessary to minimize power consumption during deep sleep.

The new state machine of power mode transition with improved sleep mode and subpower domains is applied in the SPCU, which has two key advantages: (1) there are two power modes in the OWCS, which are the active mode and the sleep mode consisting of shallow and deep sleep mode, (2) it is recursive between subpower domains and their corresponding CPU power domain. In other words, all subpower domains in the SPCU will be compulsorily switched off if CPU power supply is switched off during deep

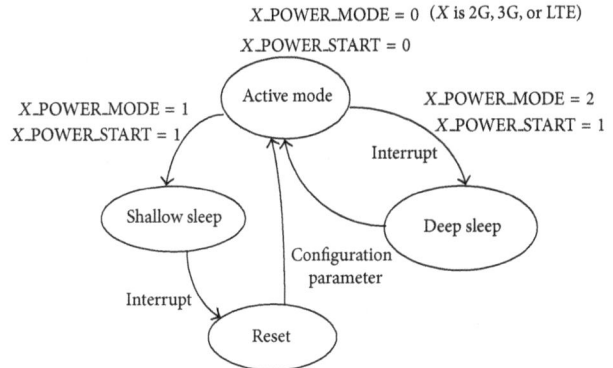

FIGURE 1: The state machine of power mode transition.

sleep. Here, we just take the 2G subsystem as an example because the power mode transition of the 3G and the LTE are similar to the 2G's power transition.

In Figure 1, with the beginning of power on, the whole OWCS normally works with active mode. At this moment, the power mode of the 2G subsystem can be indicated by the signal 2G_POWER_MODE because the value of signal 2G_POWER_MODE is zero during the active mode. In this case, depending on our needs, the voltage supply of 2G CPU may be initialized with the three-voltage option (V_{High} = 1.25 V, V_{Medium} = 1.0 V, V_{Low} = 0.80 V). In default, the voltage of 2G CPU is supplied with V_{Medium} and the clock of 2G CPU is provided with 360 MHz that is consistent with the medium voltage. As soon as the signal 2G_POWER_MODE is set to one and the signal 2G_POWER_START is set to one, the 2G subsystem will immediately enter the shallow sleep mode, in which the voltage supply of 2G CPU will not be switched off but the clock of the 2G CPU is decreased from 360 MHz to 32.768 kHz that is generated by on-chip oscillator [11]. As a result, the whole 2G subsystem reduces power consumption by slowing the clock frequency. Furthermore, the 2G subsystem will be woken up from the shallow sleep and be restored to the previous state if the interrupt happens. After the 2G subsystem is woken up, the clock frequency of 2G CPU will go back to the normal frequency. On the other hand, when the signal POWER_MODE is set to two and the signal 2G_POWER_START is set to one, the 2G subsystem will enter the deep sleep from active mode, see also Figure 1.

Especially, the power and clock of the 2G CPU is fully switched off during the deep sleep period, and its corresponding sub-power domains are also switched off. This makes the data contents of memory lost in deep sleep mode. In order to avoid losing data, the current data should be saved into the memory by setting retention mode before the deep sleep is requested. However, the power of the on-chip oscillator never be switched off in both sleep modes because it provides 32.768 kHz clock to awake circuit. The 2G subsystem will be woken up from the deep sleep and be rebooted as soon as the internal or external interrupt happened. After the 2G subsystem is rebooted, its clock frequency is back to the normal 360 MHz.

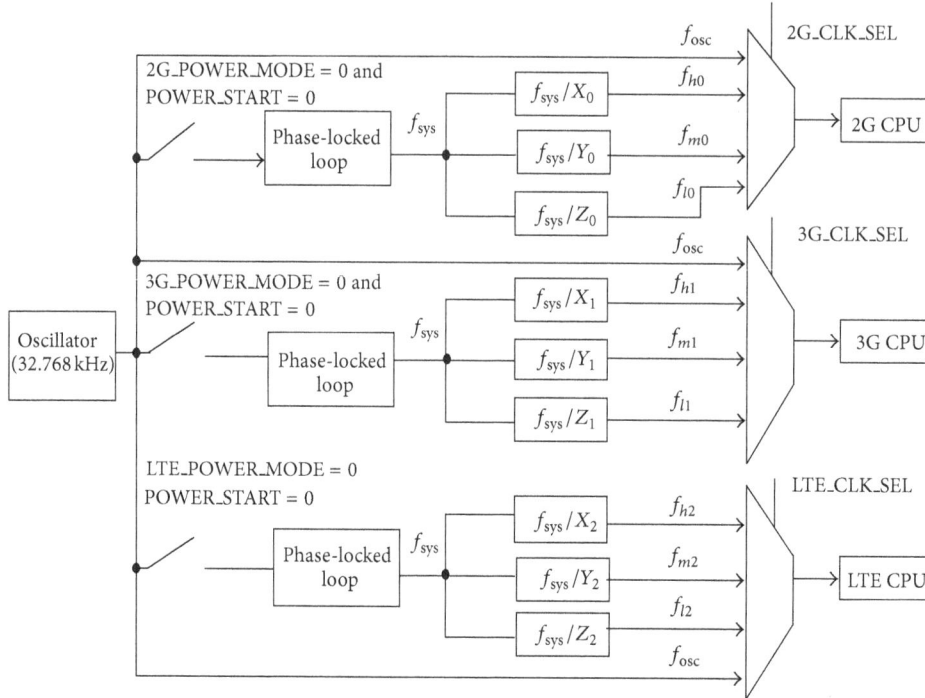

FIGURE 2: The clock controller.

3. The Clock Controller

As shown in Figure 2, the clock controller includes three phase-locked loops (PLLs) [11, 12], three clock dividers, and three digital multiplexes. In the previous clock controller, a Phase-Locked Loop (PLL) is used for 2G and 3G subsystem [13]. However, when LTE subsystem is merged into the smart OWCS, power saving has to be considered again [14, 15]. Here, we present an optimized clock controller that has three independent PLLs for DVFS. Compared with conventional design, this gives us much more flexibility to execute DVFS with three frequencies and voltages because every PLL is dedicated to 2G, 3G, or LTE subsystem. Moreover, every PLL can convert a low-frequency external clock signal that is generated by the on-chip 32.768 kHz oscillator to a high-speed internal clock for maximum.

Depending on the different frequency requirement, the clock frequency output may be configured by programming desired N, P, and K values according to (1). Normally, only if the signal POWER_MODE and the signal POWER_START in every subsystem (2G, 3G or LTE) are set to 00B at the same time, the on-chip 32.768 kHz oscillator will start to output clock to the corresponding PLL. For example, if 2G_POWER_MODE and 2G_POWER_START are set to 00B, the 32.768 kHz oscillator will output clock to the PLL of 2G subsystem. However, if the subsystem (2G, 3G or LTE) is in the deep sleep mode, the on-chip 32.768 kHz oscillator is still working but it stops providing the clock to the corresponding PLL so that no clock signal is coupled to the subsystem's CPU. Here, the generalized PLL output in three subsystems can be defined as

$$f_{sys} = f_{osc} * \frac{N}{P * K},$$ (1)

where N, P, and K are integers predefined according to the actual requirement. The CPU clock is derived from the oscillator clock (f_{osc}), multiplied by N, divided by P, and divided by K. The clock output from the 2G clock dividers can be stated as follows:

$$f_{h0} = \frac{f_{sys}}{X_0},$$

$$f_{m0} = \frac{f_{sys}}{Y_0},$$ (2)

$$f_{l0} = \frac{f_{sys}}{Z_0},$$

where X_0, Y_0, and Z_0 are the integers and $X_0 < Y_0 < Z_0$, f_{h0} represents the high input clock frequency of the 2G CPU, f_{m0} shows the medium input clock frequency of 2G CPU, f_{l0} means the low-input clock frequency of 2G CPU. Moreover, we can conclude according to (2) that $f_{h0} > f_{m0} > f_{l0}$.

The clock output from the 3G clock dividers can be written as follows:

$$f_{h1} = \frac{f_{sys}}{X_1},$$

$$f_{m1} = \frac{f_{sys}}{Y_1},$$ (3)

$$f_{l1} = \frac{f_{sys}}{Z_1},$$

where X_1, Y_1, and Z_1 are the positive integers and $X_1 < Y_1 < Z_1$, f_{h1} represents the high input clock frequency of the 3G

CPU, f_{m1} shows the medium input clock frequency of the 3G CPU, and f_{l1} stand for the low-input clock frequency of the 3G CPU. Further, we can also derive from the above Equations (3) that $f_{h1} > f_{m1} > f_{l1}$.

The clock output from the LTE clock dividers can be defined as follows:

$$f_{h2} = \frac{f_{sys}}{X_2},$$

$$f_{m2} = \frac{f_{sys}}{Y_2}, \qquad (4)$$

$$f_{l2} = \frac{f_{sys}}{Z_2},$$

where X_2, Y_2, and Z_2 are the positive integers and $X_2 < Y_2 < Z_2$, f_{h2} represents the high input clock frequency of the LTE CPU, f_{m2} shows the medium input clock frequency of the LTE CPU, and f_{l2} means the low-input clock frequency of the LTE CPU. Beyond this, we can similarly deduce on the basis of (4) that $f_{h2} > f_{m2} > f_{l2}$.

For the three digital multiplexes, the control signals (2G_CLK_SEL, 3G_CLK_SEL and LTE_CLK_SEL) can be set with three different frequency clock (f_{high}, f_{Medium}, and f_{Low}) that are absolutely synchronous to three voltage configurations (V_{High}, V_{Medium}, and V_{Low}); that is, when one kind of voltage supply is configured on the subsystem of the OWCS, the voltage configuration also can be passed to the signal X_CLK_SEL in the meantime. Also, during the shallow sleep, the clock controller works with bypassed mode, where the clock fosc can be directly bypassed to the digital multiplexes without going through the PLLs and can be chosen as the clock output of the multiplex by setting the control signal X_VSET to 11B.

The clock of the 2G, the 3G, or the LTE can be selected by the digital multiplexer. The formula of the clock output can be described as follows:

$$f_{out} = \begin{cases} f_{ln} & \text{if } X\_CLK\_SEL = 00B \\ f_{mn} & \text{if } X\_CLK\_SEL = 01B \\ f_{hn} & \text{if } X\_CLK\_SEL = 10B \\ f_{osc} & \text{others,} \end{cases} \qquad (5)$$

where X is 2G, 3G, or LTE, n is 0, 1, or 2.

4. The Sleep Timer

4.1. The Real-Time Sleep Timer. In the conventional design, sleep time is mainly controlled by external timer [15–17]. If external timer is occupied by other module during sleep, the sleep time will not be controlled again. In order to resolve this problem, the real-time sleep timer is newly introduced in our SPCU.

The main concept is that the 2G, the 3G, or the LTE subsystem has its own real-time sleep timer that can precisely control the sleep time, which gives us great flexibility when we control the sleep time of the 2G, the 3G, or the LTE subsystem. With reference to Figure 3, the real time sleep timer includes

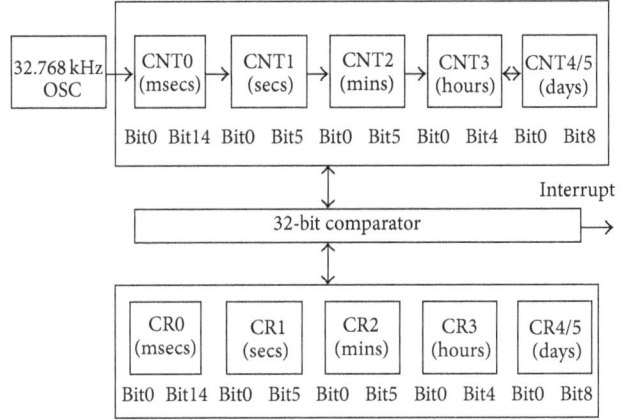

FIGURE 3: The real-time sleep timer block.

the milliseconds counter, the seconds counter, the minutes counter, the hours counters and the days counters. In default, the milliseconds counter CNT0 are 15 bits, which are set to an initial value of 000_0000_0000_0000B. One increment of the timer is then made for every cycle of the input clock. With an input clock frequency of 32.768 kHz which is from on-chip oscillator, one second will be equivalent to an overflow of a 15-bit timer. Under this condition, the register CNT5-CNT0 actually holds the real-time value in the range of milliseconds, seconds, minutes, hours, and days. Register CNT0 has 15 bits that hold the range of milliseconds of the clock. The register CNT2 and CNT1 are 6 bits, which hold the minutes and seconds of the clock. The seconds and minutes counter will be reset to zero after 60 counts. The CNT3 register holds the hours of the clock. It resets to zero after 24 counts. The registers CNT4 and CNT5 hold the days for the clock. They count for one year before it is reset to zero. One year could be 365 or 366 days. Bit TYR is used to select between a normal year and a leap year.

The real time sleep timer has to be initialized before the sleep mode start. While it is in operation, the contents registers (CR0-CR5) register will be compared to the real time sleep timer counter (CNT0-CNT5) that is programmable with any values that indicates the seconds, minutes, hours and days. An interrupt will be generated in sleep mode when the contents are equal, which will rouse the 2G, the 3G, or the LTE subsystem from the sleep mode.

4.2. The GSM Sleep Timer. The resynchronization problem of GSM system always arises in previous 2G subsystem [18], where there is time mismatch between GSM system and base station when the OWCS is woke up from sleep mode; the main reason is that GSM system [14, 15] has been reset since whole OWCS woke up. As a result, the parameter of GSM system has to be set again in order to synchronize with the base station. To cope with this shortcoming in the OWCS, the GSM sleep timer is firstly introduced in the SPCU, which is one special sleep timer that only belongs to the 2G subsystem. It is emphasized that the GSM sleep timer is basically independent from the system sleep mode, so that GSM sleep timer can be dedicated to work as a wake-up source. The time

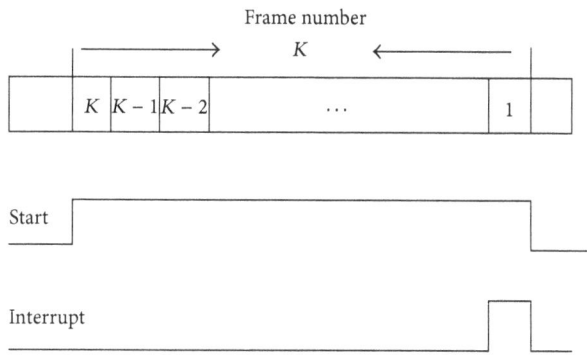

FIGURE 4: The GSM sleep timer.

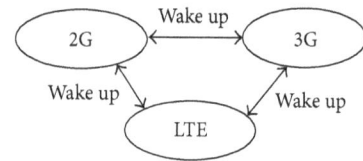

FIGURE 5: The 2G, the 3G, and the LTE wake up relationship.

unit of the GSM sleep timer is based on the time division multiple access (TDMA) frame. Every TDMA frame time is about 4.615 milliseconds that are synchronized with the GSM communication network [16–18]. The advantage of the GSM sleep timer is that it offers a convenient hardware solution to switch off the 2G subsystem for an accurate integer multiple of a TDMA frame (sleep duration). During the sleep mode, the GSM sleep timer controller is clocked with 32.768 kHz that is from on-chip oscillator. The TDMA frame number is predefined via software before the 2G subsystem enters the sleep mode. When the control signal START is set to the high level, the GSM sleep timer controller will immediately start to run, which will decrease one with every 4.615 milliseconds until the TDMA frame number is zero. In the meantime, the control signal START will become low and the SPCU will send an interrupt to the 2G CPU so that the 2G subsystem will be subsequently woken up. Figure 4 explains the exact timing of the whole GSM sleep timer between the active and the sleep mode. In the single antenna interference cancellation (SAIC) mode, the sleep duration accuracy was improved from 1.04 to 0.52 ppm compared to earlier implementations.

5. The Wake-Up and Reset

As we know, there is no LTE subsystem in the earlier OWCS [19, 20], so the old SPCU design does not cover the power system of LTE subsystem. The improvement in the new SPCU is that the LTE subsystem can be taken as an independent wake-up source, which not only can wake up the 2G or 3G subsystem but it also can be woken up from sleep by 2G or 3G subsystem. Furthermore, soft reset is specially introduced in order that subsystem can go back to the previous running state from shallow sleep. Unlike soft reset, hardware reset is dedicated to the OWCS system crash. Moreover, reset behaviour cannot be recorded by SPCU in the conventional design, so it is not convenient to check wake-up source and reset type. The advantage in new SPCU is that reset status recorder and wake-up status recorder are merged and implemented well in our design.

5.1. The Wake-Up Source.
For the 2G, the 3G, or the LTE subsystem, its wake-up source is composed of internal and external source. The internal sources are the three independent sleep timers and the DSP. The external sources are the chip pins, the 3G, and the LTE. For the 3G and the LTE, they also can be woken up by the 2G when they are in sleep mode state. The flexible wake-up relationship among the three subsystems is given in Figure 5.

After the 2G, the 3G, or the LTE subsystem is woken up and is reset, the SPCU will not be reset yet, and then the wake-up status register of the SPCU can record which source is triggered. The wake-up status register has to be cleaned before going into next sleep mode.

5.2. The Reset Type

5.2.1. The Soft Reset.
When the 2G, the 3G, or the LTE subsystem separately enters the shallow sleep mode, it can be woken up and be restored to the previous state via soft reset. Note that the soft reset is not effective to the deep sleep.

5.2.2. The Hardware Reset.
The hardware reset is a special reset in the OWCS because it can deal with the urgent case. After the 2G, the 3G, or the LTE subsystem enters the deep sleep from active mode, the whole on-chip system will be woken up and rebooted if reset key is pressed. This kind of hardware reset is not frequently used in the actual application because it resets the 2G, the 3G, and the LTE subsystem at the same time.

5.2.3. The Wake-Up Reset.
Compared to the soft reset, the wake-up reset plays an important role in the on-chip wireless communication chip, which is often used in both the shallow sleep and the deep sleep. When the 2G, the 3G, or the LTE subsystem is woken up from the shallow sleep or the deep sleep by the interrupts, the reset status register will record reset status and the wake-up status register will record which source is working.

6. The Voltage Supply Monitor

In traditional OWCS design, the system voltage supply is not monitored exactly in real time [21–23]. From the point of view of power supply security, SPCU must be sensitive to the changing voltage supply so that the OWCS system can get timely prewarning to eliminate some potential risks. In this context, the distinctive supply voltage monitor is firstly introduced in this paper in order to enhance guarding on system voltage.

In our SPCU, the basic functionality of the supply voltage monitor is primarily targeted as a measurement and analysis

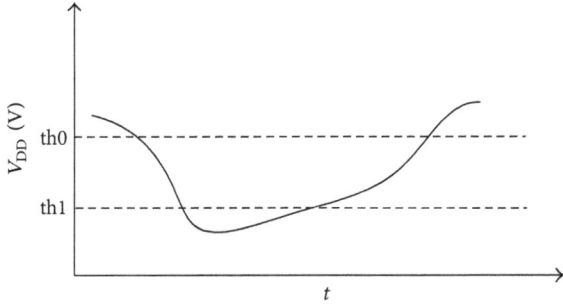

FIGURE 6: The voltage supply monitor.

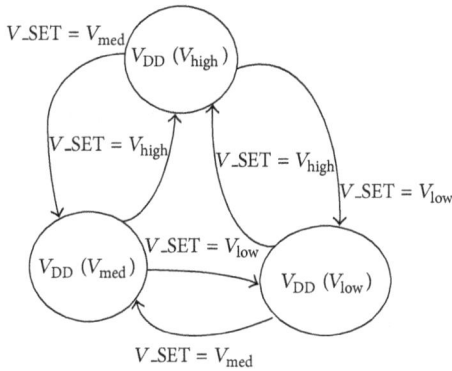

FIGURE 7: The voltage transition.

tool during normal operation of the chip. It gives a permit to measure and guard the internal subsystem CPU supply voltage on the chip for every individual CPU clock cycle. Furthermore, it is exposed to the worst operating condition that the highest on-chip voltage drops. Meanwhile, it is optionally fed up with three subsystem CPU clock frequency. The major benefit of the supply voltage monitor is that it can ensure that each subsystem CPU is supplied with the normal voltage in real time. Once the 2G, the 3G, or the LTE CPU voltage is less than a fixed threshold value in the active mode, the voltage supply monitor will send one interrupt signal to the subsystem CPU and configure the voltage again (see Figure 6). The voltage output from the voltage converter is adjusted in such a way that the supply voltage drops below the critical voltage threshold even under worst-case conditions. The supply voltage monitor has two threshold voltages, which means the supply voltage of subsystem (2G, 3G, or LTE) has double protection. For the two threshold voltages, they can be configured in advance by means of setting the voltage control register. The predefined threshold voltage formula can be stated as follows:

$$\text{Threshold} = 0.80 + 0.01 * X, \qquad (6)$$

where X is from 0 to 40. The precision of two threshold voltages can be guaranteed in real operation according to (6).

7. The Voltage Converter

The voltage converter plays an important role in supplying voltage for the OWCS. In traditional design, a voltage converter is used to provide voltage supply to 2G or 3G subsystem [23–26]. However, in practice, it sometimes leads to voltage requirement conflict because 2G and 3G subsystems work in parallel.

Unlike common voltage converter, new voltage converter has the three independent step-down converters, which are the 2G voltage converter, the 3G voltage converter, and the LTE voltage converter. Particularly, the three step-down converters can operate in two modes. One mode is the pulse frequency modulation (PFM) mode with a lower quiescent current at limited output current, the other mode is the pulse width modulation (PWM) mode with full output current but a higher quiescent current. Throughout the sleep phases, the PFM mode is required so that the current consumption of the on-chip system and the memory is very low ($\ll 1\,\text{mA}$). As a result, the power consumption is minimized. Whenever the subsystem is woken up, the mode of the step-down converter is forced to switch to PWM mode.

It is shown in Figure 7 that the voltage converter output of every subsystem (2G, 3G, LTE) is configured with three voltages (V_{Low}, V_{High}, and V_{Medium}) according to the input signal V_SET. Three voltage configuration details by the 2-bit width signal V_SET is shown in Table 1. Furthermore, the voltage converter output will be switched off in deep sleep mode. It will be powered on again when the input signal $RESET\_X$ is active by waking up or hardware reset. Hence, the voltage converter can reduce the power consumption of the on-chip system with flexible voltage configuration. Figure 8 presents the three voltage transitions for three subsystems CPU voltage.

8. The DVFS Theory

Currently, the researchers and designers are still adopting software method predicts performance requirement of CPU according to the sequence of event priority in the software scheduler [3–9]. Although it can low power in a way, it is only efficient on the case that CPU frequency is not high, it will become difficult with the increment of CPU frequency because the software method cannot correctly respond to the high frequency of CPU, so it is also failed to estimate the performance requirement of CPU in time [27, 28]. To overcome this, the hardware DVFS is fully introduced in this paper, which has very fast tracking and response speed on CPU behaviour.

For the OWCS, its dynamic power formula is stated as follows:

$$P = \alpha C V^2 f, \qquad (7)$$

where α represents the percentage of logic cell between 0 and 1 switching, C is a constant that represents the circuit load, V represents the CPU voltage, and f represents the CPU frequency.

We can easily know according to (7) that the power of CPU can be lowered by reducing the voltage and frequency.

TABLE 1: The binary voltage configuration.

Voltage configuration type	Binary value
V_{low}	00B
V_{med}	01B
V_{high}	10B

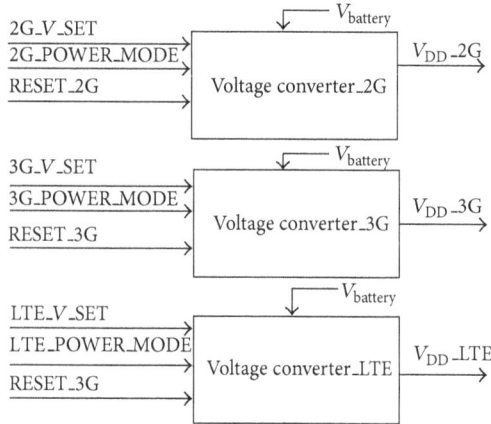

FIGURE 8: The voltage converter block.

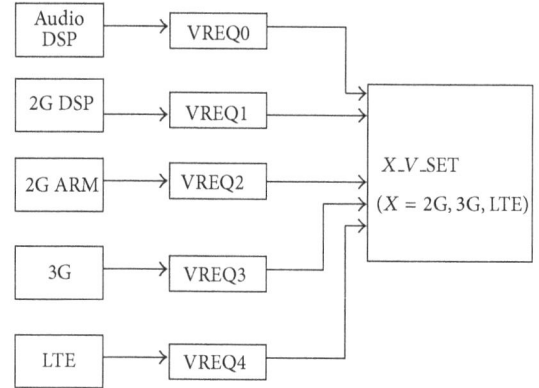

FIGURE 9: The DVFS structure.

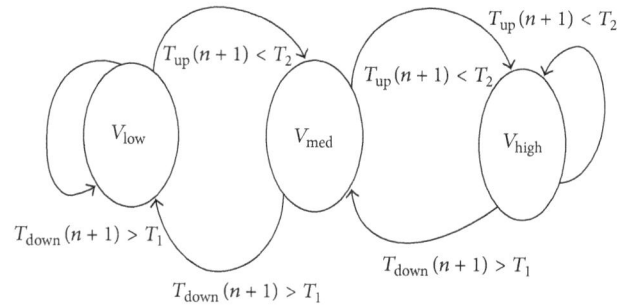

FIGURE 10: The intelligent transition for DVFS.

To some degree, it is tedious to change the dynamic voltage and frequency by means of software. In order to cut down the software overloads on the DVFS, the 2G, the 3G, the LTE, and the DSP subsystems can request the DVFS that is finished by the hardware in parallel or respectively. The structure of hardware DVFS in the active mode is given in Figure 9. With reference to Figure 9, the function of DVFS can be enabled by setting the signal VREQn (n is 0, 1, 2, 3) that is 1-bit width. Once the signal VREQn is set, the voltage of the subsystem CPU or DSP need not be adjusted by the software again and again. In comparison to the previous software design, it firstly strengthens the accuracy of voltage and frequency estimation. Secondly, it lightens the load of CPU timing tracking in a way.

In the progress of the DVFS, the SPCU can accurately predict the voltage that is needed in the next period of time according to the current CPU idle time, which supports two scaling steps (down, up). Each CPU of the three subsystems can be separately configured with the scaling down threshold value ($T1$) and the scaling up threshold value ($T2$). It is worth emphasizing that the two threshold values of the CPU idle time have to be set before the DVFS is requested. In the meantime, the moving average algorithm (MAA) is firstly adopted in the DVFS. The MAA not only tracks and samples the idle time of the every CPU with small enough intervals but also executes the accumulation and average calculation of the idle times. The MAA formula is given as follows:

$$T_{up}(n+1) = \frac{1}{N} \sum_{k=0}^{N-1} T(n-k),$$

$$T_{down}(n+1) = \frac{1}{M} \sum_{k=0}^{M-1} T(n-k), \tag{8}$$

where n, M, and N are the positive integers and usually $M > N > 0$. $T_{up}(n+1)$ stands for the average of the idle time from the sampling timing 0 to $N-1$, $T_{down}(n+1)$ stands for the average of the idle time from the sampling timing 0 to $M-1$. Based on (8), the voltage and clock of CPU scaling down step condition is fulfilled if $T_{down}(n+1) > T1$. Similarly, the voltage and clock of CPU scaling up step condition is fulfilled if $T_{up}(n+1) < T2$. Figure 10 shows the specific automatic transition for the DVFS.

The DVFS has its own timer that can be set to the expected maximum voltage settling time. For example, if a voltage ramping slew rate of 5 mV per microsecond is used in changing to the adjacent voltage, it only takes 40 microseconds to stabilize from V_{Low} (0.8 V) to V_{Medium} (1.0 V). Whenever the voltage scaling timer elapses, an interrupt can be triggered.

As shown in Table 2, it is obvious that the power consumption of each CPU is reduced with the DVFS in the actual test. Furthermore, compared with conventional software way (CSW) [29–31], the hardware DVFS has the absolute advantage in saving energy. Thus, we can clearly get a conclusion that the hardware DVFS is an efficient and smart way to save power. In Future, the hardware DVFS will be dominant in the OWCS because of the high efficiency.

9. The Aging Monitor

It is meaningful for designers to analyze the important aging data so as to optimize the power system of OWCS. But in the conventional OWCS, the aging monitor has never been

TABLE 2: Saving power with the DVFS.

CPU access to memory	DVFS and CSW are disabled	CSW is enabled	Saving power	DVFS is enabled	Saving power
2G CPU	280 mW	265 mW	5.3%	221 mW	21.1%
3G CPU	198 mW	176 mW	11.1%	157 mW	20.7%
LTE CPU	183 mW	168 mW	8.2%	141 mW	23%

used successfully because of its implement complexity [31]. To solve this problem, a novel aging monitor is exactly described in this part.

As the OWCS ages, the reliability of internal components begins to diminish. The OWCS ages in operational use during which the internal components are exposed to varying operational temperature and voltages. In fact, the effects of aging are proportional to the cumulative temperature and voltages experienced during use. So internal components which operate at higher temperatures and voltages age faster and deteriorate quicker than those components experiencing more moderate temperatures and voltages. To a certain degree, it is challenging to monitor the age of target circuit component effectively. To solve it in time, we give the particular aging monitor concept to the OWCS.

In Figure 11, the age of the target circuit component in the OWCS may be monitored by using at least one aging monitor that includes one reference oscillator circuit and one aging oscillator circuit. For the aging oscillator circuit, it comprises a ring oscillator that generates aging clock signal having an aging frequency f_{AGE} that may change over time. Enable unit A is coupled to selectively enable or disable aging oscillator circuit. The aging oscillator circuit only generates aging clock signal when enabled. One or more components in the aging oscillator circuit may degrade over time when stressed. The degradation of those components may cause the aging frequency f_{AGE} to change. The aging oscillator circuit is positioned to proximate to the target circuit component. such that oscillator circuit and target circuit may experience the similar operational stress (e.g., temperature, voltage, etc.). In a word, the aging oscillator circuit and the target circuit component are exposed to an identical operating environment. Thus, degradation in the aging oscillator circuit and the target circuit component will be correlated.

In the same way, the reference oscillator has a ring oscillator that generates reference clock signal having a reference frequency f_{REF} that may change over time. Enable unit B is coupled to selectively enable or disable reference oscillator circuit. For the reference oscillator circuit, it can be enabled for short periods of time, just long enough to compare f_{REF} of reference clock signal with f_{AGE} of aging clock signal. When reference oscillator circuit is disabled, the components in the reference oscillator circuit are not stressed and are electrically isolated from the target circuit component. Thus, when the reference oscillator circuit is disabled, it does not experience the aging effects experienced by the target circuit component. Related to the cumulative operating time of the target circuit component and the aging oscillator circuit, the

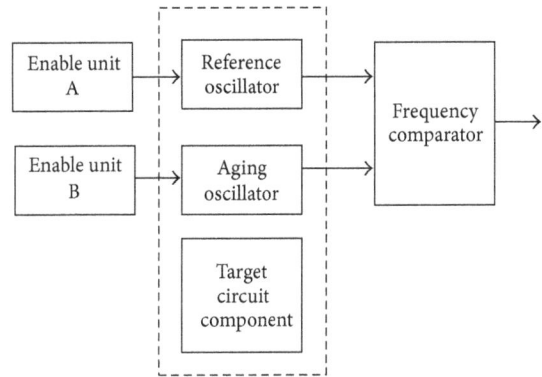

FIGURE 11: The aging monitor block.

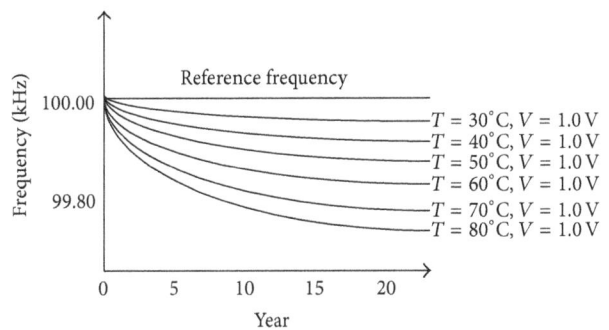

FIGURE 12: The aging progress.

reference oscillator circuit is operated for very short periods of time so that it is not stressed significantly.

In Figure 11, the frequency comparator is coupled to receive and compare reference clock signal and aging clock signal. In response, the frequency comparator generates an age signal that is proportional to the operation age of target circuit component. Age signal generated by the frequency comparator may then be analyzed. The OWCS has multiple aging monitors that track operational age of the multiple circuit components, such as memory controller, subsystem CPU. Typically, the subsystem CPU may be specially measured by multiple aging monitors.

According to Figure 12, we clearly explain that the aging effect is closely related to the operational temperature, voltage supply, and time. Assume that the environment temperature and voltage supply is not unchanged; as time goes, the aging effect becomes more apparent than before. In summary, the aging test in the OWCS is an indispensible flow that still needs to be improved in the future.

10. Conclusion

With the reliability and the flexibility of the SPCU, each subsystem of the OWCS can be easily provided with the three different voltage and clock and request the DVFS. Besides, the sleep time can be precisely performed by two kinds of the sleep timers during the sleep mode; the voltage supply of each subsystem in the OWCS can be monitored in real time. It is worth noting that the new design can be used to solve the problem on how to save the power consumption and how to overcome instability of voltage supply. In general, it is obvious that the SPCU can make the OWCS save the power and monitor the age of OWCS well, which has been successfully used in the OWCS. Considerable more work, hopefully, will be done in this area on how to achieve the lowest power consumption in the OWCS by this method provided in this paper.

Acknowledgment

This work was supported by Foundation of Chinese Academy of Sciences under Contract no. 36275128.

References

[1] Fenghuili, *Principles of Digital Communications*, The Electronics Industry Press, Beijing, China, 2012.

[2] LiXiaofeng, *Communication Theory*, Tsinghua University Press, Beijing, China, 2008.

[3] M. Peng, R. Li, and Y. Wang, "Phase-based dynamic voltage scaling algorithm," *Journal of Computer Research and Development*, vol. 45, no. 6, pp. 1093–1098, 2008.

[4] J. O. Coronel and J. E. Simó, "High performance dynamic voltage/frequency scaling algorithm for real-time dynamic load management," *Journal of Systems and Software*, vol. 85, no. 4, pp. 906–919, 2012.

[5] F. Firouzi, M. E. Salehi, F. Wang, and S. M. Fakhraie, "An accurate model for soft error rate estimation considering dynamic voltage and frequency scaling effects," *Microelectronics Reliability*, vol. 51, no. 2, pp. 460–467, 2011.

[6] M. Y. Lim, V. W. Freeh, and D. K. Lowenthal, "Adaptive, transparent CPU scaling algorithms leveraging inter-node MPI communication regions," *Parallel Computing*, vol. 37, no. 10-11, pp. 667–683, 2011.

[7] M. Li, "Approximation algorithms for variable voltage processors: min energy, max throughput and online heuristics," *Theoretical Computer Science*, vol. 412, no. 32, pp. 4074–4080, 2011.

[8] N. B. Rizvandi, J. Taheri, and A. Y. Zomaya, "Some observations on optimal frequency selection in DVFS-based energy consumption minimization," *Journal of Parallel and Distributed Computing*, vol. 71, no. 8, pp. 1154–1164, 2011.

[9] B. I. Rani, C. K. Aravind, G. S. Ilango, and C. Nagamani, "A three phase PLL with a dynamic feed forward frequency estimator for synchronization of grid connected converters under wide frequency variations," *International Journal of Electrical Power & Energy Systems*, vol. 41, no. 1, pp. 63–70, 2012.

[10] M. Weixin, *Modern Communication Network*, The People's Posts and telecommunications Press, Beijing, China, 2010.

[11] Y. Quanke, *Analog Electronics Technology Base*, High Education Press, Beijing, China, 2010.

[12] D. Jiyu, *Theory and Application of the Phase-Lock Loop*, The People's Posts and Telecommunications Press, Beijing, China, 2011.

[13] Z. Qinggui, *Theory and Application of the Phase-Lock Loop Integrated Circuit*, Shanghai Science and Technology Press, Shanghai, China, 2011.

[14] Y. Quanke, *Analog Electronics Technology Base*, High Education Press, Beijing, China, 2010.

[15] M. Dehao, *Digital Communication Base*, The People's Posts and Telecommunications Press, Beijing, China, 2010.

[16] W. Guohua, *Theory and Application of the GSM Communication Network*, The People's Posts and Telecommunications Press, Beijing, China, 2011.

[17] Guoqing, *Satellite Communication System*, The Electronic Industry Press, Beijing, China, 2010.

[18] D. Zuhui, *Digital Mobile Communication*, The Electronic Industry Press, Beijing, China, 2011.

[19] G. S. Kim, Y. H. Je, and S. Kim, "An adjustable power management for optimal power saving in LTE terminal baseband modem," *IEEE Transactions on Consumer Electronics*, vol. 55, no. 4, pp. 1847–1853, 2009.

[20] N. Chuberre, C. Nussli, P. Vincent, and J. Ntsonde, "Satellite digital multimedia broadcasting for 3G and beyond 3G systems," in *Proceedings of the 13th IST Mobile & Wireless Communication Summit 2004*, Lyon, France, 2004.

[21] M. Etinski, J. Corbalan, J. Labarta, and M. Valero, "Optimizing job performance under a given power constraint in HPC centers," in *Proceedings of the International Conference on Green Computing (Green Comp '10)*, pp. 257–267, August 2010.

[22] X. Fan, W. D. Weber, and L. A. Barroso, "Power provisioning for a ware housesized computer," in *Proceedings of the 34th Annual International Symposium on Computer Architecture (ISCA '07)*, pp. 13–23, ACM, New York, NY, USA, 2007.

[23] R. Ge and K. W. Cameron, "Power-aware speedup," in *Proceedings of the 21st International Parallel and Distributed Processing Symposium (IPDPS '07)*, p. 56, March 2007.

[24] L. Johnsson, D. Ahlin, and J. Wang, "The SNIC/KTH PRACE Prototype: achieving high energy efficiency with commodity technology without acceleration," in *Proceedings of the International Conference on Green Computing (Green Comp '10)*, pp. 87–95, August 2010.

[25] N. Kappiah, V. W. Freeh, and D. K. Lowenthal, "Just in time dynamic voltage scaling: exploiting inter-node slack to save energy in MPI programs," in *Proceedings of the ACM/IEEE 2005 Supercomputing Conference (SC '05)*, p. 33, November 2005.

[26] E. Le Sueur and G. Heiser, "Dynamic voltage and frequency scaling: the laws of diminishing returns," in *Proceedings of the 2010 Workshop on Power Aware Computing and Systems (HotPower '10)*, Vancouver, Canada, October 2010.

[27] M. Y. Lim and V. W. Freeh, "Determining the minimum energy consumption using dynamic voltage and frequency scaling," in *Proceedings of the 21st International Parallel and Distributed Processing Symposium (IPDPS '07)*, p. 348, March 2007.

[28] S. Cho and R. G. Melhem, "On the interplay of parallelization, program performance, and energy consumption," *IEEE Transactions on Parallel and Distributed Systems*, vol. 21, no. 3, pp. 342–353, 2010.

[29] N. Kamyabpour and D. B. Hoang, "Modeling overall energy consumption in wireless sensor networks," in *Proceedings of the 11th International Conference on Parallel and Distributed Computing, Applications and Technologies (PDCAT '10)*, pp. 8–11, December 2010.

[30] R. Xiaojun, X. Qin, Z. Zong et al., "An energy-efficient scheduling algorithm using dynamic voltage scaling for parallel applications on clusters," in *Proceedings of 16th International Conference on Computer Communications and Networks (ICCCN '07)*, pp. 735–740, 2007.

[31] D. Love, R. Heath, V. Lau, D. Gesbert, B. Rao, and M. Andrews, "An overview of limited feedback in wireless communication systems," *IEEE Journal on Selected Areas in Communications*, vol. 26, no. 8, pp. 1341–1365, 2008.

9

Supporting Seamless Mobility for P2P Live Streaming

Eunsam Kim,[1] Sangjin Kim,[2] and Choonhwa Lee[3]

[1] Department of Computer Engineering, Hongik University, Seoul 121-791, Republic of Korea
[2] Hyundai Autoever Corporation, 576 Sam, Uiwang, Gyeonggi 437-040, Republic of Korea
[3] Division of Computer Science and Engineering, Hanyang University, Seoul 133-791, Republic of Korea

Correspondence should be addressed to Choonhwa Lee; lee@hanyang.ac.kr

Academic Editor: Jingjing Zhou

With advent of various mobile devices with powerful networking and computing capabilities, the users' demand to enjoy live video streaming services such as IPTV with mobile devices has been increasing rapidly. However, it is challenging to get over the degradation of service quality due to data loss caused by the handover. Although many handover schemes were proposed at protocol layers below the application layer, they inherently suffer from data loss while the network is being disconnected during the handover. We therefore propose an efficient application-layer handover scheme to support seamless mobility for P2P live streaming. By simulation experiments, we show that the P2P live streaming system with our proposed handover scheme can improve the playback continuity significantly compared to that without our scheme.

1. Introduction

With the widespread deployment of high speed broadband networks such as FTTH, the IPTV services converging broadcasting and communication technologies have emerged. So far most commercial IPTV systems have employed the client/server architecture where video data are transmitted only from servers to clients. To support a huge number of IPTV subscribers at the same time, however, the client/server architecture should employ CDN (content distribution networks) structures to reduce data transmission delay. As the number of IPTV subscribers increases, the client/server architecture thus causes high expense for expanding network capacity by adding proxy servers to accommodate all the increasing subscribers [1]. The personalized IPTV services including time-shifted TV may make it even more difficult for IPTV systems to manage network traffic efficiently [2, 3].

Many research efforts have therefore been made on peer-to-peer (P2P) live streaming since it is a cost-effective and scalable alternative to client/server architectures on the Internet [4–6]. In P2P live streaming systems, peers exchange distributed video data with each other on virtual overlay networks. Thus, the better performance can be achieved as the number of participating peers increases.

On the other hand, with recent advance in wireless networks and advent of powerful mobile devices such as smart phones, the video streaming services have become feasible in mobile platforms [7–9]. Since these mobile IPTV services can provide users with the mobility and portability in wireless networks, the users' demand to enjoy IPTV services with mobile devices has been increasing rapidly. One of the most challenging issues when designing mobile IPTV systems is that users may experience the degradation of service quality due to data loss caused by the handover occurring when mobile devices are moving across APs. To get over this problem, it is thus essential to provide seamless mobility for P2P live streaming.

To minimize the transmission delay during the handover period, many schemes have been proposed at different layers of the protocol stack including data link [10, 11], network [12, 13], and transport layer [14, 15] depending on the characteristics of each layer. In the handover schemes at the layers below the application layer, however, the data loss inherently occurs while the network is being disconnected to switch APs. To avoid playback jitter in P2P live streaming systems, it is thus important to compensate for the amount of data that could not be received during the handover period. We thus need a new handover scheme at the application layer apart from that at lower layers. In fact, several application-layer handover schemes have been proposed recently but they did not consider P2P streaming structures, all based on the client/server architecture using CDN structures with low scalability and high cost.

In this paper, we therefore propose an efficient application-layer handover scheme to provide seamless mobility in the presence of handover by considering mobile peers' limited resources and the unstable characteristics of wireless networks. In our proposed scheme, to receive data from neighbor peers at a faster speed, neighbor peers transmit data to a mobile peer through a push manner for the period around the handover. To further improve the performance, an agent peer for the mobile peer is selected among stationary peers. The agent peer receives data in place of the mobile peer before and during the handover and then transmits the data through a new AP after the handover. It can transmit data at a faster speed compared to the other neighbor peers because it is selected depending on its RTT value from a new AP and the appropriateness of its buffered period from a mobile peer's perspective.

Through extensive simulations, we demonstrate the effectiveness of our proposed handover scheme. The simulation results show that the mobile P2P streaming system with our handover scheme improved the playback continuity significantly compared to that without our scheme. We also show that we can further improve the performance by adjusting the weight value used when selecting an agent peer depending on peers' current situation.

The remainder of this paper is organized as follows. Section 2 describes related work to mobile IPTV systems and handover schemes. Section 3 describes our design considerations to develop an efficient handover scheme. Section 4 proposes an efficient handover scheme for mobile P2P live streaming systems. Section 5 presents extensive simulation results. Finally, Section 6 offers conclusions.

2. Related Work

So far most streaming systems have employed the client/server architecture based on content delivery networks (CDNs) that consist of many proxy servers geographically distributed on the Internet. However, this architecture requires tremendous cost to expand the network capacity for the rapidly increasing number of IPTV subscribers. To solve this scalability problem, many P2P streaming systems have thus been proposed. They can be broadly classified into tree-push and mesh-pull structures. The tree-push structures require much overhead to rebuild tree structures whenever peers join or leave [4]. On the other hand, the mesh-pull structures provide robust structure against peers' churn while creating long startup delay and requiring a large number of data exchanges among peers [5]. Thus, several hybrid push-pull architectures such as mTreebone [6] have been proposed to offer a good tradeoff between two structures. However, these P2P structures did not consider mobile platforms in a wireless network environment, only having focused on constructing overlay networks using stationary peers in a wired network environment.

With recent bandwidth improvement in wireless networks and advent of various mobile devices, mobile IPTV systems have become feasible. So far most of research efforts on mobile P2P streaming systems have been made in MANET [16] or iMANET [17]. However, they are not suitable for large-scale P2P systems due to their characteristics of high energy consumption and low scalability caused by direct communication between peers. To provide IPTV services with mobile devices, some systems have attempted to simply add mobile devices to the existing P2P streaming structure based on wired networks via APs [18]. As a result, they did not consider the characteristics of mobile computing environment such as low bandwidth, unstable wireless signal, and peers' high mobility when designing the mobile P2P live streaming systems.

On the other hand, many handover schemes have been proposed at each protocol layer to support mobile peers' mobility. The handover schemes at the data link layer have focused on performing the fast handover between two APs according to their signal strength using RSSI (received signal strength indication) [10, 11]. At the network layer, the proposed schemes have attempted to reduce the handover period by receiving CoA (care of address) as quickly as possible [12, 13]. To do so, they predict mobile peers' moving directions when they move from one subnet to another subnet. At the transport layer, new protocols such as mSCTP and mDCCP to add mobility features to the existing TCP and UDP protocols have been proposed [14, 15]. In these handover schemes operating at the layers below the application layer, however, it is not possible to avoid data loss because the network is physically disconnected during the handover period. To compensate for such data loss at lower layers, several application layer handover schemes have therefore been developed apart from lower layer schemes [19–21]. However, those application layer handover schemes have been developed based only on CDN structures, not considering the P2P live streaming structures.

3. Design Considerations for Seamless Mobility

We describe several considerations to reflect the characteristics of mobile devices and wireless networks when designing an efficient application-layer handover scheme for P2P live streaming systems.

3.1. Data Transmission Manners for P2P Live Streaming. In a mesh-based P2P streaming architecture, the data unit for data delivery and display is a video block. Each video is divided into small blocks, which are distributed to other peers through the mesh structure. Each peer displays video blocks after buffering and sequencing received blocks in memory. Peers periodically exchange their status using buffer maps that represent the blocks' availability in peers' buffers. After obtaining buffer maps from its neighbors, a peer can determine to which neighbor peers it will request missing blocks. In such a mesh-based streaming structure, where data are transmitted in a pull manner, playback should be delayed until a peer can receive the sufficient data to start to playback a video.

On the other hand, in a tree-based P2P streaming architecture, peers receive video data from an origin server or parent peers only in a push manner. This structure enables peers to transmit data at a faster speed because they can keep transmitting data without any specific requests from their child peers once the tree structure is constructed.

In general, a mobile peer tends to experience data loss while communicating with others due to unstable wireless network environment. The mesh structure is thus more suitable for mobile P2P streaming architecture since a mobile peer can receive data more stably by requesting the retransmission of lost blocks. However, a peer has the longer delay when receiving data in a pull manner compared to that in a push manner. This is because, in a pull manner, it can receive the desired blocks from neighbor peers by specifically requesting them after exchanging buffer maps. Moreover, since a mobile peer cannot receive any data block during the handover, the transmission delay can be much longer after the handover. In our P2P live streaming system, a mobile peer therefore receives data in a push manner only for a short period around the handover to receive data at a faster speed.

3.2. Criteria for Selecting Neighbor Peers for a Mobile Peer. Even though a mobile peer is not able to receive any data only for a short period due to network condition, especially during the handover, it must continue to playback the video, keeping consuming the data that have been buffered before that period. As a result, a mobile peer may experience playback jitter unless it can quickly obtain the required data as soon as the network is available. To avoid such degradation of playback quality in our P2P live streaming system, we consider the proximity to a mobile peer as the first criterion when selecting neighbor peers. This can reduce the network latency through the shortened transmission route.

On the other hand, peers buffer the data corresponding to a specific period around their current playback positions. Since lag times between an origin server and peers are getting large as the number of peers increases, however, peers' buffering periods also become widely distributed. Furthermore, when supporting VCR operations, peers' playback positions become distributed more widely. Note that a mobile peer can receive data at a faster speed as neighbor peers are buffering more data required for its immediate playback. The other criterion is therefore how much data required by a mobile peer a candidate peer is currently buffering.

3.3. Handover Prediction. If a mobile peer cannot receive sufficient data before the handover, they may not be able to continue playing back the video due to lack of buffered data even though they receive the data at a fast speed after the handover. To prevent this situation, it is necessary to receive as much data as possible by predicting the handover before it actually happens. In our P2P live streaming system, we adopt the most common technique using signal strength of APs, that is, RSSI, to predict the handover. In other words, a mobile peer predicts that handover will occur soon when the difference of signal strength between the current and the target AP becomes smaller than the given threshold. Once the handover is predicted, neighbor peers can transmit data to the mobile peer at a faster speed by switching their transmission manner to a push one.

4. An Efficient Handover Scheme for P2P Live Streaming

In this Section, we propose a new application-layer handover scheme to provide the seamless mobility in P2P live streaming systems. We first describe our handover behavior model according to the state of each peer. We then explain our agent peer selection policy to minimize playback jitter.

4.1. Handover Behavior Model

4.1.1. State Transition for Handover Behavior. In our proposed handover scheme, there are four states for each mobile peer: normal (N), prediction (P), handover (H), and after-handover (A) state. When a mobile peer is triggered by one of several events, it changes its state after taking the corresponding action depending on its current state as follows.

(i) The N state represents the state where the signal of the current AP measured by a mobile peer is stronger than those of other adjacent APs by the threshold value for handover prediction.

(ii) The P state indicates the period between after the handover is predicted to occur and before the handover actually occurs.

(iii) The H state indicates the period when a mobile peer is switching its current AP to a new one while the network is being disconnected due to the handover.

(iv) The A state represents the period when a mobile peer is receiving data through a newly connected AP for a short period until going to N state.

Figure 1 shows a state transition diagram of a mobile peer for its handover behavior. A mobile peer can move to another state depending on the events relating to the handover. First, a mobile peer usually enters into N state when it joins. When the difference of RSSI values of the current and target AP from the mobile peer becomes smaller than the threshold, it transits to P state. When the handover actually occurs, its state moves to H state. If the signal of the target AP from the mobile is getting weaker at P state, it returns to N state. Once the mobile peer is connected to the target AP, it transits to A state. Its state is changed to N state when the following two requirements are met: one is that the amount of buffered data should reach the same buffering level as that for initial playback and the other is that the difference of RSSI values of the current and each of other adjacent APs should become larger than the threshold value. If the handover occurs successively at A state, it returns to H state.

4.1.2. Data Transmission Route and Manner. In our P2P streaming system, a mobile peer has different data transmission route and manner according to its current state. As

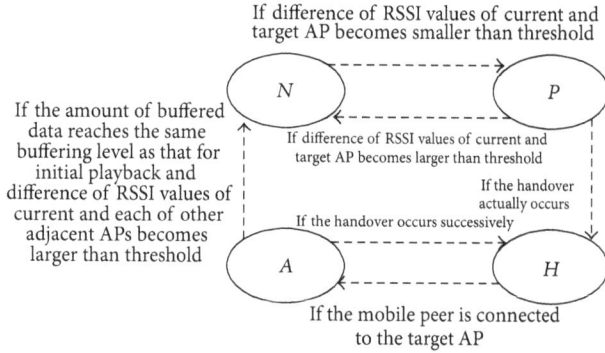

FIGURE 1: State transition diagram for handover behavior.

TABLE 1: Summary of simulation parameter values.

Parameter	Default value
Moving speed of mobile peers	5~20 Km/h
Handover latency	0.5~1 second
Bandwidth of backbone networks	100 Gbps
Bandwidth of wired networks	100 Mbps
Bandwidth of wireless networks	20 Mbps
Video playback rate	750 Kbps
Number of peers	1000
Number of neighbor peers	5
RTT values	1–200 ms
Buffering interval	3 seconds
W	0.5

shown in Figure 2(a), at N state, it receives data from neighbor peers in a pull manner ($N2M\_PL$: neighbors to mobile in a pull manner). To maximize the amount of data that can be received at P state before the handover, the mobile peer receives data from neighbor peers after switching the transmission manner to a push one ($N2M\_PS$: neighbors to mobile in a push manner) as shown in Figure 2(b). To transmit data to the mobile peer at a faster speed after the handover, an agent peer also receives as much data as possible at this state. The agent peer thus receives data from neighbor peers in a push manner ($N2A\_PS$: neighbors to agent in a push manner).

As shown in Figure 2(c), at H state, the mobile peer cannot receive any data during the handover while the agent peer maintains $N2A\_PS$ because it can still receive data from neighbor peers. The mobile peer transits to A state once it is connected to a new AP. It receives data as fast as possible from the agent peer as well as from neighbor peers in a push manner to minimize playback jitter ($N2M\_PS$, $A2M\_PS$: agent to mobile in a push manner) as shown in Figure 2(d). After returning to N state, the mobile peer receives data from newly selected neighbor peers in a pull manner ($N2M\_PL$) as shown in Figure 2(e).

4.2. Agent Peer Selection Policy. To further improve the playback continuity of a mobile peer when the handover occurs, a tracker server selects an agent peer for the mobile peer among stationary peers. In our P2P live streaming system, the agent peer plays an important role in reducing playback jitter caused by the handover. The agent peer receives data in place of the mobile peer from the moment the handover is predicted until the handover ends. It then transmits the buffered data to the mobile peer as fast as possible through a new AP so that the mobile peer cannot experience buffer starvation. To select the most suitable peer as an agent peer to perform this task, we thus consider a couple of criteria: RTT value and the appropriateness of the buffered period. The RTT value is considered to measure the transmission delay between a new AP and a candidate peer. The appropriateness of buffered period is considered to estimate how appropriate the period of the data buffered in a candidate peer is for the immediate playback of a mobile peer. In other words, it indicates how much required data from a mobile peer's perspective a candidate peer is currently

buffering. The following equation represents the criteria to select an agent peer for a mobile peer:

$$\text{MIN}\left\{W \times \text{RTT}_i + (1 - W) \times \text{ABP}_i\right\}. \qquad (1)$$

In (1), i is an index for a specific candidate peer, RTT_i and ABP_i denotes the RTT value from a new AP and the appropriateness of the buffered period of a candidate peer with an index of i, respectively, and W is the weight value between RTT_i and ABP_i. The candidate peer with a minimum value of (1) is selected as an agent peer for the corresponding mobile peer.

It is noted that W can be adjusted according to peers' current situation. If the network latency affects the performance more significantly in some situation, we need to increase W. On the contrary, in the situation where the appropriateness degree of buffered data for the playback of the mobile peer is a more important factor to improve the performance, it is necessary to decrease W.

5. Experimental Evaluation

To show the effectiveness of our proposed handover scheme for mobile P2P live streaming systems, we have performed extensive simulations using a QualNet network simulator. The default values of simulation parameters are shown in Table 1. They are used throughout our simulations unless otherwise indicated. It is assumed that mobile peers are moving at a speed of the range from 5 to 20 Km/h and the latency range of the handover is from 0.5 to 1 second. The bandwidths of backbone, wired, and wireless networks are set to 100 Gbps, 100 Mbps, and 20 Mbps, respectively. Each video has 750 Kbps playback rate. The numbers of peers are 1000 and each peer can have at most 5 neighbor peers. The RTT values between peers range from 1 to 200 ms and the average buffering interval starting from the current playback positions of peers is 3 seconds. The weight value of (1), that is, W, is set to 0.5.

5.1. Effectiveness of Our Handover Scheme. Figure 3 shows the comparison of playback continuity ratios for 7 seconds around the handover in two cases: with and without our

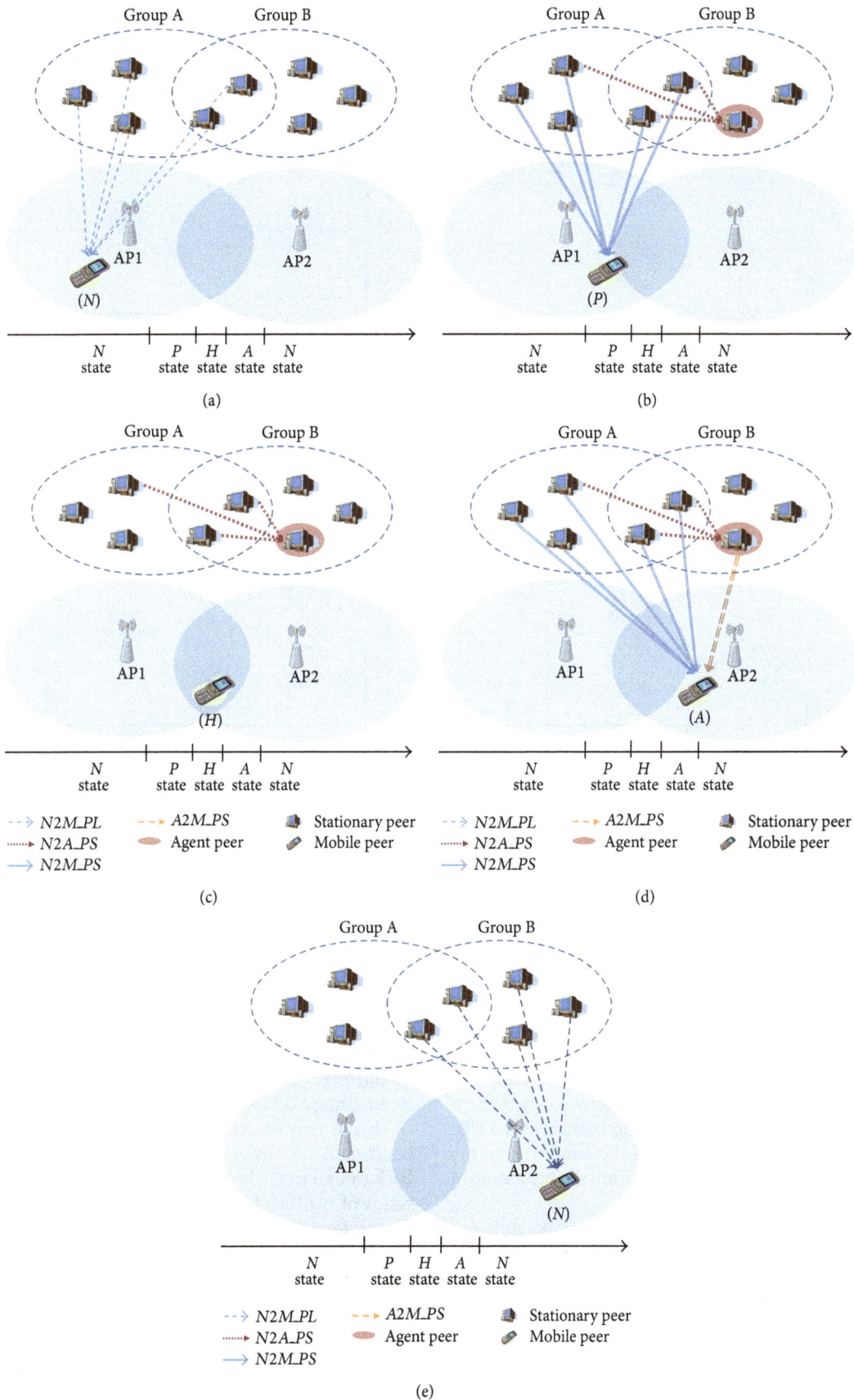

FIGURE 2: Data transmission route and manner according to a mobile peer's state: (a) N state before handover, (b) P state, (c) H state, (d) A state, and (e) N state after handover.

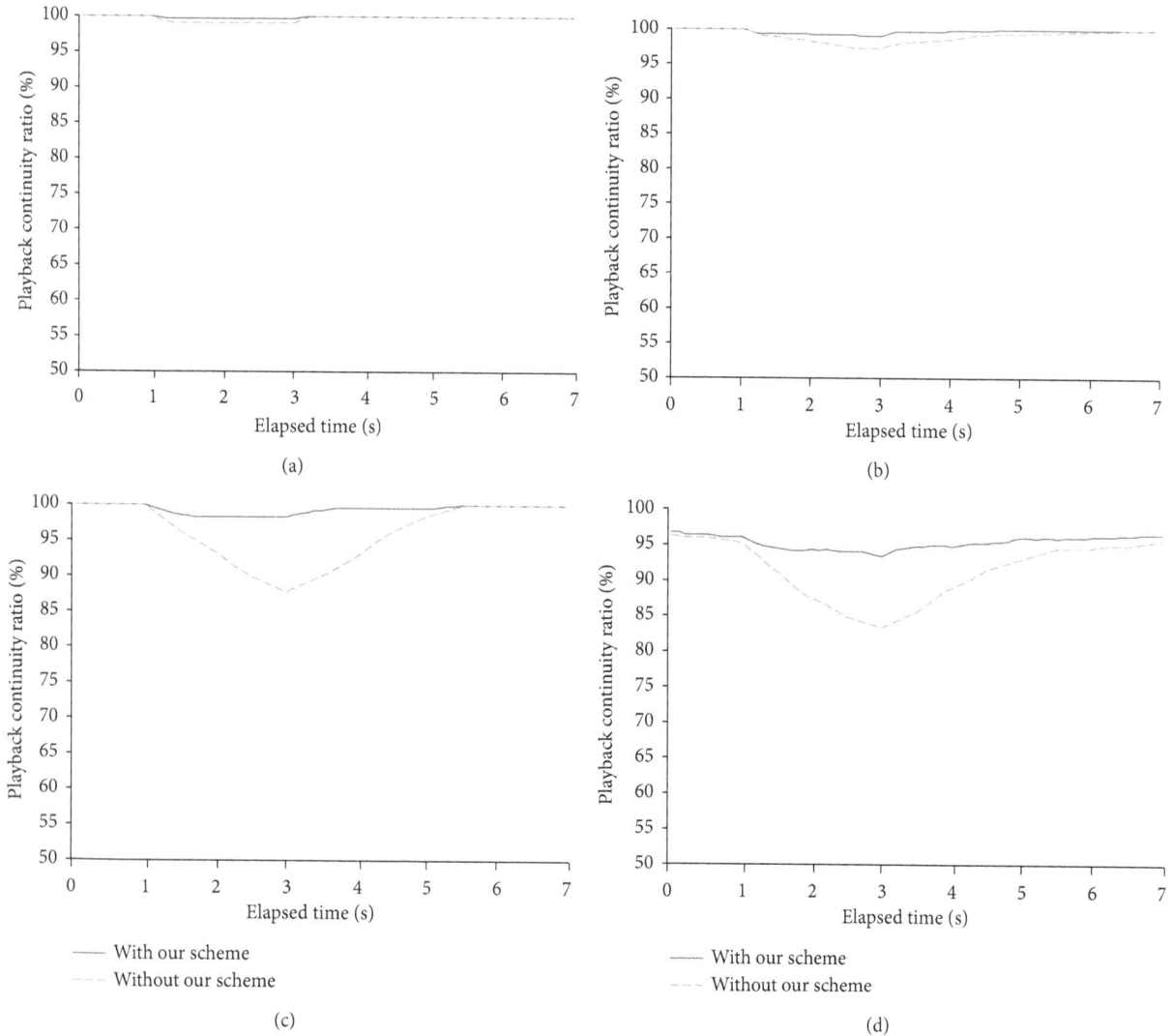

FIGURE 3: Playback continuity ratios of two cases according to effective wireless network bandwidth: (a) with effective bandwidth of 16 Mbps, (b) with effective bandwidth of 14 Mbps, (c) with effective bandwidth of 12 Mbps, and (d) with effective bandwidth of 10 Mbps.

proposed scheme. The case without our proposed scheme indicates the existing P2P live streaming systems based on mesh structure that do not perform any particular operation for the handover. To examine the impact of effective wireless network bandwidths on the performance, we varied them from 16 Mbps to 10 Mbps by generating background traffic from 4 to 10 Mbps. You can see that the handover occurs immediately after one second position from the beginning of the x-axis in Figure 3.

The experimental results show that the case with our handover scheme improved the playback continuity ratios significantly compared to the case without our scheme. Especially, in case of effective bandwidth of 10 Mbps in Figure 3(d), the difference of the minimum playback continuity ratios between two cases was 10.1%. That is, the minimum playback continuity ratio of the case with our scheme was 93.4% while that of the case without our scheme was only 83.3%. This implies that our handover scheme performs effectively no matter how much bandwidth the wireless network provides.

This is possible because our handover scheme can obtain the sufficient amount of data required by a mobile peer in advance through handover prediction and an agent peer. The mobile peer can also rush to receive data after the handover by adopting a push transmission manner.

It can also be seen from Figure 3 that, as the effective bandwidth of wireless networks decreases, that is, as the background traffic becomes heavier, the playback continuity ratios of two cases also decrease. It is noted that, however, the performance degradation degree in the case with our handover scheme is much lower than that without our scheme. As the background traffic increases from 4 Mbps to 10 Mbps, the minimum playback continuity ratio of the case with our scheme was reduced by only 6.3% while that without our scheme was reduced by 15.7%. This result indicates that our handover scheme can utilize the decreased network bandwidth efficiently. That is, a mobile peer can overcome the shortage of the network bandwidth after the handover by receiving as much data as possible before and during

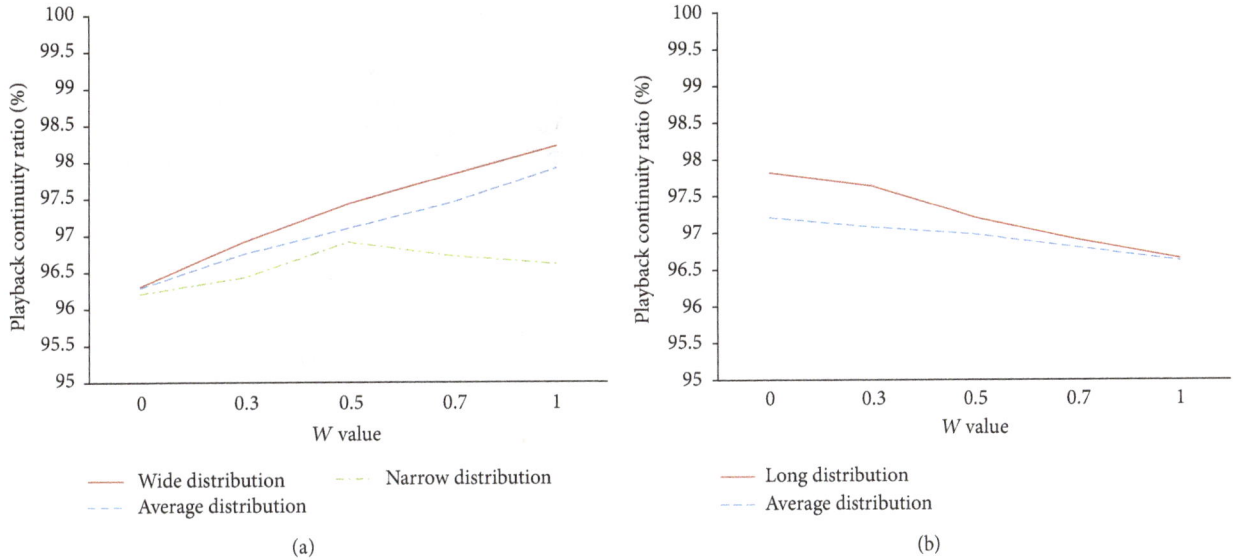

FIGURE 4: Impact of two criteria for agent peer selection on the playback continuity ratio: (a) RTT values and (b) lengths of average buffering interval.

the handover together with an agent peer. It can also reduce the transmission latency and the number of control messages considerably by receiving data in a push manner for the period around the handover.

5.2. Impact of Agent Peer Selection Criteria. To investigate the impact of two criteria in (1) on the performance when selecting an agent peer, we have made three different distributions for each criterion. The RTT values between a mobile peer and other candidate peers are distributed as follows: narrow distribution (5–100 ms), average distribution (1–200 ms), and wide distribution (0.1–1000 ms). The lengths of peers' average buffering intervals starting from the current playback positions are also distributed as follows: short distribution (1 second), average distribution (3 seconds), and long distribution (5 seconds). In our simulations, peers' RTT values and buffering interval lengths are randomly generated within the given range of each distribution. We also set the effective wireless network bandwidth to 12 Mbps for these simulations.

Figure 4(a) shows the playback continuity ratios according to the distribution degree of RTT values while fixing the distribution degree of buffering interval lengths to a short one. In case we employ wide and average distribution for buffering interval lengths, we achieved the highest playback continuity ratio when W is 1 while achieving the lowest one when W is 0. The differences between the highest and lowest ratio in case of wide and average distribution are 1.9% and 1.6%, respectively. Note that, as W increases, the playback continuity ratios also keep increasing almost linearly. This implies that, as peers are distributed more widely, that is, as the differences in RTT values are getting larger, it is more advantageous to select the peer with shorter RTT value from the mobile peer as an agent peer as indicated in (1). When using the narrow distribution for buffering interval lengths, the playback continuity ratio was highest when W is 0.5 while it was lowest when W is at both ends 0 and 1. This indicates

that the short distribution of RTT values and the narrow distribution of buffered interval lengths have similar degree from the perspective of the agent peer selection criteria. In such peers' situation, we thus need to consider two criteria to the same degree to maximize the performance.

Figure 4(b) shows the performance according to the distribution degree of buffering interval lengths when the distribution degree of RTT values is fixed to a narrow one. The simulation results show similar trends to those in Figure 4(a) except that the performance improved with the decreased value of W. That is, in case of the long and average distribution for RTT values, we achieved the highest playback continuity ratio when W is 0 while achieving the lowest one when W is 1. The improvement ratios in the long and average distribution are 1.2% and 0.6%, respectively. It can also be seen that the performance keeps improving with the decreased value of W. This indicates that, as average buffering interval lengths of peers are getting longer, it is more beneficial to select the peer that are buffering more data required by the mobile peer. In this case, we thus need to put more weight on the appropriateness of buffering period as indicated in (1).

From the simulation results in Figure 4, we have observed the impact of two criteria including RTT values and the appropriateness of the buffered periods when selecting an agent peer. Note that we can make the agent peer selection policy flexible in different peers' situation by adjusting W value, thereby further improving the performance.

6. Conclusions

We have presented an efficient application-layer handover scheme in mobile P2P live streaming systems to improve the playback continuity ratio significantly even though the handover occurs. This improvement was possible because a mobile peer can receive the sufficient amount of data required for the video playback in advance through handover prediction and an agent peer. It can also receive data at a faster speed for the period around the handover by employing a push transmission manner. As the video contents requiring the higher playback rate, such as 3D and UD (ultrahigh definition) TV, are emerging, our handover scheme is expected to be widely applied to many applications in mobile platforms to provide seamless mobility even though the handover occurs.

Conflict of Interests

The authors declare that there is no conflict of interests regarding the publication of this paper.

Acknowledgments

This research was supported by Basic Science Research Program through the National Research Foundation of Korea (NRF) funded by the Ministry of Education, Science and Technology (2013R1A1A2009913) and the MKE (The Ministry of Knowledge Economy), Korea, under the ITRC (Information Technology Research Center) support program (NIPA-2014-H0301-14-1017) supervised by the NIPA (National IT Industry Promotion Agency).

References

[1] X. Cheng and J. Liu, "NetTube: exploring social networks for peer-to-peer short video sharing," in *Proceedings of the 28th IEEE Conference on Computer Communications (INFOCOM '09)*, pp. 1152–1160, Rio de Janeiro, Brazil, April 2009.

[2] E. Kim and C. Lee, "An on-demand TV service architecture for networked home appliances," *IEEE Communications Magazine*, vol. 46, no. 12, pp. 56–63, 2008.

[3] C. Lee, E. Hwang, and D. Pyeon, "A popularity-aware prefetching scheme to support interactive P2P streaming," *IEEE Transactions on Consumer Electronics*, vol. 58, no. 2, pp. 382–388, 2012.

[4] M. Castro, P. Druschel, A. Kermarrec, A. Nandi, A. Rowstron, and A. Singh, "SplitStream: high-bandwidth multicast in cooperative environments," in *Proceedings of the 19th ACM Symposium on Operating Systems Principles (SOSP '03)*, pp. 298–313, Bolton Landing, NY, USA, October 2003.

[5] X. Hei, C. Liang, J. Liang, Y. Liu, and K. Ross, "A measurement study of a large-scale P2P IPTV system," *IEEE Transactions on Multimedia*, vol. 9, no. 8, pp. 1672–1687, 2007.

[6] F. Wang, Y. Xiong, and J. Liu, "mTreebone: a hybrid tree/mesh overlay for application-layer live video multicast," in *Proceedings of the 27th IEEE International Conference on Distributed Computing Systems (ICDCS '07)*, Toronto, Canada, June 2007.

[7] D. Kim, E. Kim, and C. Lee, "Efficient peer-to-peer overlay networks for mobile IPTV services," *IEEE Transactions on Consumer Electronics*, vol. 56, no. 4, pp. 2303–2309, 2010.

[8] Q. Qi, Y. Cao, T. Li, X. Zhu, and J. Wang, "Soft handover mechanism based on RTP parallel transmission for mobile IPTV services," *IEEE Transactions on Consumer Electronics*, vol. 56, no. 4, pp. 2276–2281, 2010.

[9] J. Koo and K. Chung, "Adaptive channel control scheme to reduce channel zapping time of mobile IPTV service," *IEEE Transactions on Consumer Electronics*, vol. 57, no. 2, pp. 357–365, 2011.

[10] M. Portoles, Z. Zhong, S. Choi, and C. Chou, "IEEE 802.11 link-layer forwarding for smooth handoff," in *Proceedings of the 14th IEEE International Symposium on Personal, Indoor and Mobile Radio Communications (PIMRC '03)*, pp. 1420–1424, Beijing, China, September 2003.

[11] P. Khadivi, T. Todd, and D. Zhao, "Handoff trigger nodes for hybrid IEEE 802.11 WLAN/cellular networks," in *Proceedings of the 1st International Conference on Quality of Service in Heterogeneous Wired/Wireless Networks (QSHINE '04)*, pp. 164–170, Dallas, Tex, USA, October 2004.

[12] S. Speicher and C. Cap, "Fast layer 3 handoffs in AODV-based IEEE 802.11 wireless mesh networks," in *Proceedings of the 3rd IEEE International Symposium on Wireless Communication Systems (ISWCS '06)*, pp. 233–237, September 2006.

[13] S. Sharma, N. Zhu, and T. Chiueh, "Low-latency mobile IP handoff for infrastructure-mode wireless LANs," *IEEE Journal on Selected Areas in Communications*, vol. 22, no. 4, pp. 643–652, 2004.

[14] K. Brown and S. Singh, "M-UDP: UDP for mobile networks," *ACM SIGCOMM Computer Communication Review*, vol. 26, no. 5, pp. 60–78, 1996.

[15] H. Elaarag, "Improving TCP performance over mobile networks," *ACM Computing Surveys*, vol. 34, no. 3, pp. 357–374, 2002.

[16] T. Hara, "Replica allocation in ad hoc networks with period data update," in *Proceedings of the 3rd International Conference on Mobile Data Management (MDM '02)*, pp. 79–86, Singapore, January 2002.

[17] F. Sailhan and V. Issarny, "Cooperative caching in ad-hoc networks," in *Proceedings of the 4th International Conference on Mobile Data Management (MDM '03)*, pp. 13–28, Melbourne, Australia, January 2003.

[18] T. Zahariadis, O. Negru, and F. Alvarez, "Scalable content delivery over P2P convergent networks," in *Proceedings of the 12th IEEE International Symposium on Consumer Electronics (ISCE '08)*, pp. 1–4, Vilamoura, Portugal, April 2008.

[19] T. Hong, K. Kang, D. Ahn, and H. Lee, "Adaptive buffering scheme for streaming service in intersystem handover between terrestrial and satellite systems," in *Proceedings of the 12th IEEE International Symposium on Consumer Electronics (ISCE '08)*, pp. 1–4, Vilamoura, Portugal, April 2008.

[20] C. Huang and C. Lee, "Layer 7 multimedia proxy handoff using anycast/multicast in mobile networks," *IEEE Transactions on Mobile Computing*, vol. 6, no. 4, pp. 411–422, 2007.

[21] B. Ciubotaru and G. M. Muntean, "SASHA—a quality-oriented handover algorithm for multimedia content delivery to mobile users," *IEEE Transactions on Broadcasting*, vol. 55, no. 2, pp. 437–450, 2009.

10

Effects of ADC Nonlinearity on the Spurious Dynamic Range Performance of Compressed Sensing

Rongzong Kang, Pengwu Tian, and Hongyi Yu

Zhengzhou Information Science and Technology Institute, Zhengzhou 450002, China

Correspondence should be addressed to Rongzong Kang; rongzongkang@outlook.com

Academic Editor: Zhongmei Zhou

Analog-to-information converter (AIC) plays an important role in the compressed sensing system; it has the potential to significantly extend the capabilities of conventional analog-to-digital converter. This paper evaluates the impact of AIC nonlinearity on the dynamic performance in practical compressed sensing system, which included the nonlinearity introduced by quantization as well as the circuit non-ideality. It presents intuitive yet quantitative insights into the harmonics of quantization output of AIC, and the effect of other AIC nonlinearity on the spurious dynamic range (SFDR) performance is also analyzed. The analysis and simulation results demonstrated that, compared with conventional ADC-based system, the measurement process decorrelates the input signal and the quantization error and alleviate the effect of other decorrelates of AIC, which results in a dramatic increase in spurious free dynamic range (SFDR).

1. Introduction

Traditional approaches to acquiring and sampling signal are based on Nyquist sampling theory, which states that the sampling rate must be at least twice the maximum frequency of the input signal. The increasing demand for ADC with both wider bandwidth and higher quantization bits seems to contradict with each other. The new theory of compressed sensing (CS) [1, 2] introduced an alternative data acquisition framework, which states that CS enables the acquisition and recover of sparse signals in some transform domains at a rate proportional to their information content that is much below the Nyquist rate.

Analog-to-information converter (AIC) is designed to acquire samples at a lower rate for compressed sensing system, and various architectures have been proposed of the recent work in this area, such as the random demodulator sampling architecture [3], the modulated-wideband converter [4], and others [5–8]. However, in the view of practical hardware implementation, the basic components constitute an AIC consists of mixer, integrator/low passed filter and ADC, and so forth. Among these components, the ADC commonly has the lowest dynamic range; an A/D converter's deviation from its ideal "linear" performance is

commonly characterized by the spurious-free dynamic range (SFDR) [9], which is defined as the difference in decibel, between the full-scale fundamental tone and the largest spurious harmonic component in the output spectrum. In order to make this notion precise, we will ignore the effects of any noise or nonlinearities from the other components except of ADC, since the SFDR of an AIC is typically dominated by the nonlinear process of ideal quantization and circuit-based (e.g., buffer, sample-and-hold) nonlinearities of ADC.

However there have been little literatures for characterizing and calculating the dynamic range performance of compressed sensing. In [10] a deterministic approach to dynamic range of a CS-based acquisition system is proposed, and the parameter of signal-to-quantization noise ratio is presented, whereas the dynamic parameter, that is, SFDR, is not considered. In [11] the quantization noise and dynamic range are considered for compressive imaging (CI) systems design and evaluate the quantization depth requirements for CI, while the quantization error and the SFDR performance of CS are still undiscussed. In [12] the impact of ADC nonlinearity in a mixed-signal CS system is studied, without considering the effect of ADC quantization error.

In this paper, we use an analytical approach couple with simulation results to formulate the SFDR performance of a compressed sensing system when considered with the quantization and nonlinearity of ADC. The background of compressed sensing is introduced firstly; then the power spectrum of quantization noise of AIC is analyzed numerically and the SFDR of ideal AIC-based system is derived. Furthermore, a detailed analysis of the other ADC nonlinear effects in SFDR performance of AIC-based system is presented. Finally, the behavioral simulations results are presented that clearly verify the accuracy of the analysis.

2. Background

2.1. Quantization-Limited SFDR of ADC-Based System. Quantization changes a sine wave from a smooth function to a staircase signal; due to this nonlinear effect, the output signal is composed of a large number of nonlinear distortion products. The most important contribution to the output distortion comes from the quantization process, because this is an inherently nonlinear process. In an ideal quantizer, suppose that there is no nonlinearity and noise exists except for the nonlinearity due to quantization. In this case, the spurious signal of ADC output is only produced by the quantization. When a sine wave is passed through the ideal quantizer, the Fourier series of the output signal leads to the closed-form expression for the magnitudes of the harmonic as [13]:

$$A_p = \delta_{p,1} A + \sum_{m=1}^{\infty} \frac{2}{m\pi} J_p (2m\pi A), \quad (1)$$

where A_p is the output amplitude of the pth harmonic, $\delta_{p,1}$ is the Kronecker delta function, A is the input amplitude, and J_p is the pth-order Bessel function of the first kind. Although the largest harmonic is always located roughly at $2\pi A$ when the quantization levels are larger than 20, we consider the third harmonic as the largest and the power of the largest harmonic as a function of the number of bits. As a result, the quantization-limited SFDR performance of an ideal ADC-based system is approximated by [13]:

$$\text{SFDR} = 8.07b + 3.29 \, \text{dB}. \quad (2)$$

2.2. Nonlinear-Limited SFDR of ADC-Based System. Besides the nonlinearity produced by quantization, the circuit imperfections such as capacitor mismatches and finite opamp DC gains are considered. These nonlinearities of ADC would also influence the SFDR performance of the system. The simplest form of a nonlinear system is the memoryless power series, which is based on normal polynomials:

$$z = \sum_{i=0}^{L} a_i y^i, \quad (3)$$

where z is the output signal, y is the input signal, and L is the order of the circuit nonlinearity. If the input signal is a single tone signal given by

$$y = A \cos (\omega t + \varphi_0), \quad (4)$$

then the amplitudes of the harmonic terms can be computed from (3). However in the case of analog circuits, the order of a polynomial expression is mostly limited to third order, polynomial coefficients of the 4th order and higher, and the nonlinearity caused by saturation at full scale are neglected.

When substituting (4) into (3), we get

$$
\begin{aligned}
z &= \sum_{i=0}^{L} a_i y^i \\
&= a_1 \left[A \cos (\omega t + \varphi_0) \right] + a_2 \left[A \cos (\omega t + \varphi_0) \right]^2 \\
&\quad + a_3 \left[A \cos (\omega t + \varphi_0) \right]^3, \quad (5)
\end{aligned}
$$

$$
\begin{aligned}
z &\cong a_1 \cos (\omega t + \varphi_0) + \frac{1}{2} a_2 A^2 \left[\cos (2\omega t + 2\varphi_0) \right] \\
&\quad + \frac{1}{4} a_3 A^2 \left[\cos (3\omega t + 3\varphi_0) \right]^2.
\end{aligned}
$$

The specific relation between polynomial coefficients and harmonic power can be expressed in general [14]. The same analysis can be done when the input signal is supposed to be two tone signals; it will cause the production of more terms, the specific terms harmonics, and intermodulation. Furthermore, the dynamic range performance of ADCs is specified in terms of one-tone and two-tone SFDR [15].

While in practice, tests which have been developed to measure the performance mostly rely on Fourier analysis using discrete Fourier transform (DFT) and the fast Fourier transform (FFT). The input analog signal is first sampled at Nyquist rate; the harmonics and intermodulation distortion are calculated through the input signal spectrum, which is estimated from the time-domain samples with nonlinear distortion via DFT. However the DFT-based method needs to avoid the leakage of the input frequency and the number of periods of the input waveform in the sample record should not be a nonprime integer submultiple of the record length, further the ADC needs to have a high resolution, which limits the maximum achievable sampling rate.

As an alternative solution to high-speed ADCs, AIC-based system enables high resolution at high frequencies while only using low frequency, sub-Nyquist ADCs [3–8]. In this work, we investigate the effect of nonlinearity induced by quantization and other circuit's nonidealities of ADC on the AIC-based system and examine the SFDR performance in the presence of these nonlinearities.

2.3. Analog-to-Information Converter (AIC). There have been many theoretical discussions on AIC system in the literature [3–8], in this work, the block diagram of a typical AIC implementation [3] called the random demodulator shown in Figure 1 is considered to compare with the conventional ADCs. In this architecture, the input signal $x(t)$ is mixed by a different pseudorandom number $p_c(t)$ waveform; then the mixer output is integrated over a time period of $1/M$. Finally, the integrator outputs are sampled and quantized, by a traditional integrate-and-dump ADC at M Hz.

Note that this AIC architecture employs sub-Nyquist rate ADCs, and the input signal is mixed with the PN sequence and sent to integrator before sampling. As a result, the spectrum of the signal sent to the ADC is relatively flat

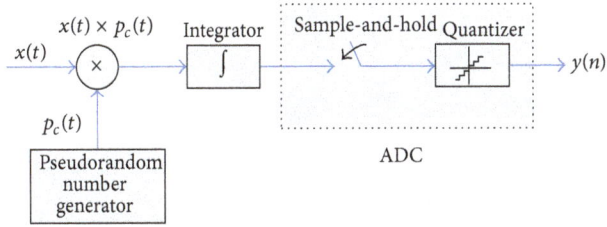

FIGURE 1: Block diagram of the random demodulator.

within the filter pass band, and then the harmonic and intermodulation energy due to the nonlinearity of ADC is spread along the signal bandwidth rather than concentrate on a few tones, which can lead to a better SFDR performance after reconstruction. In the following section we present our framework for investigating the impacts of nonlinearity caused by quantization and other circuits induced on the SFDR performance of AIC-based system.

3. SFDR Performance of AIC-Based System

3.1. Quantization-Limited SFDR of AIC-Based System. In an ideal case the dynamic range performance is mainly limited by quantization error; the spectra of the AIC quantization output is analyzed in this section.

As we know, the time-domain expression of the measuring process of AIC is given by

$$y_i = \langle x, \phi_i \rangle = \sum_{j=1}^{N} \phi_{ij} x_j, \qquad (6)$$

where ϕ_{ij} is the element of measurement matrix and x_j is the element of the input signal. Suppose that the measurement matrix is sub-Gaussian random matrix; then the element ϕ_{ij} is independent centered sub-Gaussian random variables with variance $1/M$, given $S_{i,j} = \phi_{i,j} x_j$; then

$$y_i = \sum_{j=1}^{N} S_{ij}. \qquad (7)$$

Then we can get the mean and variance of $S_{i,j}$:

$$E\left[S_{i,j}\right] = E\left[\phi_{i,j} x_j\right] = x_j E\left[\phi_{i,j}\right] = 0.$$

$$D\left[S_{i,j}\right] = E\left[S_{i,j}^2\right] = x_j^2 E\left[\phi_{i,j}^2\right] = \frac{x_j^2}{M}. \qquad (8)$$

According to the central limit theorem, when $N \to \infty$, the y_i subject to Gaussian distribution with mean 0 and variance $\sum_{j=1}^{N} (x_j^2/M) = \|X\|_2^2/M$.

As we know that when the input signal subjected to Gaussian distribution with mean 0, then the relation between autocorrelation function $R_e(m)$ of quantization error and input signal can be expressed as follows:

$$R_e(m) = \frac{\Delta^2}{2\pi^2} \sum_{k=1}^{\infty} \frac{1}{k^2} \exp\left[-4\pi^2 \frac{\sigma^2}{\Delta^2} k^2 \left(1 - r_y(m)\right)\right], \qquad (9)$$

where Δ is the quantization step size and σ^2 is the variance of the input signal. $r_y(m) = R_y(m)/R_y(0)$ represents the normalized autocorrelation function. While the autocorrelation function of the measurement value can be expressed as

$$
\begin{aligned}
R_y(m) &= E\left[y_i y_{i+m}\right] \\
&= E\left[\sum_{j=1}^{N} \phi_{i,j} x_j \sum_{k=1}^{N} \phi_{i+m,k} x_k\right] \\
&= E\left[\sum_{j=1}^{N} \sum_{k=1}^{N} \phi_{i,j} \phi_{i+m,k} x_j x_k\right] \\
&= \sum_{j=1}^{N} \sum_{k=1}^{N} E\left[\phi_{i,j} \phi_{i+m,k}\right] x_j x_k.
\end{aligned}
\qquad (10)
$$

Because the element of the measurement matrix is independent, then $R_y(0) = \|X\|_2^2/M$, when $j = k$ and $m = 0$, for others $R_y(m)$ equal to 0, so normalized autocorrelation function is

$$r_y(m) = \begin{cases} 1, & m = 0, \\ 0, & \text{else.} \end{cases} \qquad (11)$$

So, the autocorrelation function $R_e(m)$ of quantization error is

$$R_e(m) = \begin{cases} \dfrac{\Delta^2}{2\pi^2} \sum_{k=1}^{\infty} \dfrac{1}{k^2}, & m = 0, \\[3mm] \dfrac{\Delta^2}{2\pi^2} \sum_{k=1}^{\infty} \dfrac{1}{k^2} \exp\left[-4\pi^2 \dfrac{\sigma^2}{\Delta^2} k^2\right], & \text{else,} \end{cases} \qquad (12)$$

where $\sigma/\Delta \geq 1$, and when $\sigma/\Delta = 1$ and $m \neq 0$,

$$R_e(m) = \frac{\Delta^2}{2\pi^2} \left[\frac{e^{-4\pi^2}}{1} + \frac{e^{-16\pi^2}}{4} + \frac{e^{-36\pi^2}}{9} + \cdots\right]. \qquad (13)$$

For $e^{-4\pi^2} \approx 7 \times 10^{-18}$, $e^{-16\pi^2} \approx 2 \times 10^{-69}$, we get $R_e(m) \approx 0$, when $m \neq 0$, and $R_e(0) = (\Delta^2/2\pi^2) \sum_{k=1}^{\infty} (1/k^2) = (\Delta^2/12)$.

From the above analysis, we know that $R_e(m)$ is approximated to δ function, and according to the Fourier transform relationship between power spectrum and autocorrelation function, the power spectrum of quantization noise is white noise spectrum. As a result, the spurious energy due to the quantization effect of ADC is spread to the whole bandwidth, and we can get a better SFDR performance of AIC-based system compared with the conventional ADC-based system.

3.2. Nonlinear-Limited SFDR of AIC-Based System. Compared with the analysis of the conventional ADC-based system, in AIC-based system, the input signal goes through random projection, filtering, and sampling.

A signal x can be viewed as a $N \times 1$ column vector in $\mathbb{R}^N$ with elements $x[n]$, $n = 1, 2, \ldots, N$. Let the matrix $\Psi = [\psi_1, \psi_2, \ldots \psi_N]$ have columns which form a basis of vectors in $\mathbb{R}^N$. And then, any signal x can be expressed as

$$x = \sum_{i=1}^{N} s_i \psi_i \quad \text{or} \quad x = \Psi s, \quad (14)$$

where s is the $N \times 1$ column vector of weighting coefficients $s_i = \langle x, \psi_i \rangle$.

Consider a generalized linear measurement process of a signal x which is K-sparse. When we say that x is K-sparse, we mean that it is well reconstructed or approximated by a linear combination of just K basis vectors from Ψ, with $K \ll N$. That is, there are only K of the s_i in (1) that are nonzero and $(N - K)$ are zero. Let Φ be an $M \times N$ measurement matrix, $M \ll N$ where the rows of Φ are incoherent with the columns of Ψ. The incoherent measurements can be obtained by computing M inner products between x and the rows of Φ as in $y_j = \langle x, \phi_j \rangle$. It can also be expressed as

$$y = \Phi x = \Phi \Psi s = \Theta s, \quad (15)$$

where $\Theta := \Phi \Psi$ is a $M \times N$ matrix. It is proved that Φ does not depend on the signal x and it can be constructed as a random matrix such as Gaussian matrix.

Then according to the nonlinearity model of ADC, we substitute the transform-domain samples into (3), and then we can get the measurement output of the AIC with nonlinear effect as follows:

$$z = \sum_{i=0}^{L} a_i y^i = a_1 [\Phi \Psi s] + a_2 [\Phi \Psi s]^2 + a_3 [\Phi \Psi s]^3. \quad (16)$$

3.3. Reconstruction of Frequency Sparse Signal. After quantization and sampling of ADC, we get the measurement in discrete values, in order to evaluate the SFDR performance of the AIC-based system, we need to compute the spectrum of the reconstruction signal. So, in this section, we frame the reconstruction problem for the AIC-based system with the nonlinearity effect.

Furthermore, the spectrum of the input signal is estimated from the measurement value z with nonlinear distortion via solving the following optimization problem:

$$\widehat{s} = \arg\min \|s\|_1 \quad \text{s.t.} \quad \|y - \Phi \Psi s\|_2 \le \varepsilon_n + \varepsilon_d, \quad (17)$$

where ε_n is the error due to the noise and ε_d is the error due to the nonlinear distortion.

Up to now, there are many mature algorithms to resolve this convex optimization problem, including interior-point algorithms [16, 17], gradient projection [18], iterative thresholding [19, 20], and greedy approaches such as orthogonal matching pursuit (OMP) [21, 22]. Here we use the algorithm of basis pursuit with denoising [23] to resolve the reconstruction problem for evaluation of SFDR performance of AIC-based system.

4. Simulation Results

Figure 2 shows the SFDR performance of conventional ADC-based system and AIC-based system of ideal ADC with

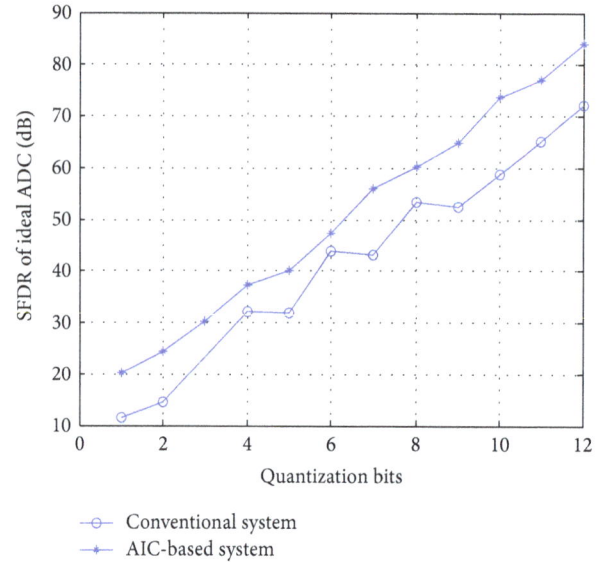

FIGURE 2: SFDR performance of AIC-based system and ADC-based with ideal ADC.

a single sinusoidal input for different quantization bits. Φ is set to an $M \times N$ Gaussian random measurement; $M = 256$, and $N = 1024$. The input signal frequency is $f_0 = 64$ Hz, $f_s = (5 * 64 - 1)$ Hz, and use BPDN [23] as the reconstruction algorithm. Every measurement was repeated 300 times to test the reproducibility.

As shown in Figure 2, the SFDR performance of AIC-based system outperforms that of conventional ADC-based system. That is because in the conventional ADC-based system, noise spectrum of sinusoid signals consists of discrete components, and the harmonic is concentrated in the odd multiple of its fundamental frequency, while in the AIC-based system the spectrum of quantization error is uniformly distributed. However the total quantization noise power represented by the area under the noise spectrum is approximately equal to $\Delta^2/12$, for AIC-based system the spurious energy is spread along the whole signal bandwidth; then each harmonic of the quantization error is thereby pulled downward into a more dense portion of the noise spectrum leading to increasing in SFDR performance. The observation from this simulation was intuitively illustrated in Figure 2.

Figure 3 shows a snapshot of the single-tone reconstructed error spectrum for conventional system and CS-based system. The second-order $a_2 = 0.1$ and third-order distortion coefficients $a_3 = 0.1$. As we can see, in the conventional ADC-based system, the spurious harmonic due to the ADC nonlinearity concentrates on the multiple of fundamental frequency, whereas, in the CS-based system, the spurious energy is spread along the whole signal bandwidth. Meanwhile amplitude of the spurious harmonic of AIC-based is lower than that of ADC-based system.

Figures 4 and 5 show the SFDR performance of ADC-based system and AIC-based system for two-tone input with quantization and different nonlinear distortion coefficients. $f_1 = 16$ Hz, $f_2 = 256$ Hz, $f_s = 1024$ Hz, and

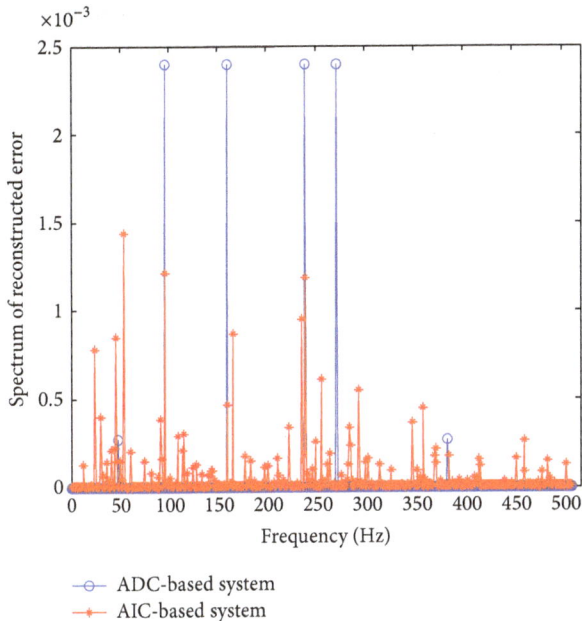

FIGURE 3: Spectrum of reconstruction error comparison between the conventional ADC-based system and the AIC-based system.

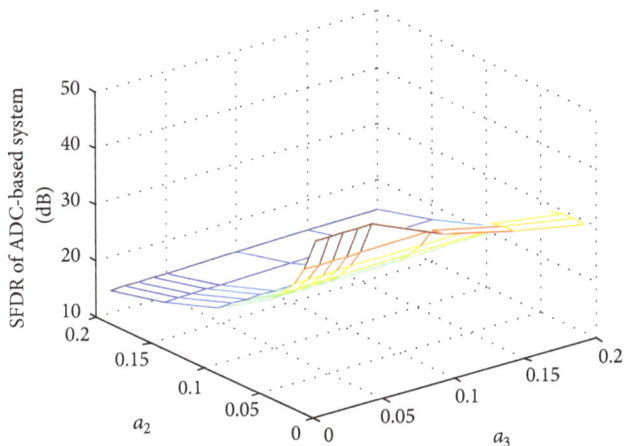

FIGURE 4: SFDR performance of conventional ADC-based system with other nonlinear effects.

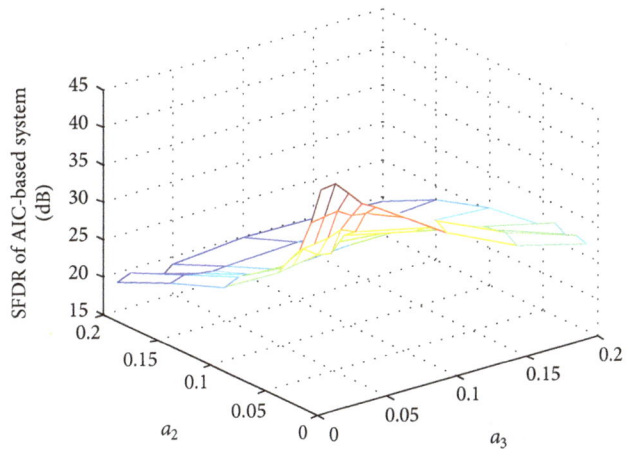

FIGURE 5: SFDR performance of for AIC-based system with other nonlinear effects.

quantization bits $N = 4$. The reconstruction algorithm and other simulation conditions are set the same as in Figure 2.

As we can see, both of the SFDR performances decrease when nonlinear distortion becomes large with the second- and third-order distortion coefficients increase. The simulation results also indicate that the second-order distortion influences the SFDR performance more seriously than that of the third-order distortion.

Comparing the results of Figure 4 with Figure 5, we can see that the SFDR performance of AIC-based system outperforms that of the ADC-based system when introducing the nonlinearity with both the quantization and circuit nonideality. This is because the randomization in AIC-based system changes the distribution of the error power from ADC nonlinear distortion; the signals sent to ADCs in the conventional Nyquist sampling architecture are original sinusoid signals, whereas those in the AIC-based system have relatively flat spectrum. As a result, by spreading the spurious energy along the signal bandwidth, the CS randomization relaxes the requirement on the ADC SFDR specification.

5. Conclusions

In this paper, we compare the SFDR performance of AIC-based system and conventional ADC-based system when considering both nonlinearity due to quantization and other circuit nonideality of ADC. We demonstrate that the quantization noise of AIC is spectrally white and uniformly distributed, and the quantization harmonics of AIC-based system is spread to the whole bandwidth, which means an improvement of SFDR performance. We show that AIC-based systems are less sensitive to the nonlinearity of ADC because of the CS randomization, which provides improvement of SFDR performance compared with conventional ADC-based system. Our results suggest that the second- and third-order distortion coefficients and quantization bits are the main factors that affect the SFDR performance of compressed sensing. The results presented in this paper can also be easily extended to the case when the signals input to AIC are multisinusoids.

Conflict of Interests

The authors declare that they have no conflict of interests regarding the publication of this paper.

References

[1] E. J. Candès and J. Romberg, "Quantitative robust uncertainty principles and optimally sparse decompositions," *Foundations of Computational Mathematics*, vol. 6, no. 2, pp. 227–254, 2006.

[2] D. L. Donoho, "Compressed sensing," *IEEE Transactions on Information Theory*, vol. 52, no. 4, pp. 1289–1306, 2006.

[3] J. A. Tropp, J. N. Laska, M. F. Duarte, J. K. Romberg, and R. G. Baraniuk, "Beyond Nyquist: efficient sampling of sparse

bandlimited signals," *IEEE Transactions on Information Theory*, vol. 56, no. 1, pp. 520–544, 2010.

[4] M. Mishali, Y. C. Eldar, O. Dounaevsky, and E. Shoshan, "Xampling: analog to digital at sub-Nyquist rates," *IET Circuits, Devices and Systems*, vol. 5, no. 1, pp. 8–20, 2011.

[5] J. N. Laska, S. Kirolos, M. F. Duarte, T. S. Ragheb, R. G. Baraniuk, and Y. Massoud, "Theory and implementation of an analog-to-information converter using random demodulation," in *Proceedings of the IEEE International Symposium on Circuits and Systems (ISCAS '07)*, pp. 1959–1962, May 2007.

[6] S. Pfetsch, T. Ragheb, J. Laska et al., "On the feasibility of hardware implementation of sub-Nyquist random-sampling based analog-to-information conversion," in *Proceedings of the IEEE International Symposium on Circuits and Systems (ISCAS '08)*, pp. 1480–1483, May 2008.

[7] S. Kirolos, J. Laska, M. Wakin et al., "Analog-to-information conversion via random demodulation," in *Proceedings of the IEEE Dallas ICAS Workshop on Design, Applications, Integration and Software (DCAS '06)*, pp. 71–74, October 2006.

[8] J. Yoo, S. Becker, and A. Emami-Neyestanak, "Design and imple-mentation of a fully integrated compressed-sensing signal acquisition system," in *Proceedings of the IEEE International Conference on Acoustics, Speech Signal Process*, pp. 5325–5328, 2012.

[9] J. Goodman, B. Miller, M. Herman, M. Vai, and P. Monticciolo, "Extending the dynamic range of RF receivers using nonlinear equalization," in *Proceedings of the IEEE International Waveform Diversity and Design Conference (WDD '09)*, pp. 224–228, February 2009.

[10] M. A. Davenport, J. N. Laska, J. R. Treichler et al., "The pros and cons of compressive sensing for wideband signal acquisition: Noise folding versus dynamic range," *IEEE Transactions on Signal Processing*, vol. 60, no. 9, pp. 4628–4642, 2012.

[11] A. Stern, Y. Zeltzer, and Y. Rivenson, "Quantization error and dynamic range considerations for compressive imaging systems design," *Journal of the Optical Society of America A*, vol. 30, no. 6, pp. 1069–1077, 2013.

[12] Z. Yu, J. Zhou, M. Ramirez, S. Hoyos, and B. M. Sadler, "The impact of ADC nonlinearity in a mixed-signal compressive sensing system for frequency-domain sparse signals," *Physical Communication*, vol. 5, no. 2, pp. 196–207, 2012.

[13] M. S. Oude Alink, A. B. J. Kokkeler, E. A. M. Klumperink, K. C. Rovers, G. J. M. Smit, and B. Nauta, "Spurious-free dynamic range of a uniform quantizer," *IEEE Transactions on Circuits and Systems II: Express Briefs*, vol. 56, no. 6, pp. 434–438, 2009.

[14] N. Björsell and P. Händel, "Achievable ADC performance by postcorrection utilizing dynamic modeling of the integral nonlinearity," *Eurasip Journal on Advances in Signal Processing*, vol. 2008, Article ID 497187, 2008.

[15] W. Kester, "Intermodulation Distortion Considerations for ADCs," Analog Devices Tutorial MT-012, 2009.

[16] S. S. Chen, D. L. Donoho, and M. A. Saunders, "Atomic decomposition by basis pursuit," *SIAM Journal on Scientific Computing*, vol. 20, no. 1, pp. 33–61, 1998.

[17] D. E. N. van Ewout Berg and M. P. Friedlander, "Probing the pareto frontier for basis pursuit solutions," *SIAM Journal on Scientific Computing*, vol. 31, no. 2, pp. 890–912, 2008.

[18] M. A. T. Figueiredo, R. D. Nowak, and S. J. Wright, "Gradient projection for sparse reconstruction: application to compressed sensing and other inverse problems," *IEEE Journal on Selected Topics in Signal Processing*, vol. 1, no. 4, pp. 586–597, 2007.

[19] I. Daubechies, M. Defrise, and C. De Mol, "An iterative thresholding algorithm for linear inverse problems with a sparsity constraint," *Communications on Pure and Applied Mathematics*, vol. 57, no. 11, pp. 1413–1457, 2004.

[20] T. Blumensath and M. E. Davies, "Iterative hard thresholding for compressed sensing," *Applied and Computational Harmonic Analysis*, vol. 27, no. 3, pp. 265–274, 2009.

[21] Y. C. Pati, R. Rezaiifar, and P. S. Krishnaprasad, "Orthogonal matching pursuit: recursive function approximation with applications to wavelet decomposition," in *Proceedings of the 27th Asilomar Conference on Signals, Systems & Computers*, pp. 40–44, November 1993.

[22] J. A. Tropp and A. C. Gilbert, "Signal recovery from random measurements via orthogonal matching pursuit," *IEEE Transactions on Information Theory*, vol. 53, no. 12, pp. 4655–4666, 2007.

[23] M. Friedlander and E. van den Berg, "Toolbox SPGL1 [EB/OL]," 2011, http://www.cs.ubc.ca/labs/scl/spgl1.

11

Efficient and Provable Secure Pairing-Free Security-Mediated Identity-Based Identification Schemes

Ji-Jian Chin,[1] **Syh-Yuan Tan,**[2] **Swee-Huay Heng,**[2] **and Raphael C.W. Phan**[1]

[1] *Faculty of Engineering, Multimedia University, 63100 Cyberjaya, Selangor, Malaysia*
[2] *Faculty of Information Science and Technology, Multimedia University, Jalan Ayer Keroh Lama,*
  75450 Bukit Beruang, Melaka, Malaysia

Correspondence should be addressed to Ji-Jian Chin; jjchin@mmu.edu.my

Academic Editor: Mirjana Ivanovic

Security-mediated cryptography was first introduced by Boneh et al. in 2001. The main motivation behind security-mediated cryptography was the capability to allow instant revocation of a user's secret key by necessitating the cooperation of a security mediator in any given transaction. Subsequently in 2003, Boneh et al. showed how to convert a RSA-based security-mediated encryption scheme from a traditional public key setting to an identity-based one, where certificates would no longer be required. Following these two pioneering papers, other cryptographic primitives that utilize a security-mediated approach began to surface. However, the security-mediated identity-based identification scheme (SM-IBI) was not introduced until Chin et al. in 2013 with a scheme built on bilinear pairings. In this paper, we improve on the efficiency results for SM-IBI schemes by proposing two schemes that are pairing-free and are based on well-studied complexity assumptions: the RSA and discrete logarithm assumptions.

1. Introduction

1.1. Background. Identification schemes allow one party, the prover, to prove itself to another party, the verifier, that it knows its secret key without revealing anything else about itself in the process. The main utilization of the identification primitive is to facilitate one-sided entity authentication and is conventionally deployed in access control mechanisms to facilitate resource control and distribution.

In traditional public key cryptography, certificates are used to ensure that a user's public key is legitimately bound to a particular user and cannot be replaced. This certificate is usually issued by a certificate authority. However, certificate management can become an issue when the number of users in a cryptosystem grows larger. One of the methods of mitigating this potentially costly problem is through the deployment of the cryptographic primitive in an identity-based setting introduced by Shamir [1].

Another issue that traditional public key cryptography faces is the revocation of user secret keys. This would necessarily involve the revocation of a user's certificate along with his public/secret key pair and is a costly operation.

Timeliness is also a factor, as the procedure to revoke a user's keys may be a (relatively) long one and creates additional load on the certificate authority, turning the revocation procedure into a potentially costly process exercise. All these compound the certificate management issue mentioned earlier.

In the identity-based setting, where certificates are not used, key revocation is conventionally done by tagging validity periods onto the user's identity-string as an extension. This also creates an issue of timeliness since revocation is only possible at the end of those validity dates. Without checking certificates, it is also difficult to check if a user is still valid or if his user secret key has been revoked already.

In [2], Boneh et al. proposed the initial groundwork for instant revocation of user keys and privileges, including the public key and certificates by introducing the concept of security-mediated cryptography. The idea of security-mediated cryptography is to necessitate the cooperation of a security mediator, a trusted third party, in any form of transaction that a user needs to use. For example, in an encryption scheme, a security mediator needs to lend his cooperation to the user in order for the user to decrypt a particular ciphertext. And for signatures, a security mediator has to agree to cooperate with a signer in order to produce a valid signature.

Specifically, this is done by separating the user's secret key into two portions during the key generation. One portion of the key is given to the mediator while the other is given to the user. Therefore, the security mediator cooperates in the form of providing his portion of the secret key to be combined with the user's portion to create the full secret key for the transaction.

1.2. Related Work. Identification schemes were first introduced by Fiat and Shamir [3], while their identity-based counterparts, namely, identity-based identification schemes, were first formalized by Bellare et al. [4] and Kurosawa and Heng [5] independently. In recent years, there have been various advances in the area of identity-based identification, such as the introduction of identity-based identification schemes in the standard model [6, 7], hierarchical model [8–10], and certificateless model [11].

Following Boneh et al.'s initial work, Ding and Tsudik expanded on security-mediated cryptography to cover identity-based encryption using the RSA assumption [12]. Shortly thereafter, Libert and Quisquater proposed the first pairing-based security-mediated identity-based encryption schemes [13]. On the signature front, Cheng et al. proposed the first security-mediated identity-based signature [14].

Chow et al. [15] and Yap et al. [16] both extended security-mediated cryptography into the certificateless setting, where the issue of key escrow was addressed. In certificateless cryptography, first proposed by Al-Riyami and Paterson [17], the key generation center only produces half of the user secret key. The user then produces the other half of the user secret key to be combined into the full user secret key, thus securing their key from even the key generation center. However, the cryptographic primitives in the certificateless setting introduce more operational cost to the scheme and are known to be difficult to be proven secure [18].

1.3. Motivations and Contribution. In 2013, Chin et al. first combined the notion of security-mediated cryptography with identifications in the identity-based setting to propose the first security-mediated identity-based identification (SM-IBI) scheme [19]. In the paper, the authors provided the first formal definitions for SM-IBIs and also proposed a concrete construction. The motivation of the authors was to allow fast revocation of keys for identification schemes in the identity-based setting.

For SM-IBI schemes, a security mediator is required to participate in the identification protocol in order for a user to authenticate himself to a verifier. The security mediator can then hold a revocation list to identify which users' secret keys have been revoked and refuse participation for those users.

However the first concrete scheme that was proposed by the authors was built based on bilinear pairings. The pairing operation is widely known by cryptographers to be a costly operation; thus, it would be beneficial to construct pairing-free alternatives to facilitate more efficient running of the cryptographic primitive.

In this paper, we propose two pairing-free SM-IBI constructs as faster alternatives to the pairing-based scheme proposed by Chin et al. Our two schemes are constructed based on the RSA assumption and the discrete logarithm assumption, respectively. The RSA-based scheme, which we name as the GQ-SM-IBI, is constructed based on the Guillou Quisquater identification scheme constructed by Bellare and Palacio [20]. On the other hand, the discrete logarithm-based scheme, which we name as the BNN-SM-IBI, is constructed based on the BNN identity-based identification scheme proposed by Bellare et al. [4]. We provide security analysis for both schemes, proving them secure against impersonation under passive attacks if the RSA assumption and the discrete logarithm assumption hold and secure against impersonation under active and concurrent attacks if the one-more RSA inversion assumption and the one-more discrete logarithm assumption hold. Lastly, we provide an efficiency analysis, both theoretically and practically, and show that both schemes are significantly faster than Chin et al.'s pairing-based SM-IBI scheme.

The rest of the paper is organized as follows. We begin with some preliminaries and review the formal definitions and security model of SM-IBI schemes in Section 2. Then we show the construction and security analysis for the GQ-SM-IBI scheme in Section 3. This is followed by the construction and security analysis for the BNN-SM-IBI scheme in Section 4. In Section 5 we show the operational costs of both schemes as well as presenting our implementation results. Finally we conclude in Section 6.

2. Preliminaries

2.1. Discrete Logarithm Assumption. Let G be a cyclic group with prime order q and let g be a generator of G. The DL problem (DL) is defined as given a number $A = g^a$ in group G, output a.

Definition 1. The discrete logarithm assumption states that there exists no polynomial-time algorithm M that is able to $(t_{DL}, \varepsilon_{DL})$-solve the discrete logarithm problem with nonnegligible probability such that

$$\Pr\left[M\left(G, q, g, A\right) = a\right] \geq \varepsilon_{DL}. \tag{1}$$

2.2. The One-More Discrete Logarithm Assumption. The one-more discrete logarithm (OMDL) problem was first introduced by Bellare and Palacio [20] in their proof against impersonation under active and concurrent attacks for the standard Schnorr identification scheme. Later work that involves proving security of identification schemes based on discrete logarithms to be secure against active and concurrent attacks for discrete logarithm also makes use of this assumption such as [4, 21].

Let G be a finite cyclic group of order q and let g be a generator of G. Define an experiment E_{DL} where an adversary M is given a challenge oracle $CHALL$ that produces a random group element $W_i \in G$ when queried and a discrete log oracle $DLOG$, which provides the discrete log $w_i \in Z_q$ corresponding to the query W_i where $g^{w_i} = W_i$. M wins if after making i queries to $CHALL_{DL}$, M is able to output solutions to all i challenges with only $i - 1$ queries to $DLOG$, meaning M has to solve at least one instance of the discrete logarithm problem without relying on the discrete log oracle. E_{DL} returns 1 if M is successful and 0 otherwise.

Definition 2. The *OMDL* assumption states that there exists no polynomial-time algorithm M that is able to $(t_{OMDL}, q_{OMDL}, \varepsilon_{OMDL})$-solve the *OMDL* problem with non-negligible probability where

$$\Pr\left[E_{DL}\left(M\left(1^k, G, g, CHALL_{DL}, DLOG\right) = 1\right)\right] \geq \varepsilon_{OMDL}. \tag{2}$$

2.3. RSA Inversion (RSAI) Assumption. Given $(N, e, X) \xleftarrow{\$} k_{RSA}(1^k)$ compute Y such that $Y = X^d \bmod N$ where $ed = 1 \bmod \phi(N)$.

Definition 3. The *RSAI* assumption states that there exists no polynomial-time algorithm M that is able to $(t_{RSAI}, \varepsilon_{RSAI})$-solve the *RSAI* problem with nonnegligible probability such that

$$\Pr\left[M\left(1^k, N, e, X\right) = Y : Y = X^d \text{ and } ed = 1 \bmod \phi(N)\right]$$
$$\geq \varepsilon_{RSAI}. \tag{3}$$

2.4. One-More RSA Inversion (OMRSAI) Assumption. This is the interactive variant of the *RSAI* problem first proposed by Bellare and Palacio [20] to prove security of the GQ identification scheme and is analogous to the *OMDL* assumption. This assumption is applied in the proof of security against active and concurrent attacks for RSA-based schemes.

Define an experiment E_{RSAI} where an adversary M is given $(N, e) \xleftarrow{\$} k_{RSA}(1^k)$ as input and access to two oracles $CHALL$ and RSA. $CHALL$ on any input returns a random point W_i, while RSA on any input h will return h^d where $ed = 1 \bmod N$. M is required to compute the *RSAI* solutions to all the target points $W_0, \ldots, W_n$ while using strictly less queries to the RSA oracle. In other words, M is required to find $W_0^d, \ldots, W_n^d$ while using the RSA oracle only $i < n$ times. E_{RSAI} returns 1 if M is successful and 0 otherwise.

Definition 4. The *OMRSAI* assumption states that there exists no polynomial-time algorithm M that is able to $(t_{OMRSAI}, q_{OMRSAI}, \varepsilon_{OMRSAI})$-win the *OMRSAI* problem with nonnegligible probability where

$$\Pr\left[E_{RSAI}\left(M\left(1^k, N, e, CHALL_{RSA}, RSA\right)\right) = 1\right] \geq \varepsilon_{RSAI}. \tag{4}$$

2.5. Definition of Security-Mediated IBI Schemes. In this section, we review the definition of SM-IBI schemes as defined by [19]. The definition follows closely to that of conventional IBI schemes, with the difference being that the prover segment is extended to encompass obtaining tokens from the security mediator. The SM-IBI scheme is defined as five probabilistic polynomial-time algorithms.

(i) *Setup.* It takes in the security parameter 1^k as input and outputs the system parameters params along with the master secret key MSK.

(ii) *Extract.* Upon receiving a user's request for a key, it takes in params MSK and a user's identity ID. Once the secret key is created, the PKG separates the key into two portions, one for the user, USK_{user}, and one for the security mediator, USK_{sem}, and returns the portions to the respective parties.

(iii) *Identification Protocol.* The identification protocol is an interactive protocol run by the 3 algorithms: *User-Prover* and *SEM-prover* on the prover side trying to authenticate himself to the *Verifier*. Both provers are used cooperatively in the interactive three-step canonical honest verifier zero knowledge proof of knowledge protocol with the verifier as follows.

(1) *User-Prover* initiates by sending his identity to ID to the *SEM-prover*. *SEM-prover* checks whether ID's keys have been revoked and stops with an error code if true. If ID is legitimate then it generates and sends SEM-COMMIT to User-Prover who then combines SEM-COMMIT with his own USER-COMMIT to form FULL-COMMIT to send to *Verifier*.

(2) *Verifier* selects a random CHALLENGE and sends it to the *User-Prover*.

(3) *User-Prover* relays CHALLENGE to *SEM-Prover* and receives SEM-RESPONSE from *SEM-Prover* which it then combines with his own USER-RESPONSE to form FULL-RESPONSE and sends to the *Verifier*. The *Verifier* will choose to either accept or reject it.

2.6. Security Model for Security-Mediated IBI. Adversaries of SM-IBI follow the description of standard identification schemes: passive and active/concurrent attackers. However, the adversary for SM-IBI is able to query additionally partial conversation components, specifically the user's prover and the security mediator's prover besides the usual full prover query.

The security of SM-IBI schemes is modelled as a game played by an adversary I against a challenger C as follows.

(i) *Setup.* C runs *Setup*, creates the system parameters params and passes them to I while keeping the master secret key MSK to itself.

(ii) *Phase 1.* This is the training phase. I is allowed to adaptively make the following queries to C.

(a) *User-Extract* (ID). C will run *Extract* but returns only the user's portion of the secret key to I.

(b) *SEM-Extract* (ID). C will run *Extract* but returns only the security mediator's portion of the secret key to I.

(c) *Full-Extract* (ID). C will run *Extract* and returns both the user's portion of the secret key and the security mediator's portion of the secret key to I.

(d) *Identification Queries* (ID). For passive adversaries, C will generate transcripts of valid conversations for I. For active/concurrent adversaries, C will act as the cheating prover, engaging I as the cheating verifier in conversations. I is able to issue any one of the following identification queries.

    (1) *SEM-Identification* (ID). C runs the security mediator's half of the prover session.

    (2) *User-Identification* (ID). C runs the user's half of the prover session.

    (3) *Full-Identification* (ID). C combines both security mediator's and user's session to generate a full and valid conversation.

(iii) *Phase 2.* I will eventually output ID^* on which it wants to be challenged on and begins its role as the cheating prover for both security mediator and user prover sessions. C on the other hand assumes the role of the verifier. I wins the game if it manages to convince C to accept with nonnegligible probability.

We say a security-mediated IBI scheme Π is (t_{SMIBI}, q_{SMIBI}, ε_{SMIBI})-secure under passive or active/concurrent attacks if for any passive or active/concurrent Type-1 impersonator I who runs in time t_{SMIBI}, $\Pr[I$ can impersonate$] < \varepsilon_{SMIBI}$, where I can make at most q_{SMIBI} full extract queries.

It is interesting to point out that extracting the security mediator's half of the secret key gives no information about the full user key and that neither security mediator's prover sessions nor user's prover sessions done alone will provide a valid conversation, but only the combined session will. This models the security requirement that any user cannot legitimately prove himself to a verifier without the security mediator's help.

3. GQ-SMIBI: RSA-Based Security-Mediated IBI Scheme

The GQ-SM-IBI scheme is derived from the GQ identification scheme proposed in [20] and is provably secure against passive attackers assuming the *RSAI* assumption and against active/concurrent attackers assuming the *OMRSAI* assumption.

The GQ-SM-IBI scheme is constructed as follows.

(1) *Setup* (1^k). It takes in the security parameter 1^k, runs the key generation algorithm for RSA, and obtains an RSA instance, $(N, e, d) \xleftarrow{\$} k_{RSA}(1^k)$. It chooses a hash function $H : \{0, 1\}^* \to \mathbb{Z}_N^*$ and publishes the system parameters $\mathsf{mpk} = \langle N, e, H \rangle$. The master secret key $\mathsf{msk} = d$ is kept secret.

(2) *Extract* ($\mathsf{mpk}, \mathsf{msk}, \text{ID}$). It takes in $\mathsf{mpk}, \mathsf{msk} = s$, and ID as input. It calculates $H(\text{ID})$ and sets

$X_{FULL} = H(\text{ID})^d$. It further chooses $d_{SEM} \xleftarrow{\$} \mathbb{Z}_{\phi(N)}^*$, sets $d_{USER} = d - d_{SEM}$, and calculates $X_{USER} = H(\text{ID})^{d_{USER}}$ and $X_{SEM} = H(\text{ID})^{d_{SEM}}$. It gives X_{USER} to the user as its partial secret key and X_{SEM} to the security mediator as its partial private key and keeps X_{FULL} secret.

(3) *Identification Protocol.* It is run by the *User-Prover* and *SEM-Prover* with *Verifier* as such.

(a) *User-Prover* sends its ID to *Sem-Prover* and chooses $y_{USER} \xleftarrow{\$} \mathbb{Z}_N^*$ and sets $Y_{USER} = y_{USER}^e$. SEM-Prover upon receiving ID checks if it is still valid and returns error if it is revoked. Otherwise, *SEM-Prover* chooses $y_{SEM} \xleftarrow{\$} \mathbb{Z}_N^*$ and sends $Y_{SEM} = y_{SEM}^e$ to *User-Prover*. *User-Prover* combines $Y_{FULL} = Y_{SEM} \times Y_{USER}$ and sends Y_{FULL} to *Verifier*.

(b) *Verifier* chooses a random challenge $c \xleftarrow{\$} Z_{2^{l(k)}}$ where $l(\cdot)$ is a super-logarithmic challenge length such that $2^{l(k)} < e$. *Verifier* then sends it to *User-Prover*. *User-Prover* relays c to *SEM-Prover*.

(c) *SEM-Prover* calculates its half of the response $z_{SEM} = y_{SEM} X_{SEM}^c$ and sends it to *User-Prover*. *User-Prover* combines it with his response $z_{USER} = y_{USER} X_{USER}^c$ to create $z_{FULL} = z_{SEM} \times z_{USER}$ and sends it as a response to *Verifier*.

Verifier checks if $z_{FULL}^e = Y_{FULL} H(\text{ID})^c$ and accepts if yes; otherwise it outputs reject.

To check for completeness,

$$
\begin{aligned}
Z_{FULL}^e &= \left[\left(y_{SEM} X_{SEM}^c \right) \left(y_{USER} X_{USER}^c \right) \right]^e \\
&= \left[y_{SEM} y_{USER} (X_{SEM} X_{USER})^c \right]^e \\
&= y_{SEM}^e y_{USER}^e \left(H(\text{ID})^{(d_{SEM}+d_{USER})ec} \right) \quad (5) \\
&= Y_{SEM} Y_{USER} \left(H(\text{ID})^{dec} \right) \\
&= Y_{FULL} \left(H(\text{ID})^c \right).
\end{aligned}
$$

3.1. Security Analysis: Impersonation under Passive Attack

Theorem 5. *The GQ-SM-IBI scheme is* (t_{SMIBI}, q_{SMIBI}, ε_{SMIBI})-*secure against impersonation under passive attacks in the random oracle if the RSA Inversion Problem is* (t_{RSAI}, ε_{RSAI})-*hard where*

$$
\varepsilon_{SMIBI} \leq \sqrt{\varepsilon_{RSAI} e \left(q_{SMIBI} + 1 \right)} + \frac{1}{2^{l(k)}}. \quad (6)
$$

Proof. Assume the GQ-SM-IBI scheme is (t_{SMIBI}, q_{SMIBI}, ε_{SMIBI})-breakable; then a simulator M that (t_{RSAI}, ε_{RSAI})-breaks the *RSAI* problem can be shown. M takes in input (N, e, y) and runs the impersonator I as a subroutine.

Without loss of generality, it can be assumed that any *SEM-Extract, User-Extract, SEM-Identification, User-Identification,* and *Full-Identification* queries are preceded by a *Create-User* query. To avoid collision and consistently respond to these queries, M maintains a list $L_H = \langle ID_i, Q_{ID_i}, f_{ID_i}, X_{FULL_{ID_i}}, X_{SEM_{ID_i}}, X_{USER_{ID_i}}, d_{SEM_{ID_i}}, d_{USER_{ID_i}} \rangle$ which is initially empty. The following shows how M simulates the environment and oracle queries for I.

(1) *Setup.* M selects a hash function $H : \{0,1\}^* \rightarrow \mathbb{Z}_N^*$ and sets the system parameters as $\mathsf{mpk} = \langle N, e, H \rangle$. It sends mpk to I.

(2) *Create-User* (ID_i). M chooses $l \in \{1, \ldots, q_H\}$ randomly and lets $ID_l = ID^*$ at this point. Whenever I makes a H query on ID_i, consider the following.

   (a) If $i = l$, M will choose $f_{ID^*} \xleftarrow{\$} \mathbb{Z}_N^*$ sets $Q_{ID^*} = y f_{ID^*}^e$ and will return Q_{ID^*} to I. It adds $\langle ID_l, Q_{ID^*}, f_{ID^*}, X_{FULL_{ID^*}} = \bot, X_{USER_{ID^*}} = \bot, X_{SEM_{ID^*}} = \bot, d_{SEM_{ID^*}} = \bot, d_{USER_{ID^*}} = bot \rangle$ to L_H.

   (b) Otherwise, for $i \neq l$, M chooses $f_{ID_i} \xleftarrow{\$} \mathbb{Z}_q^*$, sets $Q_{ID_i} = f_{ID_i}^e$, and returns Q_{ID_i} to I. M also sets $X_{FULL_{ID_i}} = f_{ID_i}$, chooses $d_{USER_{ID^*}} \xleftarrow{\$} \mathbb{Z}_{\phi(N)}^*$, and sets $X_{SEM_{ID_i}} = Q_{ID_i}^{d_{SEM_{ID_i}}}$ as the SEM portion of the user private key and $X_{USER_{ID_i}} = f_{ID_i} Q_{ID_i}^{-d_{USER_{ID_i}}}$ as the user portion of the user private key. It can be seen that

$$X_{SEM_{ID_i}} \times X_{USER_{ID_i}}$$
$$= Q_{ID_i}^{d_{SEM_{ID_i}}} \times f_{ID_i} Q_{ID_i}^{-d_{SEM_{ID_i}}} \qquad (7)$$
$$= f_{ID_i}$$
$$= X_{FULL_{ID_i}}.$$

M adds $\langle ID_i, Q_{ID_i}, f_{ID_i}, X_{FULL_{ID_i}}, X_{USER_{ID_i}}, X_{SEM_{ID_i}}, d_{SEM_{ID_i}}, d_{USER_{ID_i}} \rangle$ to L_H. M returns Q_{ID_i} to I.

(3) *SEM-Extract* (ID_i). If $ID_i = ID^*$ and *USER-Extract* (ID^*) has been queried before, M aborts. If $ID_i = ID^*$, but *USER-Extract* (ID^*) has not been queried before, M chooses $d_{SEM_{ID^*}} \xleftarrow{\$} \mathbb{Z}_{\phi(N)}^*$ and sets $X_{SEM_{ID^*}} = Q_{ID^*}^{d_{SEM_{ID^*}}}$ as the SEM portion of the user private key, saves it in L_H, and returns it to I. Otherwise for all other $ID_i \neq ID^*$, M just retrieves $X_{SEM_{ID_i}}$ in L_H and returns it to I.

(4) *USER-Extract* (ID_i). If $ID_i = ID^*$ and *USER-Extract* (ID^*) has been queried before, M aborts. If $ID_i = ID^*$, but *SEM-Extract* (ID^*) has not been queried before, M chooses $d_{USER_{ID^*}} \xleftarrow{\$} \mathbb{Z}_{\phi(N)}^*$ and sets $X_{USER_{ID^*}} = Q_{ID^*}^{d_{USER_{ID^*}}}$ as the USER portion of the user

private key, saves it in L_H, and returns it to I. Otherwise for all other $ID_i \neq ID^*$, M just retrieves $X_{USER_{ID_i}}$ in L_H and returns it to I.

(5) *Identification* $(ID_j, session)$. I will act as the cheating verifier to learn information from valid conversation transcripts from M. If $ID_j \neq ID^*$ M just retrieves ID_j's entry in L_H and runs the identification protocol for either the SEM's interactions, the user's interactions, or both combined as in the protocol.

Otherwise, for $ID_j = ID^*$, M then creates a full valid transcript for each mth query by picking $z_{m,USER_{ID^*}}, z_{m,SEM_{ID^*}} \xleftarrow{\$} \mathbb{Z}_q^*$, and $c_{m,ID^*} \xleftarrow{\$} \mathbb{Z}_q^*$ and sets $Z_{m,FULL_{ID^*}} = z_{m,USER_{ID^*}} \times z_{m,SEM_{ID^*}}$. M also sets $Y_{m,USER_{ID^*}} = z_{m,USER_{ID^*}}^e Q_{ID^*}^{-(c_{m,ID^*}/2)}$, $Y_{m,SEM_{ID^*}} = z_{m,SEM_{ID^*}}^e Q_{ID^*}^{-(c_{m,ID^*}/2)}$, and $Y_{m,FULL_{ID^*}} = z_{m,FULL_{ID^*}}^e Q_{ID^*}^{-c_{m,ID^*}}$. M tags the session with $session = m$. If I queries

   (i) *SEM-Identification* (ID^*, m), M will return $\langle Y_{m,SEM_{ID^*}}, c_{m,ID^*}, z_{m,SEM_{ID^*}} \rangle$;

   (ii) *User-Identification* (ID^*, m), M will return $\langle Y_{m,USER_{ID^*}}, c_{m,ID^*}, z_{m,USER_{ID^*}} \rangle$;

   (iii) *Full-Identification* (ID^*, m), M will return $\langle Y_{m,FULL_{ID^*}}, c_{m,ID^*}, z_{m,FULL_{ID^*}} \rangle$.

If no session is specified for the query, M just returns the next session in sequence. One can see that *SEM-Identification* (ID^*, m) and *User-Identification* (ID^*, m) can be combined to create a full and valid transcript on session m:

$$Y_{m,FULL_{ID^*}} H(ID^*)^{c_{m,ID^*}}$$
$$= Y_{m,SEM_{ID^*}} Y_{m,USER_{ID^*}} Q_{ID^*}^{c_{m,ID^*}}$$
$$= z_{m,SEM_{ID^*}}^e Q_{ID^*}^{-(c_{m,ID^*}/2)} z_{m,USER_{ID^*}}^e Q_{ID^*}^{-(c_{m,ID^*}/2)} Q_{ID^*}^{c_{m,ID^*}}$$
$$= \left[z_{m,SEM_{ID^*}} z_{m,USER_{ID^*}} \right]^e Q_{ID^*}^{-c_{m,ID^*}} Q_{ID^*}^{c_{m,ID^*}} \qquad (8)$$
$$= z_{m,SEM_{ID^*}}^e z_{m,USER_{ID^*}}^e$$
$$= z_{m,FULL_{ID^*}}^e.$$

Eventually, I stops Phase 1 and outputs the challenge ID, ID^* on which it wishes to be challenged on. M checks if $ID^* = ID_l$ from L_H and aborts if not. Otherwise, M runs I now as a cheating prover on ID^* by obtaining its commitment, $Y_{FULL_{ID^*}}$, selecting a challenge $c_1 \xleftarrow{\$} \mathbb{Z}_q^*$ and obtaining the response $z_{FULL_{ID^*},1}$ from I. M then resets I to the step whereby I just sent $Y_{FULL_{ID^*}}$, selects a second challenge $c_2 \xleftarrow{\$} \mathbb{Z}_q^*$, and receives $z_{FULL_{ID^*},2}$ as response. It should hold that $z_{FULL_{ID^*},1}^e = Y_{FULL_{ID^*}} H(ID)^{c_1}$ and $z_{FULL_{ID^*},2}^e = Y_{FULL_{ID^*}} H(ID)^{c_2}$. One can then obtain $(z_{FULL_{ID^*},1}/z_{FULL_{ID^*},2})^e = H(ID)^{c_1-c_2}$. Since e is a prime and

$-e < c_1 - c_2 < e$, $GCD(e, c_1 - c_2) = 1$. Use the extended Euclidean algorithm to obtain integers S and T such that $eS + (c_1 - c_2)T = 1$. It follows that

$$
\begin{aligned}
H(\text{ID}^*) &= Q_{\text{ID}^*}^{eS+(c_1-c_2)T} \bmod N \\
&= \left(yf_{\text{ID}^*}^e\right)^{eS+(c_1-c_2)T} \bmod N \\
&= \left(yf_{\text{ID}^*}^{e^2 S}\right)\left(yf_{\text{ID}^*}^{e(c_1-c_2)T}\right) \bmod N \\
&= \left(yf_{\text{ID}^*}^{e^2 S}\right)\left(\frac{z_{\text{FULL}_{\text{ID}^*},1}}{z_{\text{FULL}_{\text{ID}^*},2}}\right)^{eT} \bmod N \\
&= \left[\left(yf_{\text{ID}^*}^{eS}\right)\left(\frac{z_{\text{FULL}_{\text{ID}^*},1}}{z_{\text{FULL}_{\text{ID}^*},2}}\right)^{T}\right]^e \bmod N.
\end{aligned}
\tag{9}
$$

M then calculates the solution to the $RSAI$ problem as follows:

$$
\begin{aligned}
yf_{\text{ID}^*}^e &= \left[\left(yf_{\text{ID}^*}^{eS}\right)\left(\frac{z_{\text{FULL}_{\text{ID}^*},1}}{z_{\text{FULL}_{\text{ID}^*},2}}\right)^{T}\right]^e \bmod N, \\
y^d f_{\text{ID}^*} &= \left(yf_{\text{ID}^*}^{eS}\right)\left(\frac{z_{\text{FULL}_{\text{ID}^*},1}}{z_{\text{FULL}_{\text{ID}^*},2}}\right)^{T} \bmod N, \\
y^d &= \frac{\left(yf_{\text{ID}^*}^{eS}\right)\left(z_{\text{FULL}_{\text{ID}^*},1}/z_{\text{FULL}_{\text{ID}^*},2}\right)^{T}}{f_{\text{ID}^*}} \bmod N, \\
y^d &= \left(yf_{\text{ID}^*}^{eS-1}\right)\left(\frac{z_{\text{FULL}_{\text{ID}^*},1}}{z_{\text{FULL}_{\text{ID}^*},2}}\right)^{T} \bmod N.
\end{aligned}
\tag{10}
$$

It remains to calculate the probability of M solving the $RSAI$ problem and winning the game. The probability of M successfully extracting two valid conversation transcripts from I is bounded by $(\varepsilon_{IBI} - (1/2^{l(k)}))^2$ as given by the reset lemma [20]:

$$
\begin{aligned}
&\Pr\left[M \text{ wins } RSAI\right] \\
&= \Pr\left[M \text{ computes } g^{ab} \bigwedge \neg\text{abort}\right] \\
&= \Pr\left[M \text{ computes } g^{ab} \mid \neg\text{abort}\right]\Pr\left[\neg\text{abort}\right] \\
&\varepsilon_{RSAI} \geq \left(\varepsilon_{SMIBI} - \frac{1}{2^{l(k)}}\right)^2 \Pr\left[\neg\text{abort}\right].
\end{aligned}
\tag{11}
$$

Finally calculate $\Pr[\neg\text{abort}]$. Let δ be the probability that I issues both a *SEM-Extract* and a *USER-Extract* query on ID^* and that I makes a total of q_{SMIBI} of such queries. The probability of M answering all the extraction queries is δ_{SMIBI}^q. In Phase 2, the probability of M not aborting is if I outputs the challenge identity ID^* that was not queried before. This is given by the probability $1 - \delta$. Compiling them, the probability of M not aborting is $\delta^{q_{SMIBI}}(1 - \delta)$. This probability is maximised at $\delta_{opt} = 1 - (1/(q_{SMIBI} + 1))$. Using δ_{opt}, the probability that M does not abort is at least $1/(2^{l(k)}(q_{SMIBI}+1))$ because the value $(1-(1/(q_{SMIBI}+1)))^{q_{SMIBI}}$

approaches $1/e$ for large q_{SMIBI}. Therefore, the advantage of M, ε_{RSAI}, and the bound of the simulation are given as follows:

$$
\begin{aligned}
\varepsilon_{RSAI} &\geq \left(\varepsilon_{SMIBI} - \frac{1}{2^{l(k)}}\right)^2 \frac{1}{(q_{SMIBI}+1)} \\
\varepsilon_{RSAI}(q_{SMIBI}+1) &\geq \left(\varepsilon_{SMIBI} - \frac{1}{2^{l(k)}}\right)^2 \\
\varepsilon_{SMIBI} &\leq \sqrt{\varepsilon_{RSAI}(q_{SMIBI}+1)} + \frac{1}{2^{l(k)}}.
\end{aligned}
\tag{12}
$$

$\square$

3.2. Security Analysis: Impersonation under Active/Concurrent Attack

Theorem 6. *The GQ-SM-IBI scheme is $(t_{SMIBI}, q_{SMIBI}, \varepsilon_{SMIBI})$-secure against impersonation under active/concurrent attacks in the random oracle if the OMCDH Problem is $(t_{OMRSAI}, q_{OMRSAI}, \varepsilon_{OMRSAI})$-hard where*

$$
\varepsilon_{SMIBI} \leq \sqrt{\varepsilon_{OMRSAI}\, e\,(q_{SMIBI}+1)} + \frac{1}{2^{l(k)}}.
\tag{13}
$$

Proof. Assume the GQ-SM-IBI scheme is $(t_{SMIBI}, q_{SMIBI}, \varepsilon_{SMIBI})$-breakable; then a simulator M that $(t_{OMRSAI}, q_{OMRSAI}, \varepsilon_{OMRSAI})$-breaks the *OMRSAI* problem can be shown. M takes in input (N, e), is given access to *CHALL* and *RSA* oracles, and runs the impersonator I as a subroutine. Any *SEM-Extract*, *User-Extract*, *SEM-Identification*, *User-Identification*, and *Full-Identification* queries are preceded by a *Create-User* query. M maintains a list $L_H = \langle \text{ID}_i, Q_{\text{ID}_i}, f_{\text{ID}_i}, X_{FULL_{\text{ID}_i}}, X_{SEM_{\text{ID}_i}}, X_{USER_{\text{ID}_i}}, d_{SEM_{\text{ID}_i}}, d_{USER_{\text{ID}_i}}\rangle$ which is initially empty. The following shows how M simulates the environment and oracle queries for I.

(1) *Create-User* (ID_i). M chooses $l \in \{1, \ldots, q_H\}$ randomly and lets $\text{ID}_l = \text{ID}^*$ at this point. Whenever I makes a H query on ID_i, consider the following.

    (i) If $i = l$, M will choose $f_{\text{ID}^*} \xleftarrow{\$} \mathbb{Z}_N^*$, queries *CHALL* for w_0, and sets $Q_{\text{ID}^*} = w_0 f_{\text{ID}^*}^e$ and will return Q_{ID^*} to I. It adds $\langle \text{ID}_l, Q_{\text{ID}^*}, f_{\text{ID}^*}, X_{FULL_{\text{ID}^*}} = \perp, X_{USER_{\text{ID}^*}} = \perp, X_{SEM_{\text{ID}^*}} = \perp, d_{SEM_{\text{ID}^*}} = \perp, d_{USER_{\text{ID}^*}} = \perp\rangle$ to L_H.

    (ii) Otherwise, for $i \neq l$, M chooses $f_{\text{ID}_i} \xleftarrow{\$} \mathbb{Z}_q^*$, sets $Q_{\text{ID}_i} = f_{\text{ID}_i}^e$, and returns Q_{ID_i} to I. M also sets $X_{FULL_{\text{ID}_i}} = f_{\text{ID}_i}$, chooses $d_{USER_{\text{ID}^*}} \xleftarrow{\$} \mathbb{Z}_{\phi(N)}^*$, and sets $X_{SEM_{\text{ID}_i}} = Q_{\text{ID}_i}^{d_{SEM_{\text{ID}_i}}}$ as the SEM portion of the user private key and $X_{USER_{\text{ID}_i}} = f_{\text{ID}_i} Q_{\text{ID}_i}^{-d_{USER_{\text{ID}_i}}}$ as the user portion of the user private key. It can be seen that

$$X_{SEM_{ID_i}} \times X_{USER_{ID_i}}$$

$$= Q_{ID_i}^{d_{SEM_{ID_i}}} \times f_{ID_i} Q_{ID_i}^{-d_{SEM_{ID_i}}} \quad (14)$$

$$= f_{ID_i}$$

$$= X_{FULL_{ID_i}}.$$

M adds $\langle ID_i, Q_{ID_i}, f_{ID_i}, X_{FULL_{ID_i}}, X_{USER_{ID_i}}, X_{SEM_{ID_i}}, d_{SEM_{ID_i}}, d_{USER_{ID_i}} \rangle$ to L_H. M returns Q_{ID_i} to I.

(2) *SEM-Extract* (ID_i). If $ID_i = ID^*$ and *USER-Extract* (ID^*) has been queried before, M aborts. If $ID_i = ID^*$, but *USER-Extract* (ID^*) has not been queried before, M chooses $d_{SEM_{ID^*}} \xleftarrow{\$} \mathbb{Z}_{\phi(N)}^*$ and sets $X_{SEM_{ID^*}} = Q_{ID^*}^{d_{SEM_{ID^*}}}$ as the SEM portion of the user private key, saves it in L_H, and returns it to I. Otherwise for all other $ID_i \neq ID^*$, M just retrieves $X_{SEM_{ID_i}}$ in L_H and returns it to I.

(3) *USER-Extract* (ID_i). If $ID_i = ID^*$ and *SEM-Extract* (ID^*) has been queried before, M aborts. If $ID_i = ID^*$, but *SEM-Extract* (ID^*) has not been queried before, M chooses $d_{USER_{ID^*}} \xleftarrow{\$} \mathbb{Z}_{\phi(N)}^*$ and sets $X_{USER_{ID^*}} = Q_{ID^*}^{d_{USER_{ID^*}}}$ as the USER portion of the user private key, saves it in L_H, and returns it to I. Otherwise for all other $ID_i \neq ID^*$, M just retrieves $X_{USER_{ID_i}}$ in L_H and returns it to I.

(4) *Identification* $(ID_i, session)$. I will act as the cheating verifier to learn information from valid conversation interactions from M. If $ID_i \neq ID^*$ M just retrieves ID_i's entry in L_H and runs the identification protocol for either the SEM's interactions, the user's interactions, or both combined as in the protocol.

Otherwise, for $ID^* = ID^*$, M then creates a full valid conversation for each mth query by querying *CHALL* for W_m and setting $Y_{FULL_{ID^*}} = W_m$. Additionally, M chooses $Y_{m,SEM_{ID^*}} \xleftarrow{\$} \mathbb{Z}_N^*$ and sets $Y_{m,User_{ID^*}} = W_m/Y_{m,SEM_{ID^*}}$. Upon receiving c_{m,ID^*} from I, M will query $RSA(W_m(W_0 f_{ID^*}^e)^{c_{m,ID^*}})$ to receive $V_m = (W_m(W_0 f_{ID^*}^e)^{c_{m,ID^*}})^d$ and sets $z_{FULL_{ID^*}} = V_m$. M then chooses $z_{m,SEM_{ID^*}} \xleftarrow{\$} \mathbb{Z}_N^*$ and then sets $z_{m,USER_{ID^*}} = V_m/z_{m,SEM_{ID^*}}$. M tags the session with $session = m$.

  (i) If I queries *SEM-Identification* (ID^*, m), M will commit $Y_{m,SEM_{ID^*}}$ and respond with $z_{m,SEM_{ID^*}}$ upon receiving c_{m,ID^*}.
  (ii) If I queries *User-Identification* (ID^*, m), M will commit $Y_{m,USER_{ID^*}}$ and respond with $z_{m,USER_{ID^*}}$ upon receiving c_{m,ID^*}.
  (iii) If I queries *Full-Identification* (ID^*, m), M will commit $Y_{m,FULL_{ID^*}}$ and respond with $z_{m,FULL_{ID^*}}$ upon receiving c_{m,ID^*}.

If no session is specified for the query, M just returns the next session in sequence. One can see that *SEM-Identification* (ID^*, m) and *User-Identification* (ID^*, m) can be combined to create a full and valid conversation on session m:

$$Y_{m,FULL_{ID^*}} H(ID^*)^{c_{m,ID^*}}$$

$$= Y_{m,SEM_{ID^*}} Y_{m,USER_{ID^*}} Q_{ID^*}^{c_{m,ID^*}}$$

$$= Y_{m,SEM_{ID^*}} \frac{W_m}{Y_{m,SEM_{ID^*}}} (W_0 f_{ID^*}^e)^{c_{m,ID^*}}$$

$$= W_m(W_0 f_{ID^*}^e)^{c_{m,ID^*}}$$

$$= \left[W_m(W_0 f_{ID^*}^e)^{c_{m,ID^*}} \right]^{de}$$

$$= \left[\left[W_m(W_0 f_{ID^*}^e)^{c_{m,ID^*}} \right]^d \right]^e \quad (15)$$

$$= \left[V_m \right]^e$$

$$= \left(\frac{V_m}{z_{m,SEM_{ID^*}}} \right)^e \left(z_{m,SEM_{ID^*}} \right)^e$$

$$= z_{m,SEM_{ID^*}}^e z_{m,USER_{ID^*}}^e$$

$$= z_{m,FULL_{ID^*}}^e.$$

Eventually, I stops Phase 1 and outputs the challenge ID, ID^* on which it wishes to be challenged on. M checks if $ID^* = ID_l$ from L_H and aborts if not. Otherwise, M runs I now as a cheating prover on ID^*. M runs by obtaining its commitment, $Y_{FULL_{ID^*}}$, selects a challenge $c_1 \xleftarrow{\$} \mathbb{Z}_q^*$, and obtains the response $z_{FULL_{ID^*},1}$ from I. M then resets I to the step whereby I just sent $Y_{FULL_{ID^*}}$, selects a second challenge $c_2 \xleftarrow{\$} \mathbb{Z}_q^*$, and receives $z_{FULL_{ID^*},2}$ as response. It should hold that $z_{FULL_{ID^*},1}^e = Y_{FULL_{ID^*}} H(ID)^{c_1}$ and $z_{FULL_{ID^*},2}^e = Y_{FULL_{ID^*}} H(ID)^{c_2}$. One can then obtain $(z_{FULL_{ID^*},1}/z_{FULL_{ID^*},2})^e = H(ID)^{c_1-c_2}$. Since e is a prime and $-e < c_1 - c_2 < e$, $GCD(e, c_1 - c_2) = 1$. Use the extended Euclidean algorithm to obtain integers S and T such that $eS + (c_1 - c_2)T = 1$. It follows that

$$H(ID^*) = Q_{ID^*}^{eS+(c_1-c_2)T} \bmod N$$

$$= (w_0 f_{ID^*}^e)^{eS+(c_1-c_2)T} \bmod N$$

$$= \left(w_0 f_{ID^*}^{e^2 S} \right) \left(w_0 f_{ID^*}^{e(c_1-c_2)T} \right) \bmod N$$

$$= \left(w_0 f_{\mathrm{ID}^*}^{e^2 S}\right)\left(\frac{z_{\mathrm{FULL}_{\mathrm{ID}^*},1}}{z_{\mathrm{FULL}_{\mathrm{ID}^*},2}}\right)^{eT} \bmod N$$

$$= \left[\left(w_0 f_{\mathrm{ID}^*}^{eS}\right)\left(\frac{z_{\mathrm{FULL}_{\mathrm{ID}^*},1}}{z_{\mathrm{FULL}_{\mathrm{ID}^*},2}}\right)^{T}\right]^{e} \bmod N.$$

$$(16)$$

M then calculates the solution to the *RSAI* problem as follows:

$$w_0 f_{\mathrm{ID}^*}^{e} = \left[\left(w_0 f_{\mathrm{ID}^*}^{eS}\right)\left(\frac{z_{\mathrm{FULL}_{\mathrm{ID}^*},1}}{z_{\mathrm{FULL}_{\mathrm{ID}^*},2}}\right)^{T}\right]^{e} \bmod N,$$

$$w_0^d f_{\mathrm{ID}^*} = \left(w_0 f_{\mathrm{ID}^*}^{eS}\right)\left(\frac{z_{\mathrm{FULL}_{\mathrm{ID}^*},1}}{z_{\mathrm{FULL}_{\mathrm{ID}^*},2}}\right)^{T} \bmod N,$$

$$(17)$$

$$w_0^d = \frac{\left(w_0 f_{\mathrm{ID}^*}^{eS}\right)\left(z_{\mathrm{FULL}_{\mathrm{ID}^*},1}/z_{\mathrm{FULL}_{\mathrm{ID}^*},2}\right)^{T}}{f_{\mathrm{ID}^*}} \bmod N,$$

$$w_0^d = \left(w_0 f_{\mathrm{ID}^*}^{eS-1}\right)\left(\frac{z_{\mathrm{FULL}_{\mathrm{ID}^*},1}}{z_{\mathrm{FULL}_{\mathrm{ID}^*},2}}\right)^{T} \bmod N.$$

M proceeds to calculate the solutions to the other challenges as follows:

$$w_j = V_j(w_0 f_{\mathrm{ID}^*})^{-cm,\mathrm{ID}^*}. \qquad (18)$$

The probability study for the simulation above is similar to that of the impersonation under passive attack game and is therefore omitted. $\square$

4. BNN-SM-IBI: An DL-Based Security-Mediated IBI Scheme

The BNN-SM-IBI scheme is derived from the BNN-IBI scheme proposed in the work of [4] and is provably secure against passive attackers assuming the DL assumption and against active/concurrent attackers assuming the *OMDL* assumption.

The BNN-SM-IBI scheme is constructed as follows.

(1) *Setup* (1^k). It takes in the security parameter 1^k. It randomly selects $x \xleftarrow{\$} \mathbb{Z}_q$, a generator $g \xleftarrow{\$} G$ and computes $X = g^x$. Setup also chooses $H : \{0,1\}^* \to \mathbb{Z}_q$ and publishes the system parameters $\mathsf{mpk} = \langle G, q, g, X, H \rangle$. The master private key $\mathsf{msk} = x$ is kept secret.

(2) *Extract* $(\mathsf{mpk}, \mathsf{msk}, \mathsf{ID})$. It takes in $\mathsf{mpk}, \mathsf{msk} = s$, and ID as input. It calculates $H(\mathsf{ID})$, picks $r_{\mathrm{FULL}}, r_{\mathrm{SEM}}, s_{\mathrm{SEM}} \xleftarrow{\$} \mathbb{Z}_q$, calculates $s_{\mathrm{FULL}} = r_{\mathrm{FULL}} + xH(\mathsf{ID})$ and $R_{\mathrm{FULL}} = g^{r_{\mathrm{FULL}}}$, and sets the full user private key as $\mathsf{usk}_{\mathrm{FULL}} = \langle s_{\mathrm{FULL}}, R_{\mathrm{FULL}} \rangle$. Additionally, it sets $s_{\mathrm{USER}} = s_{\mathrm{FULL}} - s_{\mathrm{SEM}}$, $r_{\mathrm{USER}} = r_{\mathrm{FULL}} - r_{\mathrm{SEM}}$, $R_{\mathrm{USER}} = g^{r_{\mathrm{USER}}}$, $R_{\mathrm{SEM}} = g^{r_{\mathrm{SEM}}}$. It then sets $\mathsf{usk}_{\mathrm{USER}} = \langle s_{\mathrm{USER}}, R_{\mathrm{USER}} \rangle$ and $\mathsf{usk}_{\mathrm{SEM}} = \langle s_{\mathrm{SEM}}, R_{\mathrm{SEM}} \rangle$. It gives

$\mathsf{usk}_{\mathrm{USER}}$ to the user as its partial private key and $\mathsf{usk}_{\mathrm{SEM}}$ to the security mediator as its partial private key and keeps $\mathsf{usk}_{\mathrm{FULL}}$ secret.

(3) *Identification Protocol*. It is run by the *User-Prover* and *SEM-Prover* with *Verifier* as such.

(a) *User-Prover* sends its ID to *Sem-Prover*, chooses $y_{\mathrm{USER}} \xleftarrow{\$} \mathbb{Z}_q$, and sets $Y_{\mathrm{USER}} = g^{y_{\mathrm{USER}}}$ and $S_{\mathrm{User}} = g^{s_{\mathrm{USER}}}$. *SEM-Prover* upon receiving ID checks if it is still valid and returns error if it is revoked. Otherwise, *SEM-Prover* chooses $y_{\mathrm{SEM}} \xleftarrow{\$} \mathbb{Z}_q$ and sends $Y_{\mathrm{SEM}} = g^{y_{\mathrm{SEM}}}$, $S_{\mathrm{SEM}} = g^{s_{\mathrm{SEM}}}$, and R_{SEM} to *User-Prover*. *User-Prover* combines $Y_{\mathrm{FULL}} = Y_{\mathrm{SEM}} \times Y_{\mathrm{USER}}$, $S_{\mathrm{FULL}} = S_{\mathrm{SEM}} \times S_{\mathrm{USER}}$, $R_{\mathrm{FULL}} = R_{\mathrm{SEM}} \times R_{\mathrm{USER}}$ and sends $Y_{\mathrm{FULL}}, S_{\mathrm{FULL}}$, and R_{FULL} to *Verifier*.

(b) *Verifier* chooses a random challenge $c \xleftarrow{\$} \mathbb{Z}_q$ and sends it to *User-Prover*. *User-Prover* relays c to *SEM-Prover*.

(c) *SEM-Prover* calculates its half of the response $z_{\mathrm{SEM}} = y_{\mathrm{SEM}} + cs_{\mathrm{SEM}}$ and sends it to *User-Prover*. *User-Prover* combines it with his response $z_{\mathrm{USER}} = y_{\mathrm{USER}} + cs_{\mathrm{USER}}$ to create $z_{\mathrm{FULL}} = z_{\mathrm{SEM}} + z_{\mathrm{USER}}$ and sends it as a response to *Verifier*.

Verifier checks if $g^{z_{\mathrm{FULL}}} = Y_{\mathrm{FULL}} R_{\mathrm{FULL}}^c X^{H(\mathrm{ID})c}$ and accepts if yes; otherwise it outputs reject.

To check for completeness,

$$\begin{aligned}
g^{z_{\mathrm{FULL}}} &= g^{z_{\mathrm{SEM}} + z_{\mathrm{USER}}} \\
&= g^{y_{\mathrm{SEM}} + cs_{\mathrm{SEM}} + y_{\mathrm{USER}} + cs_{\mathrm{USER}}} \\
&= g^{y_{\mathrm{SEM}} + y_{\mathrm{USER}} + c(s_{\mathrm{SEM}} + s_{\mathrm{USER}})} \\
&= g^{y_{\mathrm{FULL}} + c(s_{\mathrm{FULL}})} \\
&= g^{y_{\mathrm{FULL}} + c(r_{\mathrm{FULL}} + xH(\mathrm{ID}))} \\
&= g^{y_{\mathrm{FULL}}} g^{r_{\mathrm{FULL}} c} g^{xH(\mathrm{ID})c} \\
&= Y_{\mathrm{FULL}} R_{\mathrm{FULL}}^c X^{H(\mathrm{ID})c}.
\end{aligned} \qquad (19)$$

4.1. Security Analysis: Impersonation under Passive Attack

Theorem 7. *The BNN-SM-IBI scheme is* $(t_{SMIBI}, q_{SMIBI}, \varepsilon_{SMIBI})$-*secure against impersonation under passive attacks in the random oracle if the DL problem is* $(t_{DL}, \varepsilon_{DL})$-*hard where*

$$\varepsilon_{SMIBI} \leq \sqrt{\varepsilon_{DL} e\left(q_{SMIBI} + 1\right)} + \frac{1}{q}. \qquad (20)$$

Proof. Assume the BNN-SM-IBI scheme is $(t_{SMIBI}, q_{SMIBI}, \varepsilon_{SMIBI})$-breakable; then a simulator M that $(t_{DL}, \varepsilon_{DL})$-breaks the DL problem can be shown. M takes in input

$(g, A = g^a)$ and runs the impersonator I as a subroutine. Without loss of generality, it can be assumed that any *SEM-Extract*, *User-Extract*, *SEM-Identification*, *User-Identification*, and *Full-Identification* queries are preceded by a *Create-User* query. To avoid collision and consistently respond to these queries, M maintains a list $L_H = \langle \text{ID}_i, Q_{\text{ID}_i}, S_{FULL_{\text{ID}_i}}, S_{SEM_{\text{ID}_i}}, S_{USER_{\text{ID}_i}}, s_{FULL_{\text{ID}_i}}, s_{SEM_{\text{ID}_i}}, s_{USER_{\text{ID}_i}}, r_{FULL_{\text{ID}_i}}, r_{SEM_{\text{ID}_i}}, r_{USER_{\text{ID}_i}}, R_{FULL_{\text{ID}_i}}, R_{SEM_{\text{ID}_i}}, R_{USER_{\text{ID}_i}} \rangle$ which is initially empty. The following shows how M simulates the environment and oracle queries for I.

(1) *Setup.* M sets the system parameters as $\mathsf{mpk} = \langle G, q, g, X = g^x, H \rangle$ and keeps master private key x secret. It sends mpk to I.

(2) *Create-User* (ID_i). M chooses $l \in \{1, \ldots, q_H\}$ randomly and lets $\text{ID}_l = \text{ID}^*$ at this point. Whenever I makes a H query on ID_i, consider the following.

    (i) If $i = l$, M chooses $Q_{\text{ID}^*} \xleftarrow{\$} \mathbb{Z}_q$ and sets $S_{FULL_{\text{ID}^*}} = A$ and $R_{FULL_{\text{ID}^*}} = AX^{Q_{\text{ID}^*}}$ and returns Q_{ID^*} to I. M adds $\langle \text{ID}_l, Q_{\text{ID}^*}, S_{FULL_{\text{ID}^*}}, S_{SEM_{\text{ID}^*}} = \bot, S_{USER_{\text{ID}^*}}, s_{FULL_{\text{ID}^*}} = \bot, s_{SEM_{\text{ID}^*}} = \bot, s_{USER_{\text{ID}^*}} = \bot, r_{FULL_{\text{ID}^*}} = \bot, r_{SEM_{\text{ID}^*}} = \bot, r_{USER_{\text{ID}^*}} = \bot, R_{FULL_{\text{ID}^*}}, R_{SEM_{\text{ID}^*}} = \bot, R_{USER_{\text{ID}^*}} = \bot \rangle$ to L_H.

    (ii) Otherwise, for $i \neq l$, M chooses $Q_{\text{ID}_i} \xleftarrow{\$} \mathbb{Z}_q^*$ and returns Q_{ID_i}. It then chooses $r_{FULL_{\text{ID}_i}}, r_{SEM_{\text{ID}_i}}, s_{SEM_{\text{ID}_i}} \xleftarrow{\$} \mathbb{Z}_q^*$, calculates $s_{FULL_{\text{ID}_i}} = r_{FULL_{\text{ID}_i}} + xQ$, and sets $r_{USER_{\text{ID}_i}} = r_{FULL_{\text{ID}_i}} - r_{SEM_{\text{ID}_i}}$ and $s_{USER_{\text{ID}_i}} = s_{FULL_{\text{ID}_i}} - s_{SEM_{\text{ID}_i}}$. M adds $\langle \text{ID}_i, Q_{\text{ID}_i}, S_{FULL_{\text{ID}_i}}, S_{SEM_{\text{ID}_i}}, S_{USER_{\text{ID}_i}}, s_{FULL_{\text{ID}_i}}, s_{SEM_{\text{ID}_i}}, s_{USER_{\text{ID}_i}}, r_{FULL_{\text{ID}_i}}, r_{SEM_{\text{ID}_i}}, r_{USER_{\text{ID}_i}}, R_{FULL_{\text{ID}_i}}, R_{SEM_{\text{ID}_i}}, R_{USER_{\text{ID}_i}} \rangle$ to L_H.

(3) *SEM-Extract* (ID_i). If $\text{ID}_i = \text{ID}^*$ and *USER-Extract* (ID^*) has been queried before, M aborts. If $\text{ID}_i = \text{ID}^*$, but *USER-Extract* (ID^*) has not been queried before, M will choose $r_{SEM_{\text{ID}^*}}, s_{SEM_{\text{ID}^*}} \xleftarrow{\$} \mathbb{Z}_q$ and sets $R_{SEM_{\text{ID}^*}} = g^{r_{SEM_{\text{ID}^*}}}$ and $S_{SEM_{\text{ID}^*}} = g^{s_{SEM_{\text{ID}^*}}}$. M also sets $S_{USER_{\text{ID}^*}} = S_{FULL_{\text{ID}^*}} / S_{SEM_{\text{ID}^*}}$ and $R_{USER_{\text{ID}^*}} = R_{FULL_{\text{ID}^*}} / R_{SEM_{\text{ID}^*}}$, saves these values in L_H, and returns them to I. These values will be used in the event of an *Identification* query later on. Otherwise for all other $\text{ID}_i \neq \text{ID}^*$, M just retrieves $\langle s_{SEM_{\text{ID}_i}}, R_{SEM_{\text{ID}_i}} \rangle$ in L_H and returns it to I.

(4) *USER-Extract* (ID_i). If $\text{ID}_i = \text{ID}^*$ and *SEM-Extract* (ID^*) has been queried before, M aborts. If $\text{ID}_i = \text{ID}^*$, but *SEM-Extract* (ID^*) has not been queried before, M will choose $r_{USER_{\text{ID}^*}}, s_{USER_{\text{ID}^*}} \xleftarrow{\$} \mathbb{Z}_q$ and sets $R_{USER_{\text{ID}^*}} = g^{r_{USER_{\text{ID}^*}}}$ and $S_{USER_{\text{ID}^*}} = g^{s_{USER_{\text{ID}^*}}}$. M also sets $S_{SEM_{\text{ID}^*}} = S_{FULL_{\text{ID}^*}} / S_{USER_{\text{ID}^*}}$ and $R_{SEM_{\text{ID}^*}} = R_{FULL_{\text{ID}^*}} / R_{USER_{\text{ID}^*}}$, saves these values in L_H, and returns them to I. These values will be

used in the event of an *Identification* query later on. Otherwise for all other $\text{ID}_i \neq \text{ID}^*$, M just retrieves $\langle s_{USER_{\text{ID}_i}}, R_{USER_{\text{ID}_i}} \rangle$ in L_H and returns it to I.

(5) *Identification* $(\text{ID}_j, session)$. I will act as the cheating verifier to learn information from valid conversation transcripts from M. If $\text{ID}_j \neq \text{ID}^*$, M just retrieves ID_j's entry in L_H and runs the identification protocol for either the SEM's interactions, the user's interactions, or both combined as in the protocol.

Otherwise, for $\text{ID}_j = \text{ID}^*$, M then creates a full valid transcript for each mth query by picking $z_{m,FULL_{\text{ID}^*}}, z_{m,SEM_{\text{ID}^*}}, c_{m,\text{ID}^*} \xleftarrow{\$} \mathbb{Z}_q$. M retrieves $S_{FULL_{\text{ID}^*}}, R_{FULL_{\text{ID}^*}}$ and sets $Y_{m,FULL_{\text{ID}^*}} = g^{z_{m,FULL_{\text{ID}^*}}} S_{FULL_{\text{ID}^*}}^{-c_{m,\text{ID}^*}}$. Additionally, M chooses $z_{m,SEM_{\text{ID}^*}} \xleftarrow{\$} \mathbb{Z}_q$, calculates $z_{m,USER_{\text{ID}^*}} = z_{m,FULL_{\text{ID}^*}} - z_{m,SEM_{\text{ID}^*}}$, and retrieves $S_{SEM_{\text{ID}^*}}, R_{SEM_{\text{ID}^*}}$ and $S_{USER_{\text{ID}^*}}, R_{USER_{\text{ID}^*}}$ from L_H as well. M then sets $Y_{m,SEM_{\text{ID}^*}} = g^{z_{m,SEM_{\text{ID}^*}}} S_{SEM_{\text{ID}^*}}^{c_{m,\text{ID}^*}}$ and $Y_{m,USER_{\text{ID}^*}} = g^{z_{m,USER_{\text{ID}^*}}} S_{USER_{\text{ID}^*}}^{c_{m,\text{ID}^*}}$. M tags the session with $session = m$. If I queries

    (i) *SEM-Identification* (ID^*, m), M will return $\langle Y_{m,SEM_{\text{ID}^*}}, S_{SEM_{\text{ID}^*}}, R_{SEM_{\text{ID}^*}}, c_{m,\text{ID}^*}, z_{m,SEM_{\text{ID}^*}} \rangle$;

    (ii) *User-Identification* (ID^*, m), M will return $\langle Y_{m,USER_{\text{ID}^*}}, S_{USER_{\text{ID}^*}}, R_{USER_{\text{ID}^*}}, c_{m,\text{ID}^*} z_{m,USER_{\text{ID}^*}} \rangle$;

    (iii) *Full-Identification* (ID^*, m), M will return $\langle Y_{m,FULL_{\text{ID}^*}}, S_{FULL_{\text{ID}^*}}, R_{FULL_{\text{ID}^*}}, c_{m,\text{ID}^*} z_{m,FULL_{\text{ID}^*}} \rangle$.

If no session is specified for the query, M just returns the next session in sequence. One can see that *SEM-Identification* (ID^*, m) and *User-Identification* (ID^*, m) can be combined to create a full and valid transcript on session m:

$$Y_{m,FULL_{\text{ID}^*}} R_{FULL_{\text{ID}^*}}^{c_{m,\text{ID}^*}} X^{H(\text{ID}^*)c_{m,\text{ID}^*}}$$

$$= Y_{m,USER_{\text{ID}^*}} Y_{m,SEM_{\text{ID}^*}} R_{USER_{\text{ID}^*}}^{c_{m,\text{ID}^*}} R_{SEM_{\text{ID}^*}}^{c_{m,\text{ID}^*}}$$
$$\times X^{H(\text{ID}^*)c_{m,\text{ID}^*}}$$

$$= \left(g^{z_{m,USER_{\text{ID}^*}}} S_{USER_{\text{ID}^*}}^{-c_{m,\text{ID}^*}} g^{z_{m,SEM_{\text{ID}^*}}} \right) S_{SEM_{\text{ID}^*}}^{-c_{m,\text{ID}^*}}$$
$$\times \left(R_{USER_{\text{ID}^*}} R_{SEM_{\text{ID}^*}} \right)^{c_{m,\text{ID}^*}} X^{H(\text{ID}^*)c_{m,\text{ID}^*}}$$

$$= g^{z_{m,USER_{\text{ID}^*}}} g^{z_{m,SEM_{\text{ID}^*}}} \left(S_{USER_{\text{ID}^*}} S_{SEM_{\text{ID}^*}} \right)^{-c_{m,\text{ID}^*}}$$
$$\times \left(R_{USER_{\text{ID}^*}} R_{SEM_{\text{ID}^*}} \right)^{c_{m,\text{ID}^*}} X^{H(\text{ID}^*)c_{m,\text{ID}^*}}$$

$$= g^{z_{m,USER_{\text{ID}^*}}} g^{z_{m,SEM_{\text{ID}^*}}} \left(\frac{S_{FULL_{\text{ID}^*}}}{S_{SEM_{\text{ID}^*}}} S_{SEM_{\text{ID}^*}} \right)^{-c_{m,\text{ID}^*}}$$
$$\times \left(\frac{R_{FULL_{\text{ID}^*}}}{R_{SEM_{\text{ID}^*}}} R_{SEM_{\text{ID}^*}} \right)^{c_{m,\text{ID}^*}} X^{H(\text{ID}^*)c_{m,\text{ID}^*}}$$

$$= g^{z_{m,FULL_{ID^*}}} A^{-c_{m,ID^*}} A^{c_{m,ID^*}} X^{-H(ID^*)c_{m,ID^*}}$$

$$\times X^{H(ID^*)c_{m,ID^*}}$$

$$= g^{z_{m,FULL_{ID^*}}}.$$

$$(21)$$

Eventually, I stops Phase 1 and outputs the challenge ID, ID^* on which it wishes to be challenged on. M checks if $ID = ID_l$ from L_H and aborts if not. Otherwise, M runs I now as a cheating prover on ID^* by obtaining its commitment, $Y_{FULL_{ID^*}}$, $S_{FULL_{ID^*}}$, $R_{FULL_{ID^*}}$, selecting a challenge $c_1 \overset{\$}{\leftarrow} \mathbb{Z}_q^*$ and obtaining the response $z_{FULL_{ID^*},1}$ from I. M then resets I to the step whereby I just sent $Y_{FULL_{ID^*}}$, $S_{FULL_{ID^*}}$, $R_{FULL_{ID^*}}$, selects a second challenge $c_2 \overset{\$}{\leftarrow} \mathbb{Z}_q^*$, and receives $z_{FULL_{ID^*},2}$ as response. M is then able to extract the full user private key as follows:

$$\frac{z_{FULL_{ID^*},1} - z_{FULL_{ID^*},2}}{c_1 - c_2}$$

$$= \frac{y_{FULL_{ID^*}} + c_1 a - y_{FULL_{ID^*}} + c_2 a}{c_1 - c_2}$$

$$= \frac{(c_1 - c_2) a}{c_1 - c_2}$$

$$= a.$$

$$(22)$$

M outputs a as the discrete log solution.

It remains to calculate the probability of M solving the CDH problem and winning the game. The probability of M successfully extracting two valid conversation transcripts from I is bounded by $(\varepsilon_{SMIBI} - (1/q))^2$ as given by the reset lemma [20]:

$$\Pr\left[M \text{ wins DL}\right]$$

$$= \Pr\left[M \text{ computes } a \bigwedge \neg\text{abort}\right]$$

$$= \Pr\left[M \text{ computes } a \mid \neg\text{abort}\right] \Pr\left[\neg\text{abort}\right]$$

$$(23)$$

$$\varepsilon_{DL} \geq \left(\varepsilon_{SMIBI} - \frac{1}{q}\right)^2 \Pr\left[\neg\text{abort}\right].$$

Finally, let δ be the probability that I issues a *SEM-Extract* and a *USER-Extract* query on ID^* and that I makes a total of q_{SMIBI} of such queries. The probability of M answering all the extraction queries is δ_I^q. In Phase 2, the probability of M not aborting is if I outputs the challenge identity ID^* that was not queried before. This is given by the probability $1 - \delta$. Compiling them, the probability of M not aborting is $\delta^{q_{SMIBI}}(1-\delta)$. This value is maximised at $\delta_{opt} = 1-(1/(q_{SMIBI}+1))$. Using δ_{opt}, The probability that M does not abort is at least $1/e(q_{SMIBI}+1)$ because the value $(1-(1/(q_{SMIBI}+1)))^{q_{SMIBI}}$

approaches $1/e$ for large q_{SMIBI}. Therefore, the advantage of M, ε_{DL}, and the bound of the simulation is given as follows:

$$\varepsilon_{DL} \geq \left(\varepsilon_{SMIBI} - \frac{1}{q}\right)^2 \frac{1}{e(q_{SMIBI} + 1)}$$

$$\varepsilon_{DL} e(q_{SMIBI} + 1) \geq \left(\varepsilon_{SMIBI} - \frac{1}{q}\right)^2$$

$$(24)$$

$$\varepsilon_{SMIBI} \leq \sqrt{\varepsilon_{DL} e(q_{SMIBI} + 1)} + \frac{1}{q}.$$

$\square$

4.2. Security Analysis: Impersonation under Active/Concurrent Attack

Theorem 8. *The BNN-SM-IBI scheme is* $(t_{SMIBI}, q_{SMIBI}, \varepsilon_{SMIBI})$*-secure against impersonation under active/concurrent attacks in the random oracle if the OMCDH Problem is* $(t_{OMCDH}, q_{OMCDH}, \varepsilon_{OMCDH})$*-hard where*

$$\varepsilon_{SMIBI} \leq \sqrt{\varepsilon_{OMCDH} e(q_{SMIBI} + 1)} + \frac{1}{q}. \qquad (25)$$

Proof. Assume the BNN-SM-IBI scheme is $(t_{SMIBI}, q_{SMIBI}, \varepsilon_{SMIBI})$-breakable; then a simulator M that $(t_{OMCDH}, q_{OMCDH}, \varepsilon_{OMCDH})$-breaks the *OMCDH* Problem can be shown. M takes in input (g, g^a), is given access to *CHALL* and *DLOG* oracles, and runs the impersonator I as a subroutine. Without loss of generality, it can be assumed that any *SEM-Extract*, *User-Extract*, *SEM-Identification*, *User-Identification*, and *Full-Identification* queries are preceded by a *Create-User* query. To avoid collision and consistently respond to these queries, M maintains a list $L_H = \langle ID_i, Q_{ID_i}, S_{FULL_{ID_i}}, S_{SEM_{ID_i}}, S_{USER_{ID_i}}, s_{FULL_{ID_i}}, s_{SEM_{ID_i}}, s_{USER_{ID_i}}, r_{FULL_{ID_i}}, r_{SEM_{ID_i}}, r_{USER_{ID_i}}, R_{FULL_{ID_i}}, R_{SEM_{ID_i}}, R_{USER_{ID_i}} \rangle$ which is initially empty. The following shows how M simulates the environment and oracle queries for I.

(1) *Setup.* M sets the system parameters as $\mathsf{mpk} = \langle G, q, g, X = g^x, H \rangle$ and keeps master private key x secret. It sends mpk to I.

(2) *Create-User* (ID_i). M chooses $l \in \{1, \ldots, q_H\}$ randomly and lets $ID_l = ID^*$ at this point. Whenever I makes a H query on ID_i, consider the following.

   (i) If $i = l$, M chooses $Q_{ID^*} \overset{\$}{\leftarrow} \mathbb{Z}_q$, queries *CHALL* for W_0, and sets $S_{FULL_{ID^*}} = W_0$ and $R_{FULL_{ID^*}} = W_0 X^{Q_{ID^*}}$. M returns Q_{ID^*} to I. M adds $L_H = \langle ID_l, Q_{ID^*}, S_{FULL_{ID^*}}, S_{SEM_{ID^*}} = \perp, S_{USER_{ID^*}} = \perp, s_{FULL_{ID^*}} = \perp, s_{SEM_{ID^*}} = \perp, s_{USER_{ID^*}} = \perp, r_{FULL_{ID^*}} = \perp, r_{SEM_{ID^*}} = \perp, r_{USER_{ID^*}} = \perp, R_{FULL_{ID^*}}, R_{SEM_{ID^*}} = \perp, R_{USER_{ID^*}} = \perp \rangle$ to L_H.

   (ii) Otherwise, for $i \neq l$, M chooses $Q_{ID_i} \overset{\$}{\leftarrow} \mathbb{Z}_q^*$ and returns Q_{ID_i}. It then chooses $r_{FULL_{ID_i}}, r_{SEM_{ID_i}}, s_{SEM_{ID_i}} \overset{\$}{\leftarrow} \mathbb{Z}_q^*$, calculates

$s_{FULL_{ID_i}} = r_{FULL_{ID_i}} + xQ$, and sets $r_{USER_{ID_i}} = r_{FULL_{ID_i}} - r_{SEM_{ID_i}}$ and $s_{USER_{ID_i}} = s_{FULL_{ID_i}} - s_{SEM_{ID_i}}$. M adds $L_H = \langle ID_i, Q_{ID_i}, S_{FULL_{ID_i}}, S_{SEM_{ID_i}}, S_{USER_{ID_i}}, s_{FULL_{ID_i}}, s_{SEM_{ID_i}}, s_{USER_{ID_i}}, r_{FULL_{ID_i}}, r_{SEM_{ID_i}}, r_{USER_{ID_i}}, R_{FULL_{ID_i}}, R_{SEM_{ID_i}}, R_{USER_{ID_i}} \rangle$ to L_H.

(3) *SEM-Extract* (ID_i). If $ID_i = ID^*$ and *USER-Extract* (ID^*) has been queried before, M aborts. If $ID_i = ID^*$, but *USER-Extract* (ID^*) has not been queried before, M will choose $r_{SEM_{ID^*}}, s_{SEM_{ID^*}} \overset{\$}{\leftarrow} \mathbb{Z}_q$ and sets $R_{SEM_{ID^*}} = g^{r_{SEM_{ID^*}}}$ and $S_{SEM_{ID^*}} = g^{s_{SEM_{ID^*}}}$. M also sets $S_{USER_{ID^*}} = S_{FULL_{ID^*}}/S_{SEM_{ID^*}}$ and $R_{USER_{ID^*}} = R_{FULL_{ID^*}}/R_{SEM_{ID^*}}$, saves these values in L_H, and returns them to I. These values will be used in the event of an *Identification* query later on. Otherwise for all other $ID_i \neq ID^*$, M just retrieves $\langle s_{SEM_{ID_i}}, R_{SEM_{ID_i}} \rangle$ in L_H and returns it to I.

(4) *USER-Extract* (ID_i). If $ID_i = ID^*$ and *SEM-Extract* (ID^*) has been queried before, M aborts. If $ID_i = ID^*$, but *SEM-Extract* (ID^*) has not been queried before, M will choose $r_{USER_{ID^*}}, s_{USER_{ID^*}} \overset{\$}{\leftarrow} \mathbb{Z}_q$ and sets $R_{USER_{ID^*}} = g^{r_{USER_{ID^*}}}$ and $S_{USER_{ID^*}} = g^{s_{USER_{ID^*}}}$. M also sets $S_{SEM_{ID^*}} = S_{FULL_{ID^*}}/S_{USER_{ID^*}}$ and $R_{SEM_{ID^*}} = R_{FULL_{ID^*}}/R_{USER_{ID^*}}$, saves these values in L_H, and returns them to I. These values will be used in the event of an Identification query later on. Otherwise for all other $ID_i \neq ID^*$, M just retrieves $\langle s_{USER_{ID_i}}, R_{USER_{ID_i}} \rangle$ in L_H and returns it to I.

(5) *Identification* (ID_j, session). I will act as the cheating verifier to learn information from valid conversations from M. If $ID_j \neq ID^*$, M just retrieves ID_j's entry in L_H and runs the identification protocol for either the SEM's interactions, the user's interactions, or both combined as in the protocol.

Otherwise, for $ID_i = ID^*$, M then creates a full valid conversation for each mth query by querying *CHALL* for W_m and setting $Y_{m,FULL_{ID^*}} = W_m$. M also retrieves. Additionally, M chooses $Y_{m,SEM_{ID_j}} \overset{\$}{\leftarrow} G$ and sets $Y_{m,User_{ID_j}} = W_m/Y_{m,SEM_{ID_j}}$. M also retrieves $S_{m,FULL_{ID_j}}, S_{m,SEM_{ID_j}}, S_{m,USER_{ID_j}}, R_{m,FULL_{ID_j}}, R_{m,SEM_{ID_j}}$, and $R_{m,USER_{ID_j}}$ from L_H. Upon receiving c_{m,ID_j} from I, M will query $z_{m,FULL_{ID_j}} = DLOG(W_m W_0^{c_{m,ID_j}})$, selects $z_{m,SEM_{ID_j}} \overset{\$}{\leftarrow} \mathbb{Z}_q$, and sets $z_{m,USER_{ID_j}} = z_{m,FULL_{ID_j}} - z_{m,SEM_{ID_j}}$. M tags the session with *session* $= m$.

(a) If I queries *SEM-Identification* (ID_j, m), M will commit $Y_{m,SEM_{ID_j}}, S_{SEM_{ID_j}}, R_{SEM_{ID_j}}$ and respond with $z_{m,SEM_{ID_j}}$ upon receiving c_{m,ID_j}.

(b) If I queries *User-Identification* (ID_j, m), M will commit $Y_{m,USER_{ID_j}}, S_{USER_{ID_j}}, R_{USER_{ID_j}}$ and respond with $z_{m,USER_{ID_j}}$ upon receiving c_{m,ID_j}.

(c) If I queries *Full-Identification* (ID_j, m), M will commit $Y_{m,FULL_{ID_j}}, S_{FULL_{ID_j}}, R_{FULL_{ID_j}}$ and respond with $z_{m,FULL_{ID_j}}$ upon receiving c_{m,ID_j}.

If no session is specified for the query, M just returns the next session in sequence. One can see that *SEM-Identification* (ID_j, m) and *User-Identification* (ID^*, m) can be combined to create a full and valid conversation on session m:

$$Y_{m,FULL_{ID_j}} R_{FULL_{ID_j}}^{c_{m,ID_j}} X^{H(ID_j)c_{m,ID_j}}$$

$$= Y_{m,USER_{ID_j}} Y_{m,SEM_{ID_j}} R_{USER_{ID_j}}^{c_{m,ID_j}} R_{SEM_{ID_j}}^{c_{m,ID_j}} X^{H(ID_j)c_{m,ID_j}}$$

$$= \frac{W_m}{Y_{m,SEM_{ID_j}}} Y_{m,SEM_{ID_j}} \left(\frac{W_0 X^{-H(ID)}}{R_{SEM_{ID_j}}} \right)^{c_{m,ID_j}}$$

$$\times R_{SEM_{ID_j}}^{c_{m,ID_j}} X^{H(ID_j)c_{m,ID_j}} \tag{26}$$

$$= W_m \left(W_0^{c_{m,ID_j}} \right) X^{-H(ID)c_{m,ID_j}} X^{H(ID_j)c_{m,ID_j}}$$

$$= W_m \left(W_0^{c_{m,ID_j}} \right)$$

$$= g^{DLOG(W_m(W_0^{c_{m,ID_j}}))}$$

$$= g^{z_{m,FULL_{ID_j}}}.$$

Eventually, I stops Phase 1 and outputs the challenge ID, ID^* on which it wishes to be challenged on. M checks if $ID^* = ID_l$ from L_H and aborts if not. Otherwise, M runs I now as a cheating prover on ID^*. M runs by obtaining its commitment, $Y_{FULL_{ID^*}}, S_{FULL_{ID^*}}, R_{FULL_{ID^*}}$, selects a challenge $c_1 \overset{\$}{\leftarrow} \mathbb{Z}_q^*$, and obtains the response $z_{FULL_{ID^*},1}$ from I. M then resets I to the step whereby I just sent $Y_{FULL_{ID^*}}, S_{FULL_{ID^*}}, R_{FULL_{ID^*}}$, selects a second challenge $c_2 \overset{\$}{\leftarrow} \mathbb{Z}_q^*$, and receives $z_{FULL_{ID^*},2}$ as response. M is then able to extract the full user private key as follows:

$$\frac{z_{FULL_{ID^*},1} - z_{FULL_{ID^*},2}}{c_1 - c_2}$$

$$= \frac{y_{FULL_{ID^*}} + c_1 w_0 - y_{FULL_{ID^*}} + c_2 w_0}{c_1 - c_2} \tag{27}$$

$$= \frac{(c_1 - c_2) w_0}{c_1 - c_2}$$

$$= w_0.$$

M outputs w_0 as the initial discrete log challenge solution.

TABLE 1: Operation costs for the GQ-SM-IBI scheme.

Algorithm	A	$M1$	$M2$	E
Setup	0	0	0	0
Extract	1	0	0	3
SEM-Prove	0	0	1	2
User-Prove	0	0	3	2
Verify	0	0	1	2

A: addition in $\mathbb{Z}_{\phi(N)}$, $M1$: multiplication in $\mathbb{Z}_{\phi(N)}$, $M2$: multiplication in $\mathbb{Z}_N^*$, and E: exponentiation in $\mathbb{Z}_N^*$.

TABLE 2: Communication costs between algorithms for the GQ-SM-IBI scheme.

Algorithms	Bitstring $|ID|$	Elements of $|e - 1|$	Elements in $\mathbb{Z}_N^*$
TA-User (Extract)	1	0	1
TA-SEM (Extract)	1	0	1
SEM-User (Identification)	1	1	2
User-Verifier (Identification)	1	1	2

TABLE 3: Operation costs for the BNN-SM-IBI scheme.

Algorithm	A	M	GM	E
Setup	0	0	0	1
Extract	3	1	0	3
SEM-Prover	1	1	0	2
User-Prover	2	1	3	2
Verifier	0	1	2	3

H: hash operation, A: addition mod q, M: multiplication mod q, GM: group multiplication, and E: exponentiation mod q.

TABLE 4: Communication costs between algorithms for the BNN-SM-IBI scheme.

Algorithms	Bitstring	Elements in G_1	Elements in $\mathbb{Z}_q$
TA-User (Extract)	1	1	1
TA-SEM (Extract)	1	1	1
SEM-User (Identification)	1	3	2
User-Verifier (Identification)	1	3	2

TABLE 5: Comparison of average simulated runtimes for SM-IBI schemes.

	Identification (ns)
GQ-SM-IBI (1024-bits)	2,158,747
BNN-SM-IBI (1024-bits)	14,064,652
BLS-SM-IBI (512-bits)	104,664,826
GQ-SM-IBI (2048-bits)	7,706,801
BNN-SM-IBI (2048-bits)	82,456,452
BLS-SM-IBI (1024-bits)	487,682,172
GQ-SM-IBI (3072-bits)	17,027,899
BNN-SM-IBI (3072-bits)	182,153,886
BLS-SM-IBI (1536-bits)	1,188,023,987

M proceeds to calculate the solutions to the other challenges as follows:

$$
\begin{aligned}
z_{m,FULL_{ID_j}} &- w_0 c_{m,ID_j} \\
&= DLOG\left(W_m W_0^{c_{m,ID_j}}\right) - w_0 c_{m,ID_j} \\
&= \left[w_m + w_0 c_{m,ID_j}\right] - w_0 c_{m,ID_j} \\
&= w_m.
\end{aligned}
\tag{28}
$$

The probability study for the simulation above is similar to that of the impersonation under passive attack game and is therefore omitted.   □

5. Efficiency Analysis

In this section we provide the breakdown of operation costs for both the GQ-SM-IBI scheme and the BNN-SM-IBI scheme.

We measure the operation costs of the GQ-SM-IBI scheme in terms of addition operations in $\mathbb{Z}_{\phi(N)}$, multiplication operations in $\mathbb{Z}_{\phi(N)}$ and $\mathbb{Z}_N^*$, and exponentiation operations in $\mathbb{Z}_N^*$. Overall, the operational costs of the GQ-SM-IBI scheme are given in Table 1.

We also provide the communication costs of the GQ-SM-IBI scheme in Table 2.

As for the BNN-SM-IBI scheme, we measure the operational costs in terms of group operational costs of addition modulo q, multiplication modulo q, multiplication in group G, and exponentiations modulo q. Overall, the operational costs of the BNN-SM-IBI scheme are given in Table 3.

We also provide the communication costs of the BNN-SM-IBI scheme in Table 4.

However, it is difficult to compare all these operational costs since the schemes are using operations defined in different fields and groups. Therefore, we built a simulator to generate running time results using a common platform.

We compare the running time of only the identification protocol, since the protocol is invoked every time a prover wishes to perform an interaction with a verifier. We compare the running times of the GQ-SM-IBI scheme and BNN-SM-IBI scheme as well as the original pairing-based scheme by [19]. We name the scheme from [19] as BLS-SM-IBI since its design follows the BLS signature scheme by Boneh et al. from [22]. For GQ-SM-IBI and BNN-SM-IBI the level of security used was 1024-bits, 2048-bits, and 3072-bits. We also compared the results with the equivalent level of security for BLS-SM-IBI pairing-based scheme of 512-bits, 1024-bits, and 1536-bits, respectively.

The simulation was run on a i7-2630QM platform with 4 GB RAM running 64-bit Windows 7. The library used was Java Cryptography Extension. We ran the simulation 100 iterations for each algorithm and took the average running time, measured in nanoseconds as presented in Table 5.

From the results, one can see that the GQ-SM-IBI scheme has the fastest identification protocol running time with the BNN-SM-IBI scheme trailing behind. However, both pairing-free schemes vastly outperform the BLS-SM-IBI that is pairing-based with at least an approximate factor of 6 to 7 for all three security levels. Therefore we obtain faster SM-IBI schemes from pairing-free alternatives.

6. Conclusion

We proposed two SM-IBI schemes that have an instant revocation feature and are very efficient. Our schemes outperform the only pairing-based SM-IBI currently known and are provably secure in the random oracle model against both passive and active/concurrent attackers.

Conflict of Interests

The authors declare that there is no conflict of interests regarding the publication of this paper.

Acknowledgment

The authors would like to acknowledge the Ministry of Education, Malaysia, for financially aiding this research through the Fundamental Research Grant Scheme FRGS/2/2013/ICT07/MMU/03/5.

References

[1] A. Shamir, "Identity-based cryptosystems and signature schemes," in *Advances in Cryptology*, G. R. Blakley and D. Chaum, Eds., vol. 196 of *Lecture Notes in Computer Science*, pp. 47–53, Springer, Berlin, Germany, 1984.

[2] D. Boneh, X. Ding, G. Tsudik, and C. M. Wong, "A method for fast revocation of public key certificates and security capabilities," in *Proceedings of the 10th Conference on USENIX Security Symposium (SSYM '01)*, D. S. Wallach, Ed., vol. 10, p. 22, Berkeley, Calif, USA, 2001.

[3] A. Fiat and A. Shamir, "How to prove yourself: practical solutions to identification and signature problems," in *Advances in Cryptology-CRYPTO '86*, A. M. Odlyzko, Ed., vol. 263 of *Lecture Notes in Computer Science*, pp. 186–194, Springer, Berlin, Germany, 1986.

[4] M. Bellare, C. Namprempre, and G. Neven, "Security proofs for identity-based identification and signature schemes," in *Advances in Cryptology-EUROCRYPT 2004*, C. Cachin and J. Camenisch, Eds., vol. 3027 of *Lecture Notes in Computer Science*, pp. 268–286, Springer, Berlin, Germany, 2004.

[5] K. Kurosawa and S. H. Heng, "From digital signature to ID-based identification/signature," in *Public Key Cryptography-PKC 2004*, F. Bao, R. H. Deng, and J. Zhou, Eds., vol. 2947 of *Lecture Notes in Computer Science*, pp. 248–261, Springer, Berlin, Germany, 2004.

[6] J. J. Chin, S. H. Heng, and B. M. Goi, "An efficient and provable secure identity-based identification scheme in the standard model," in *Public Key Infrastructure*, S. F. Mjølsne, S. Mauw,

and S. K. Katsikas, Eds., vol. 5057 of *Lecture Notes in Computer Science*, pp. 60–73, Springer, Berlin, Germany, 2008.

[7] J. J. Chin and S. H. Heng, "An adaptive-secure k-resilient identity-based identification scheme in the standard model," in *Proceedings of the FTRA-ACSA Summer*, Vancouver, Canada, June 2012.

[8] J. J. Chin, S. H. Heng, and B. M. Goi, "Hierarchical identity-based identification schemes," in *Security Technology*, D. Slezak, T. H. Kim, W. C. Fang, and K. P. Arnett, Eds., vol. 58 of *Communications in Computer and Information Science*, pp. 93–99, Springer, Berlin, Germany, 2009.

[9] A. Fujioka, T. Saito, and K. Xagawa, "Secure hierarchical identity-based identification without random oracles," in *Information Security*, D. Gollmann and F. C. Freiling, Eds., vol. 7483 of *Lecture Notes in Computer Science*, pp. 258–273, Springer, Berlin, Germany, 2012.

[10] A. Fujioka, T. Saito, and K. Xagawa, "Applicability of OR-proof techniques to hierarchical identity-based identification," in *Cryptology and Network Security*, J. Pieprzyk, A. R. Sadeghi, and M. Manulis, Eds., vol. 7712 of *Lecture Notes in Computer Science*, pp. 169–184, Springer, Berlin, Germany, 2012.

[11] J. J. Chin, R. C. W. Phan, R. Behnia, and S. H. Heng, "An efficient and provably secure certificateless identification scheme," in *SECRYPT 2013*, P. Samarati, Ed., pp. 371–378, SciTePress, 2013.

[12] X. Ding and G. Tsudik, "Simple identity-based cryptography with mediated RSA," in *Topics in Cryptology-CT-RSA 2003*, M. Joye, Ed., vol. 2612 of *Lecture Notes in Computer Science*, pp. 193–210, Springer, Berlin, Germany, 2003.

[13] B. Libert and J. J. Quisquater, "Efficient revocation and threshold pairing based cryptosystems," in *Proceedings of the 22nd Annual Symposium on Principles of Distributed Computing (PODC '03)*, E. Borowsky and S. Rajsbaum, Eds., pp. 163–171, ACM, Boston, Mass, USA, July 2003.

[14] X. Cheng, L. Guo, and X. Wang, "An identity-based mediated signature scheme from bilinear pairing," *International Journal of Network Security*, vol. 2, no. 1, pp. 29–33, 2006.

[15] S. S. M. Chow, C. Boyd, and J. M. G. Nieto, "Security-mediated certificateless cryptography," in *Public Key Cryptography-PKC 2006*, M. Yung, Y. Dodis, A. Kiayias, and T. Malkin, Eds., vol. 3958 of *Lecture Notes in Computer Science*, pp. 508–524, Springer, Berlin, Germany, 2006.

[16] W. S. Yap, S. S. M. Chow, S. H. Heng, and B. M. Goi, "Security mediated certificateless signatures," in *Applied Cryptography and Network Security*, J. Katz and M. Yung, Eds., vol. 4521 of *Lecture Notes in Computer Science*, pp. 459–477, Springer, Berlin, Germany, 2007.

[17] S. S. Al-Riyami and K. G. Paterson, "Certificateless public key cryptography," in *Advances in Cryptology-ASIACRYPT 2003*, C.-S. Laih, Ed., vol. 2894 of *Lecture Notes in Computer Science*, pp. 452–473, Springer, Berlin, Germany, 2003.

[18] J. J. Chin, R. Behnia, S. H. Heng, and R. C. W. Phan, "Cryptanalysis of a certificateless identification scheme," *Security and Communication Networks*, 2014.

[19] J. J. Chin, R. Behnia, S. H. Heng, and R. C. Phan, "An efficient and provable secure security-mediated identity-based identification scheme," in *Proceedings of the 8th Asia Joint Conference on Information Security (Asia JCIS '13)*, pp. 27–32, IEEE, Seoul, South Korea, 2013.

[20] M. Bellare and A. Palacio, "GQ and Schnorr identification schemes: proofs of security against impersonation under active and concurrent attacks," in *Advances in Cryptology-CRYPTO*

2002, M. Yung, Ed., vol. 2442 of *Lecture Notes in Computer Science*, pp. 162–177, Springer, Berlin, Germany, 2002.

[21] S. Y. Tan, S. H. Heng, R. C. W. Phan, and B. M. Goi, "A variant of Schnorr identity-based identification scheme with tight reduction," in *Future Generation Information Technology*, T. H. Kim, H. Adeli, D. Slezak et al., Eds., vol. 7105 of *Lecture Notes in Computer Science*, pp. 361–370, Springer, Berlin, Germany, 2011.

[22] D. Boneh, B. Lynn, and H. Shacham, "Short signatures from the weil pairing," *Journal of Cryptology*, vol. 17, no. 4, pp. 297–319, 2004.

12

A Social Diffusion Model with an Application on Election Simulation

Jing-Kai Lou,[1,2] **Fu-Min Wang,**[3] **Chin-Hua Tsai,**[3] **San-Chuan Hung,**[3] **Perng-Hwa Kung,**[3] **Shou-De Lin,**[3] **Kuan-Ta Chen,**[1] **and Chin-Laung Lei**[2]

[1] *Institute of information Science, Academia Sinica, 128 Academia Road, Section 2, Nankang, Taipei 115, Taiwan*

[2] *Department of Electrical Engineering, National Taiwan University, No. 1, Section 4, Roosevelt Road, Taipei 10617, Taiwan*

[3] *Department of Computer Science and Information Engineering, National Taiwan University, No. 1, Section 4, Roosevelt Road, Taipei 10617, Taiwan*

Correspondence should be addressed to Jing-Kai Lou; kaeaura@gmail.com

Academic Editors: H. R. Karimi, Z. Yu, and W. Zhang

Issues about opinion diffusion have been studied for decades. It has so far no empirical approach to model the interflow and formation of crowd's opinion in elections due to two reasons. First, unlike the spread of information or flu, individuals have their intrinsic attitudes to election candidates in advance. Second, opinions are generally simply assumed as single values in most diffusion models. However, in this case, an opinion should represent preference toward multiple candidates. Previously done models thus may not intuitively interpret such scenario. This work is to design a diffusion model which is capable of managing the aforementioned scenario. To demonstrate the usefulness of our model, we simulate the diffusion on the network built based on a publicly available bibliography dataset. We compare the proposed model with other well-known models such as independent cascade. It turns out that our model consistently outperforms other models. We additionally investigate electoral issues with our model simulator.

1. Introduction

Huge success of viral marketing nowadays clearly shows that acquaintances indeed greatly influence people adopting a new or different opinion. This implicates that people, in a way, attempt to plant their intrinsic ideas, opinions, or preferences in others' minds through exchanging opinions over and over in different circumstances. One interesting and long-discussed scenario is election. Elections in the modern world are an essential mechanism to aggregate the opinions of the masses and to make joint decisions for a variety of purposes. People share thoughts and even attempt to convince others to adopt their attitudes during the election season.

As social media such as Facebook are widely utilized, it becomes quite convenient for people to manifest themselves. Social media exposure grants people a hitherto wide range to deliver their views. Social media extremely accelerates and facilitates such opinion-exchange interactions among individuals. As opinions interflow, the intrinsic opinions of an irresolute person could eventually be assimilated to those of the determined ones. Then a consensus or a public opinion appears.

From a research aspect, understanding the progress of human negotiation benefits the real world applications. For instance, social scientists would wonder to what extent the opinions' exchange among friends can affect each other's viewpoints. Campaign companies would inquire how to promote a candidate with limited budgets. Such questions are not easy to answer via a user study, particularly when the number of participants becomes huge.

Opinion diffusion on social networks has been studied for decades. Unfortunately, many previous models, such as the Independent Cascade Model, Linear Threshold Model, SIR/SIS model, and heat diffusion model, cannot manage the election scenario intuitively due to the following two reasons. First, people have their intrinsic opinions more or less, which is absent in the aforementioned models. In such a manner, people may not serve as neutral relays. People amplify the opinions they stand for and deamplify opinions they stand against. Second, opinions could be multidimensional, for example, a viewpoint for multiple electoral candidates. Most diffusion models adopt single values to represent the opinions for simplicity. A single value clearly cannot directly represent the views about multiple candidates. Our goal is to design a suitable diffusion model which is capable of managing the propagation of viewpoints. Up to date we have not yet seen too many computational approaches with systematic and quantifiable studies on this issue.

Inspired by the real world phenomena, we have realized several preferable properties to manage the information diffusion like opinion polling. The properties are *high-dimensional media*, *input dependence*, *deterministic convergence*, and *consensus*. A summary of the properties is as follows. First, we prefer the media (which represents preference toward candidates) propagated throughout the process being a unit vector because, democratically, individuals (or nodes) have equal rights in casting votes. Second, the preference distribution should be significantly affected by the initial intrinsic preference as well as the neighbors through social network. Finally, we hope the propagation converges eventually, and a common trend appears after numerous interactions [1]. In this paper, we show that our model is the only one satisfying all properties among the existing models.

The novelty and contributions of this paper can be viewed from several different angles.

(1) This work strategically demonstrates a plausible process to answer a set of real world problems.

    (a) We start by designing a preference negotiation model (with theoretical guarantees) to manage high-dimensional information. We assess the quality of this model by proving its convergence and several other important properties.

    (b) We conduct an experiment to demonstrate the validity of our model in predicting the change of citation preference among authors through collaboration networks.

    (c) To build the diffusion simulation framework, we further devise suite of satellite algorithms for preference profile sampling, deployment, and seed-voter selection.

(2) Our case studies, in practice, have provided the concrete solutions to the real world problems of concern.

    (a) The simulation shows that election outcomes can be significantly affected by the social factor. We find that each individual preference profile extremely changes after a preference diffusion. Using Kendal τ coefficient to measure the similarity before and after the diffusion, it usually results in a coefficient below 0.5.

    (b) With the simulation, we additionally examine several well-known voting schemas to verify

their vulnerability to vote-buying. Among them, Borda Count voting schema performs best to resist vote-buying. Plurality voting is the most vulnerable to manipulation.

2. Preliminary

We review the previously done works related to the information diffusion and electoral issues.

2.1. Diffusion Model. To develop models for diffusion simulation or prediction, researchers unearth the underlying mechanisms or the inherent patterns of information diffusion from real word phenomena and utilize these findings.

The Linear Threshold Model (LT model for short) [2, 3] and Independent Cascade Model (IC model for short) [3, 4] are the most well-known and fundamental ones to describe how the information propagates step by step in a network. Inspired by the ideas of the two models, various models have been proposed later for more specific scenarios.

The LT model at first intends to describe the process of shutdowns due to chain effect of energy overload in a power grid. The concept is then adopted for simulating the information diffusion. In the LT model, nodes in a network are the containers of energy (information) and the amounts carried are represented as real values. Each node has a predesignated carrying capacity and initially carries no energy. Once the simulation proceeds, some nodes are assigned as the early adopters, the first groups gaining energy (information), to carry energy, and the carried amount increases progressively. Once the amount of carried energy exceeds their capacity, the nodes become active (overload) and pass excessive energy to other linked nodes. This leads to a propagation of power overload.

With an operation similar to the LT model, the IC model further simplifies the carried information as a binary value. Nodes become active once they receive the information passed from the linked neighbors. There is no predetermined capacity for nodes in IC model. Instead, each edge is associated with a real number representing the probability of successful information pass along it. During the diffusion, the active nodes continually attempt to send out the information through edges until all linked neighbors become active.

Kempe et al. (2003) [3] generalized the IC model by introducing a General Cascade Model. Gruhl et al. (2004) [5] and Leskovec et al. (2006) [6] proposed generative model to simulate blog essay generation based on the IC Model. These models assume nodes can turn from inactive to active given a certain probability for cascading. Based on the LT model and the IC model, Saito et al. (2010) [7] proposed the Asynchronous Linear Threshold Model and Asynchronous Independent Cascade Model.

Another influential line of research, following the success of the PageRank algorithm, puts the propagation process in an explicit recursive mathematical form. Heat diffusion [8, 9] is a physics phenomenon describing heat flows from high temperature positions to low temperature positions. Inspired by the heat diffusion, Ma et al. (2008) [8] proposed a model to analyze candidate selection strategies for market promotion. The process is formulated as

$$\frac{f_i(t + \Delta t) - f_i(t)}{\Delta t} = \alpha \sum_{j:(v_j, v_i) \in E} \left(f_j(t) - f_i(t) \right), \quad (1)$$

where $f_i(t)$ is the heat of node i at time t and α is the thermal conductivity, namely, the heat diffusion coefficient.

In heat diffusion process, each vertex receives heat from its neighbors, which is similar to our model. The major difference, which will be discussed in detail in the following section, is that heat diffusion model lacks a normalization phase (since it considers only the propagation of one value) and a fusion phase (because the heat itself can disappear after diffusion, so there is no need to fuse on heat diffusion model).

Inspired by these previous works, our model takes the strong points of these approaches, namely, their focus on mimicking social interaction traits such as forming consensus as well as their incorporation of structural information into the propagation process, and blends them into a more coherent framework that could be used to answer real world problems mentioned in the introduction.

2.2. Electoral Issues. In 1992, Bartholdi III et al. [10] first studied the complexity of the process to determine needed actions by organizer to add or remove candidates to manipulate election results (where it is recognized as the classical social choice theory). However, they did not propose any model for the interactions between voters. Gibbard [11] and Satterthwaite [12] showed that every election scheme with at least three possible outcomes is subject to individual manipulation. This means the minority has a chance to manipulate the group decision to secure a preferred outcome. Gibbard and Satterthwaite also addressed the computational difficulty in minority manipulation. However, their model assumes the independence of voters, which does not consider nor discuss the effect from other voters on voter's preference. Existing studies in this direction mainly focus on the complexity and feasibility issues, which is very different from our goal.

Liu (2009) [13] attempted to check whether the preference distribution changes if the number of political experts in a communication network increases. They use an agent-based model for simulation. Each agent in the model maintains a binary value toward a candidate (instead of a real value or ranking) and simply disseminates the values to other agents in the nearby 3 by 3 matrix.

Yoo et al. (2009) [14] proposed semisupervised importance propagation model. Their idea is, to some extent, similar to our "fusion phase" by adding the original score into the accumulated score obtained from the neighbor. The difference between their model and ours is that theirs deals with a single value instead of a vector, and therefore they do not perform the normalization over candidate scores like we do.

The election manipulation is a long-discussed issue. Nevertheless, the social factor is absent in these works. Here we bring a marriage between the social network analysis and the electoral issues.

3. The Proposed Model

We here propose the diffusion model to unearth how the communications affect the individual decisions. Abbreviations section lists the notations used in the rest of the paper.

3.1. Preference Propagation Model. We first define a preference profile p_v of an individual v, which is a k-dimensional vector that represents v's preference toward k different candidates. The jth element in p_v is an integer in $[1, k]$ indicating this individual's preference for candidate j (smaller numbers denote higher ranks). To facilitate the operation of the preference profiles, we translate p_v into a score vector s_v, for all v, using the following equation:

$$s_v[i] = \frac{(k - p_v[i] + 1)}{T}, \quad \forall i \in 1, 2, \ldots, k, \quad (2)$$

where $T = k(k + 1)/2$. This transformation can be regarded as a normalization process as in s_v not only does the preferred candidate receive higher score but also the sum of all elements equals 1. Using the score vector of each individual, we can create an n by k matrix $S = (s_{v_1}, s_{v_1}, \ldots, s_{v_n})^t$ denoted by the preference matrix. We denote the preference matrix of a given time stamp t since the propagation process starts as $S(t)$.

The information propagates one iteration after the other in our model, and each iteration consists of three phases: *propagation*, *normalization*, and *fusion*.

In the propagation phase, each node v synchronically propagates the preference score vector s_v to the neighboring nodes. To describe such operation mathematically, we define an $n \times n$ forward transition matrix F such that the multiplication of F and $S(t)$ represents the score of each node obtained from all neighbors after this phase. We denoted it by $S_p(t)$.

We assume the edge directions in a network G reveal the direction of influence. Therefore, $F = (KA)^t$, where K is a diagonal matrix with the inverse of degree of each node in the diagonal and A is the adjacency matrix of G. Note that F is identical to the forwarding matrix of a random walk algorithm. The only difference is that F in a random walk algorithm is multiplied by a vector instead of by a matrix S.

In S_p, each row represents the neighbors' accumulated preference scores toward each candidate. Unlike S, the elements in each row of S_p do not add up to one. To ensure that every individual has equal influence while casting votes, we normalize each row of S_p so that its elements add up to one. Therefore, in the second phase, S_p is multiplied by an $n \times n$ diagonal normalization matrix N, where each element in the diagonal of N is equal to the sum of all elements in the corresponding row of S_p. After the second phase, we will obtain a new scoring matrix $S_n(t) = NFS(t)$.

The major difference between our propagation model and the diffusion models for electricity/heat (see Section 2 for more details) lies in the intrinsic difference of the media that are propagated. Electricity or heat flows from one place to another (that is, a flow from node A to node B implies the material does not exist in A anymore). Opinions, by contrast, do not vanish after propagation (that is, A's inclination towards a candidate does not disappear even after bringing his or her opinions to B). Therefore we add a third phase to include a fusion model that integrates an individual's own opinions $S(t)$ with the opinion $S_n(t)$ gathered from its neighbors.

In the fusion phase, we introduce a parameter for each individual: the susceptible ratio, a real number $\epsilon \in [0, 1]$ that represents how easily an individual can be affected by others. Given a susceptibility parameter for each individual, we can then create a susceptible matrix E, an $n \times n$ diagonal matrix with the ϵ value of each individual in the diagonal. If E is equal to the identity matrix I, which would imply all individuals are equally and highly susceptible to one another, then $S(t + 1)$ should be equivalent to its neighbors' opinion $S_n(t)$. On the opposite side, if E is equal to the zero matrix, implying all individuals are impervious to one another, then $S(t+1)$ should be identical to $S(t)$. Thus, after one iteration of propagation, the preference score matrix can be represented as

$$S(t + 1) = (I - E) S(t) + ENFS(t) = ((I - E) + ENF) S(t). \tag{3}$$

Note that we assume that E does not change over time, and neither does F (which is only dependent upon topology). Interestingly, at first glance one might assume that N changes iteratively; it actually does not. Because the sum of each column in F equals 1 and the scores are always normalized for all candidates, it is not hard to prove that

$$N_{ij} = \begin{cases} \left(\sum_{j=1}^{n} F_{i,j}\right)^{-1} & \text{when } i = j \\ 0 & \text{otherwise,} \end{cases} \tag{4}$$

which depends only on F. Therefore, we can rewrite $S(t + 1)$ as $\mathcal{X}S(t)$, where $\mathcal{X}$ is a time-independent matrix, which becomes an important feature for the proof of convergence in the next section.

The above concludes one iteration of propagation. In the next iteration, $S(t + 1)$ becomes the initial preference score for the individuals and the same process can be executed to obtain another round of propagation results $S(t + 2)$. Algorithm 1 is the algorithm for our model.

3.2. Proof of Convergence and Consensus. In this section, we show the convergent property of our proposed scheme. The score matrix becomes invariant after a sufficient number of propagations. Moreover, we show that given certain conditions all rows in the converged score matrix are identical. In other words, a consensus within a community will eventually be reached through information propagations in our model.

Let $\mathcal{X}$ denote the overall preference propagation operation of all three phases explicitly laid down in the previous section,

$$S(t + 1) = \mathcal{X}S(t) = [(I - E) + ENF] S(t). \tag{5}$$

To provide intuition for the forthcoming deductions and to borrow results of the properties of $\mathcal{X}$ from Section 3.1, we start by pointing out the similarities as well as the differences between $\mathcal{X}$ and the PageRank matrix $\mathcal{G}$. First, the entity $\mathcal{X}$ acting on $S(t)$ is actually a matrix consisting of the vectors of

probabilities instead of a simple vector of probabilities. As a result, the columns of $\mathcal{X}$ do not add up to 1 (only the rows do) and therefore it is not a stochastic matrix. Furthermore, a social personal relationship network is intrinsically more localized compared to the World Wide Web, and, as such, the favorable positive definite property enjoyed by $\mathcal{G}$ does not necessarily hold for $\mathcal{S}$. That said, these complexities, while no doubt complicating the theoretical treatment of our algorithm, are in fact a natural manifestation of the increased richness of our target of research in hand—social networks.

We start our deduction of the convergence of $\mathcal{X}$ by enlisting the Perron-Frobenius theorem [15] which states that an irreducible, acyclic matrix has a single eigenvalue that is strictly larger than the others. Under the assumption that the graph being induced by $\mathcal{X}$, $G_{\mathcal{X}}$, is strongly connected and that the weights matrix E has entries smaller than one but not all zeros, $\mathcal{X}$ is irreducible and acyclic and thus applies to the Perron-Frobenius theorem. We denote the dominant real positive eigenvalue of $\mathcal{X}$ by r. Armed with this fact, we are able to transform $\mathcal{X}$ into its Jordan canonical form

$$\mathcal{X} = P^{-1} J_{\mathcal{X}} P, \quad J_{\mathcal{X}} = \begin{pmatrix} J_{\mathcal{X}_1} & 0 & \cdots \\ 0 & J_{\mathcal{X}_2} & \cdots \\ \vdots & \vdots & \ddots \end{pmatrix}, \tag{6}$$

by which the leading block $J_{\mathcal{X}_1}$ is a 1×1 matrix $[r]$, and other $J_{\mathcal{X}_i}$'s correspond to their strictly smaller eigenvalues $\lambda_{\mathcal{X}_i}$. Since, by the rules of matrix multiplication, the effects of $\mathcal{X}$ on $S(t)$ can be analyzed one by one with respect to $S(t)$'s column vectors without loss of generality, we will proceed with our proof of $S(t)$'s convergence by concentrating on $S(t)$'s column vectors which we denote by lower case $s(t)$. Decomposing $s(0)$ into the sum of $\mathcal{X}$'s eigenvectors, $c_1 v_1 + c_2 v_2 + \cdots$, we obtain the general form of the time evolution of $s(t)$,

$$s(t) = J_{\mathcal{X}}^t (c_1 v_1 + c_2 v_2 + \cdots) = r^t (c_1 v_1 + b_t), \tag{7}$$

where

$$\|b_t\| = \frac{1}{r^t} \left\| J_{\mathcal{X}_2}^t c_2 v_2 + \cdots \right\|$$

$$\leq \sum_{i=2}^{|V|} \left(\frac{|\lambda x_i|}{r}\right)^t \|c_i v_i\| \longrightarrow 0, \quad \text{as } t \longrightarrow \infty. \tag{8}$$

The above shows that $\|b_t\|$ converges to zero when t is large, and therefore $S(t)$ converges to $r^t(c_1 v_1)$. To get an intuition for the speed of this convergence, we turn to a special case where the susceptible ratios are identical; that is, E is a scalar ϵ. In this case, we apply the Perron-Frobenius theorem again on NF and we again obtain NF's Jordan form

$$NF = P^{-1} J_{NF} P, \quad J_{NF} = \begin{pmatrix} J_{NF_1} & 0 & \cdots \\ 0 & J_{NF_2} & \cdots \\ \vdots & \vdots & \ddots \end{pmatrix}. \tag{9}$$

However, note that since it needs not to be acyclic, it is strictly larger than the other. Now, using this basis we find that $\mathcal{X}$ is equal to

> R: iteration number; P: initial preference profiles
> E: susceptible matrix; F: forwarding matrix
> N: normalization matrix
> $S(0) = PreferenceToScore(P)$
> **for** $t = 0$ **to** R **do**
> $S_p(t) = FS(t)$
> $S_n(t) = NS_p(t)$
> $S(t + 1) = ES(t) + (I - E)S_n(t)$
> **end for**
> **return** $S(R)$

ALGORITHM 1: Preference propagation model.

$$\epsilon \begin{pmatrix} \ddots & & & & & & 0 \\ & \ddots & & & & & \\ & \dfrac{\left(1 - \epsilon + \epsilon\lambda_{NF_i}\right)}{\epsilon} & 1 & & & & \\ & & \dfrac{\left(1 - \epsilon + \epsilon\lambda_{NF_i}\right)}{\epsilon} & 1 & & & \\ & & & \dfrac{\left(1 - \epsilon + \epsilon\lambda_{NF_i}\right)}{\epsilon} & \ddots & & \\ 0 & & & & & \ddots & \end{pmatrix}. \tag{10}$$

Since a Jordan canonical form is unique, we obtain $\lambda_{\mathscr{X}_i} = (1 - \epsilon + \epsilon\lambda_{NF_i})/\epsilon$. From this result, we confirm that when $\epsilon = 0$, $\mathscr{X}$ degenerates to the trivial diagonal case and that as ϵ approaches 1, the rate of convergence is geometrically proportional to ϵ/r.

We are now one step away from the final proof of S's convergence. Recalling that $s(t) \rightarrow r^t c_1 v_1$, once $r \leq 1$ is established, $S(t)$ converges. To prove this, we take advantage of the Collatz-Wielandt theorem which gives the following formula for r: $r = \max_{x \in N} f(x)$, where $f(x) = \min_{1 \leq i \leq n; x_i \neq 0}[\mathscr{X}x]/x_i$ and $N = \{x \mid x \geq 0 \text{ with } x \neq 0\}$.

We begin by asserting that the upper bound of $f(x)$ is 1. To prove this, we suppose the opposite holds, that means there exists x such that $f(x) = \min_{1 \leq i \leq n; x_i \neq 0}[\mathscr{X}x]/x_i = \alpha > 1$. This implies the following list of equations:

$$1 < \alpha \leq \frac{1}{x_1}\left(\mathscr{X}_{11}x_1 + \mathscr{X}_{12}x_2 + \cdots + \mathscr{X}_{1n}x_n\right)$$

$$\vdots \tag{11}$$

$$1 < \alpha \leq \frac{1}{x_n}\left(\mathscr{X}_{n1}x_1 + \mathscr{X}_{n2}x_2 + \cdots + \mathscr{X}_{nn}x_n\right).$$

Note that $\sum_{j=1}^{n}\mathscr{X}_{ij} = 1, \forall i$. Thus, the above list of equations can be arranged into

$$\mathscr{X}_{12}\left(\frac{x_2}{x_1} - 1\right) + \mathscr{X}_{13}\left(\frac{x_3}{x_1} - 1\right) + \cdots + \mathscr{X}_{1n}\left(\frac{x_n}{x_1} - 1\right)$$

$$> 0$$

$$\vdots$$

$$\mathscr{X}_{n1}\left(\frac{x_1}{x_n} - 1\right) + \mathscr{X}_{n2}\left(\frac{x_2}{x_n} - 1\right) + \cdots + \mathscr{X}_{n(n-1)}\left(\frac{x_{n-1}}{x_n} - 1\right)$$

$$> 0. \tag{12}$$

However, by denoting i by the subscript that has $x_i = \max_{1 \leq j \leq n}x_j$ and remembering that $\mathscr{X}$ is a nonnegative matrix, one of the above equations would not hold:

$$\mathscr{X}_{i1}\left(\frac{x_1}{x_i} - 1\right) + \mathscr{X}_{i1}\left(\frac{x_2}{x_i} - 1\right) + \cdots + \mathscr{X}_{in}\left(\frac{x_n}{x_i} - 1\right) > 0. \tag{13}$$

This justifies the assertion that $f(x) \leq 1$. Combining this result with the observation that the trivial vector $(1, 1, \ldots)$ yields $f(x) = 1$, we conclude that $\max_{x \in N} f(x) = 1$. Therefore, $r = 1$, and $S(t)$ converges to $c_1 v_1$.

For networks that are not strongly connected we can always find the SCCs in linear time, and the problem reduces to the smaller "source SCCs" of the network since the matrices of all the other SCCs have a Perron root smaller than 1 and their elements eventually vanish. For the remaining source SCCs, since no vertices have susceptibility ratios equal to 1, according to the above results they all converge. The net effect is exemplified by the stark difference between the individuals belonging to the various source SCCs and the rest. Whereas source SCC vertices will converge to their own respective common values, the others may converge to different values and act as followers in terms of aligning their own preferences to the weighted average of those belonging to the sources. Figure 1 gives an example of such phenomenon. Let the initial preference matrix of all the nodes in Figure 1 be

$$
\begin{pmatrix}
s_A & s_B & s_C & s_D & s_E \\
s_A' & s_B' & s_C' & s_D' & s_E' \\
s_A'' & s_B'' & s_C'' & s_D'' & s_E''
\end{pmatrix}
\begin{matrix}
\text{candidate1} \\
\text{candidate2} \\
\text{candidate3,}
\end{matrix}
\qquad (14)
$$

where each row in the preference matrix denotes each node's preference for candidates 1, 2, and 3, respectively. Then after infinite number of propagations, the preference matrix will become

$$
\begin{pmatrix}
s(\infty)_{AB} & s(\infty)_{AB} & s(\infty)_C & s(\infty)_D & s(\infty)_E \\
s(\infty)_{AB}' & s(\infty)_{AB}' & s(\infty)_C' & s(\infty)_D' & s(\infty)_E' \\
s(\infty)_{AB}'' & s(\infty)_{AB}'' & s(\infty)_C'' & s(\infty)_D'' & s(\infty)_E''
\end{pmatrix}, \qquad (15)
$$

in which the preferences of nodes A and B in Figure 1 for candidate 1 converge to the common value $s(\infty)_{AB}$, for candidate 2 converge to the common value $s(\infty)_{AB}'$, and for candidate 3 converge to the common value $s(\infty)_{AB}''$. However, for nodes D and E, given that the SCC composed by them $\{D, E\}$ is under the influences of both opinion leaders SCC $\{A, B\}$ and $\{C\}$, their eventual preferences instead of aligning themselves to a common value become a linear combination of the preferences of $\{A, B\}$ and $\{C\}$. The exact details of this combination depend on the structure of the network.

The preference propagation model simulates this unique behavior of people by projecting the preferences' vector onto the leading uniform eigenvector denoting equilibrium. In addition, it also attempts to mimic the real world by distinguishing the opinion leaders from the followers. As with its real world counterpart, this process is solely determined by the initial preferences of every individual and the structure of the embedding social network.

Another example is shown in Figure 2, time evolution of preferences held by nodes in a social network demonstrating the effects of opinion leaders, creating their own consensus, and passing it down to opinion followers in a cascading manner. We see that the opinion follower SCC composed by nodes 15 to 20 is colored with various shades of gray depending on its distance to the two opinion leader SCCs composed by nodes 1 to 3 and 4 to 7. We observe that the preference of the opinion leader SCCs 1 to 3 is first passed to the opinion follower SCCs 8 to 11 (in 10th propagation round) and then subsequently passed to the opinion follower SCCs 12 to 14 through the efforts of SCCs 8 to 11 in a cascaded manner.

This simple example demonstrates that the strongly connected source components form the opinion leader groups, while each follower node is affected by (i.e., linear combination) the opinions of its surrounding opinion leader groups. Our framework models the real world observation about how less-convinced personnel are affected by the mass opinions they encountered.

3.3. Comparison with Other Models.

We here discuss what the most salient characteristics of a successful social model are

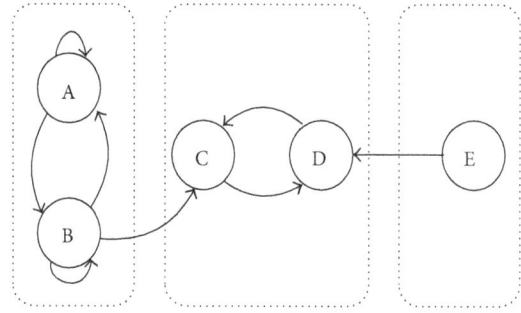

FIGURE 1: Nodes A, B form an opinion leader SCC, while node C by itself is another opinion leader SCC. Nodes E and D form an opinion follower SCC.

based on common observations and beliefs, in an attempt to contrast the most distinguishing features of our model with the other previously proposed frameworks.

High Dimension Media. Since a personal preference describes the order of preference of all possible candidates, the media in an ideal model should be represented as ordered lists instead of as a single value. Most of the propagation models such as Linear Threshold Model, Cascade Independent Model, or DiffusionRank model, unfortunately, only handle binary or real value in propagation.

Topology Dependence and Input Dependence. The word of mouth is the main strategy for a person to affect others. The real world process of guiding friends toward the adoption of self-preference goes mutually and simultaneously. To state such phenomenon, the outgoing persuasions of a person should ideally become a combination of self-preference and the incoming preferences. An ideal model should take into account both network structure and initial personal preference. Moreover, we would like a model's way of incorporating these two factors to be as natural as possible, instead of relying on ad hoc stopping designs or simply restricting the number of time nodes or individuals' interactions.

Deterministic Convergence. Of course an ideal model should converge or end eventually, or else it would be difficult for the modeler to interpret the results. As far as we know, there are currently two kinds of designs to achieve such a convergence. The first one, such as LT model and IC model, attaches a binary status to each node in a network to determine whether it is visited. The *inactive* status means the node is not yet visited while the *active* status means the node is visited. With such design, preference propagation to inactive nodes can be easily monitored. Moreover, the propagation converges in such model when none of the existing nodes can change the status anymore.

Following the success of the PageRank algorithm, the second popular approach is building the convergence mechanism into a model inherently, so that after sufficient iterations the model converges and produces a definite result.

To make results easily analyzable, convergent models that can generate identical results, given both the same initial preferences of nodes and network structure, are preferred.

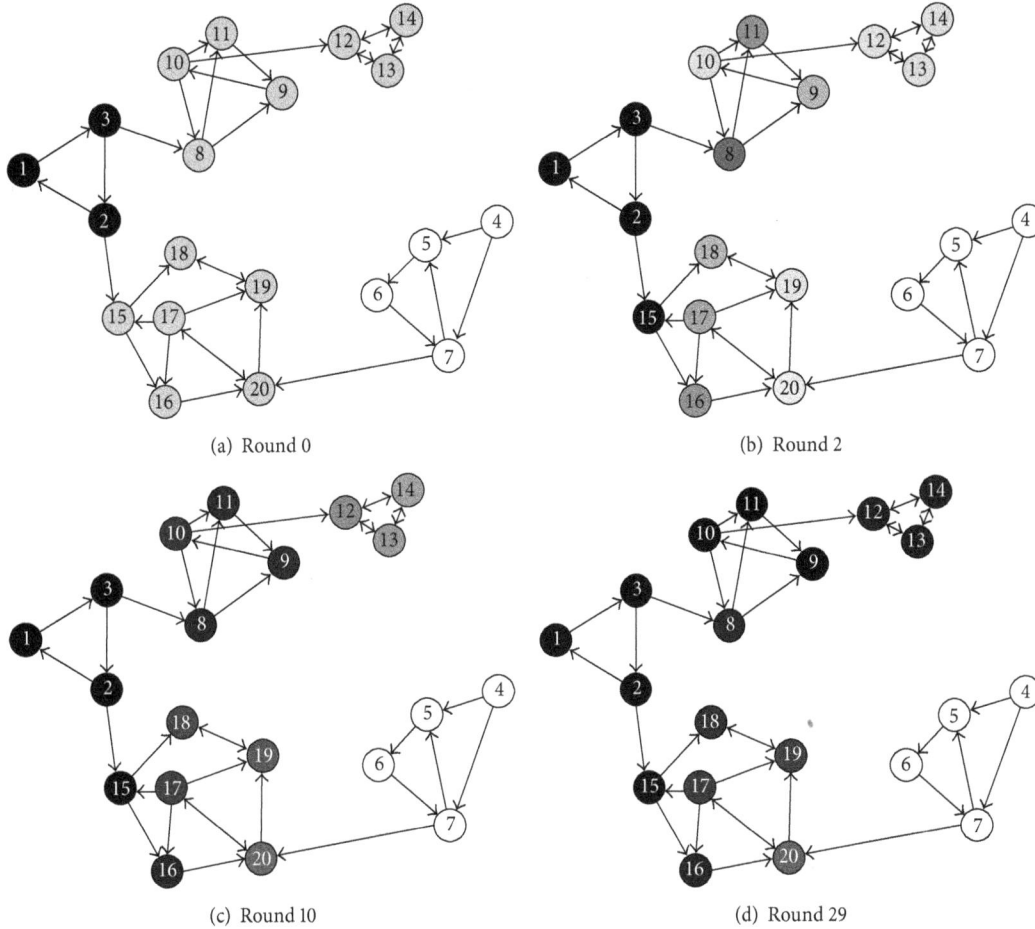

(a) Round 0

(b) Round 2

(c) Round 10

(d) Round 29

FIGURE 2: Time evolution of preferences held by nodes in a social network, demonstrating the effects of opinion leaders creating their own consensus and passing it down to opinion followers in a cascading manner.

Consensus. The problem of reaching a consensus among agents has been studied since around 1970 [16, 17] with simulation models such as the voter model [1]. Mossel et al. gave a theoretical proof that the consensus could be reached with the voter model [18]. Thus an ideal model should be able to reflect specific common traits. In particular, we observe that one such universal trait is people in the same community (i.e., SCC) having the tendency to align their preference after sufficient exchanges. This translates into the fact that an ideal model should contain some kind of homogeneity inside a group.

To see how our model and other proposed frameworks capture the above characteristics of real world social interactions, we conducted several experiments and recorded their results in Table 1 for ease of comparison. We particularly chose models that are most representative in their own stance, namely, the Linear Threshold Model, Independent Cascade Model, PageRank model, and DiffusionRank model, for comparison. Note that since the propagating media in these models are not a vector of preference, we made the following enhancements for each of them to handle such cases. For the LT and IC models, we assume that each vertex initially held approval for its top k preferred candidates (nonapproval for

the others), and thus, for every candidate, we can obtain a list of seeds as inputs into the LT and IC models. We then execute the model separately on each candidate, gather their results, and normalize them to form the final preference of each vertex. For the PageRank and DiffusionRank models, given that they can take real values as inputs, we simply executed these models separately for each candidate in the preference list and then integrated the results to be a vector of real numbers.

As shown in Table 1, we see that our model is the only model that operates directly on a list of preferences, whereas other models work restrictively on single boolean or real values and have to be executed separately to obtain a joint preference, which fail to consider the correlation of the preference score among candidates. We note that all models provide convergent results. Besides, since the IC model carries a random component, it does not deliver repeatable final preference results.

To examine whether these models can give a kind of consensus to nodes that belong to a strongly connected network, we execute all models on a strongly connected graph until they naturally stop or converge. It turns out that, except for our model, none showed signs of reaching consensus

TABLE 1: Comparison of models on the abilities to capture characteristics of real world social networks interactions.

	Convergence	Repeatability of final state	Consensus	Input dependence	Media space
Proposed model	√	√	√ (if SCC)	√	R^k
LT model	√	√		√	Boolean
IC model	√	√		√	Boolean
PageRank	√	√			R
DiffusionRank	√	√		√	R

among the final output preferences. Note that our model does not produce consensus given non-SCC components.

To see whether these models take into account the initial preferences held by nodes, we fed all models with six different initial preferences and see whether they give six different results. It is not surprising that the PageRank model returns identical results regardless of the input, indicating that it takes into account only the structure of the network but ignores the initial preferences held by each node or individual. In conclusion, our model is the only framework that supports all five criteria set by observations from real world social networks.

4. Experiment

In this section, we compare our model with well-known diffusion models to evaluate the performance. We examine whether all the aforementioned algorithms, including ours, can capture the preference transition in social networks to a certain extent. To conduct such validation, we require the information such as the network structure and the node preferences over time.

4.1. Preference Data. The citations of scientific research papers implicitly reveal the research interests of the authors. In other words, we believe that the acts such as citing or submitting to the journals or the conferences would be an indicator of the authors' interests. By utilizing this fact, we can infer the researchers' preference from their corresponding top frequently cited conferences and journals. We have further realized that one author's preference could be influenced by the other coauthors. It is particularly correct for advisor-student relationship since the advisors and students usually affect each other's research interests. We therefore have designed an experiment to model how researchers' preferences can be affected by the collaborators.

We use KDD Cup 2003 ArXiv HEP-TH (High Energy Physics Theory) citation network [19] with the corresponding paper meta information as our evaluation dataset. This dataset contains the citations from 1992 to 2003. We select the top 16 journals that possess most papers as the candidates to construct the preference lists and construct the yearly preference lists for all authors. A preference list consists of the citation count of the corresponding journals within one year. Thus, for each author, we have 12 lists representing their preferences.

The reason we use the citations rather than the publications of authors is that the publications imply not only

preference but also capability. To fairly present the interests, we use the citations. In addition, we construct a collaborative network from this dataset as the underlying social preference diffusion backbone. To easily perceive the changes in interests, we remove the authors who had fewer than 5 publications in the dataset, which results in a network with 2683 nodes.

4.2. Model Comparison. Since we already have all the required information including network structure and preference transition, the next step is to study which diffusion model predicts the preference transition better. We assume a good diffusion model could capture the progression of the authors' research interests through collaborations. To do so, we initially set up the node preference according to the actual data in year x and then compare the predicting results with the actual preference in year $x + k$. The following issues are noted in the experiment.

High Dimension Media. To represent the order in preference toward all candidates, the media in an ideal model ought to be an ordered list instead of a single value. Nonetheless, most well-known diffusion models, such as LT, IC, and DiffusionRank, only treat the media as boolean or real number. For comparison, we exploit these models in our problem by executing them independently for each candidate. We evaluate the candidate rank based on each independent diffusion result.

Determinism of the Final State. Except for the IC model, outcome of all the models mentioned above is deterministic. Because the parameter (i.e., diffusion probability) in IC model is a nondeterministic factor, we execute the experiment 20 times and average the results.

Initialization. Because the media in LT and IC models are not native for high dimension, we singly process the propagation for each candidate. That means, in our experiment, the active mode of top 1% authors to a specific publisher is initially set active in LT and IC models, while the rest of publishers are set inactive. We further set the diffusion probability of each edge as $1/N$, where N is the degree of its source node in IC model. In LT model, we assign links with identical weight and nodes with the same threshold. The parameters in LT and IC are then tuned to find the optimal outcome. The propagation process is executed multiple times with different thresholds and the performance is averaged. For DiffusionRank model, we use the parameter settings suggested by the authors of [9].

TABLE 2: Compare the result after one round for each model with the ground truth of years 1997, 1998, and 1999.

	Kendall's tau			Top 3 Jaccard coefficients		
Year	1997	1998	1999	1997	1998	1999
Independent Cascade	0.007	0.012	0.015	0.011	0.014	0.015
Linear Threshold	0.172	0.167	0.167	0.171	0.195	0.212
DiffusionRank	0.221	0.181	0.160	0.216	0.222	0.213
Proposed(0.00)	0.240	0.204	0.178	0.242	0.243	0.225
Proposed(0.25)	0.243	0.206	0.180	0.248	0.244	0.226
Proposed(0.50)	0.243	0.206	0.180	0.247	0.243	0.227
Proposed(0.75)	0.243	0.206	0.180	0.246	0.243	0.226
Proposed(1.00)	0.230	0.190	0.163	0.204	0.179	0.156

TABLE 3: Consider the result after $k \times R$ rounds for each model, and compare it with the ground truth of year $1996 + k$. The table shows the average of the similarity scores for 1997, 1998, and 1999.

	Kendall's tau					Top3 Jaccard coefficients				
Round	1	2	3	4	5	1	2	3	4	5
Independent Cascade	0.011	0.011	0.011	0.011	0.011	0.013	0.013	0.013	0.013	0.013
Linear Threshold	0.168	0.168	0.168	0.168	0.168	0.192	0.192	0.192	0.192	0.192
DiffusionRank	0.186	0.186	0.186	0.186	0.186	0.217	0.217	0.217	0.217	0.217
Proposed(0.00)	0.208	0.209	0.207	0.206	0.205	0.238	0.240	0.237	0.236	0.234
Proposed(0.25)	**0.210**	0.209	0.208	0.207	0.207	0.241	0.241	0.240	0.239	0.238
Proposed(0.50)	0.209	0.210	0.209	0.209	0.208	0.240	0.242	0.242	0.241	0.240
Proposed(0.75)	0.209	0.209	0.209	0.209	0.209	0.239	0.241	0.242	**0.242**	0.241
Proposed(1.00)	0.194	0.194	0.194	0.194	0.194	0.179	0.179	0.179	0.179	0.179

4.3. Experiment Result. Diffusion models are evaluated by comparing their predictions about preference in 1997, 1998, and 1999, while using the real preference during the period from 1993 to 1996 as initial status. To measure the similarity between predicting and real results, we adopt Kendall's tau coefficient [20] and the Jaccard coefficient. We individually measure the similarity for each author, each node in the network, and then average them as a performance indicator. Because Kendall's tau coefficient is not well defined with tie scores, we manually set Kendall's tau score as 0 when there is a tie on all 16 publishers. Furthermore, we calculate the Jaccard coefficient performing on the top 3 highest scored publishers.

Firstly, for the sake of knowing the correspondence between the extent of changes in iterations and in years, we execute one-iteration propagation in each model and then compare the results with the ground truth in 1997, 1998, and 1999, respectively. We also try different susceptible ratio ϵ in our model, as $\epsilon = 1.0$ implies the authors stick to their own preferences without considering the effect from the neighbors. Table 2 shows the results, which we find quite suitable to take one iteration as a period of a year. The results demonstrate that our model consistently outperforms the 2nd best model DiffusionRank, regardless of which susceptible ratio is used as long as it is not 1.0.

Secondly, we execute the diffusion algorithms for multiple rounds and compare them with the ground truth of years 1997–1999. Table 3 shows the average of the scores for 1997, 1998, and 1999. Note that LT and IC models stop when there is no possible activation (regarded as one round), which implies

that authors are not affected by their neighbors after the first round is completed. Tables 2 and 3 additionally show that the impervious preferences ($\epsilon = 0$) reach a performance similar to the best result, which might reveal the slowly changing nature. Nevertheless, the results show that our model can faithfully capture the trait of the social influence even if the authors' interests change slowly.

5. A Social-Based Simulation Framework for Election Behavior

Based on the preference negotiation model, we implement a simulation framework, shown in Figure 4, granting us to analyze the social impact to elections.

5.1. System Architecture. Voters possess their own preference profiles to each candidate in the early stage of an election. A faithful simulation ought to produce preference profiles that are similar to the practical cases. Thus, we produce profiles satisfying certain distributions according to the data collected from a historical election.

Although we have the voter's preference data, unfortunately, there is no information telling us the relationship between the voters. To deal with this, we propose several plausible scenarios to deploy the profile on a given social network. Because people generally become friends due to similar tastes and thoughts, the profiles should not be distributed randomly. We design several plausible ways to distribute preference profiles on a social network.

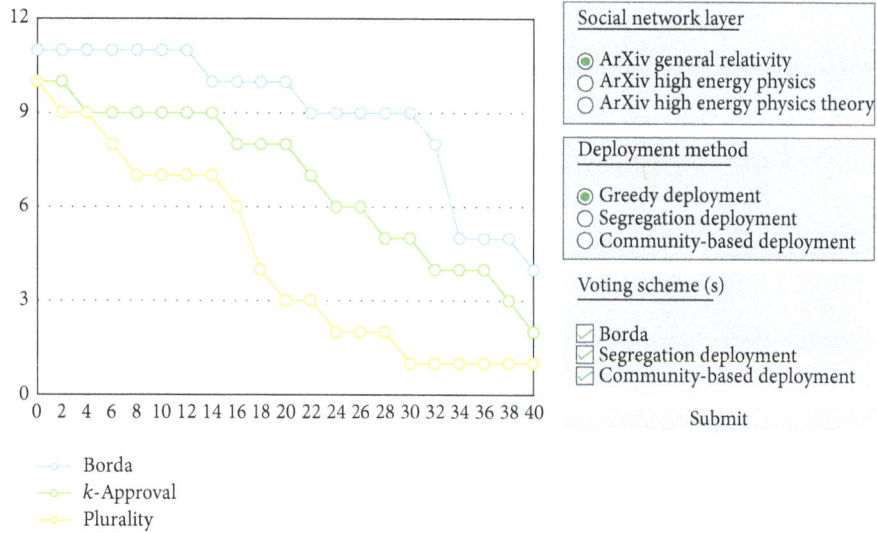

FIGURE 3: Our user-interface for the convenience of doing experiments.

Once the preference profiles are assigned to each node, the system starts executing the preference negotiation model. Nodes in the network start to persuade others and be persuaded. It is not necessary for a simulation to run until convergence, because in the real world there exist some elections which leave insufficient time for people to negotiate, debate, or exchange opinions before having to cast their votes.

Finally, since in each stage there are several parameters we can adjust, we have designed a user-interface that allows the easy execution for experiments (see Figure 3).

5.2. Preference Profile Generation. In the Preference Profile Generation stage, we create preference profiles based on historical election data obtained from OpenSTV, an online voting record database. We choose to use the "Melbourne City Council Victoria Australia 2008—Lord Mayor Leadership team" data set because it is by far the most complete dataset we have found. This dataset was recorded in November 29, 2008, and has the ballot size of 57,961 and 11 candidates. It consists of the preference lists for all voters. A preference list is a sorted list of candidates revealing the preference order of this particular ballot.

We propose a ranking-preserved sampling method to produce the preference profiles based on the historical data, with the aim of preserving the rank of each candidate. Given our historical data, we first learn a $k \times k$ matrix M_r, where k is equal to the total number of candidates. The (i, j)th elements of M_r encode the probability that the jth position in ranking belongs to candidate i according to the historical dataset, $\text{Prob}(C_i \mid \text{Rank} = j)$. Each column of M_r yields a probability distribution for each candidate of a given rank. Given M_r, we can iteratively sample candidates in each rank (from higher to lower) based on the distribution (with the natural restriction of prohibiting the same candidate in different positions in a single ballot).

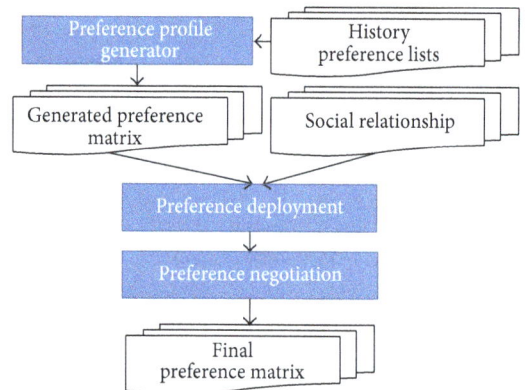

FIGURE 4: The flow chart of the proposed framework.

5.3. Preference Deployment. As the old saying goes "birds of a feather flock together," we presume that the people of similar preference profiles have a higher likelihood of being close to each other in the network. Below are three algorithms to realize such idea.

Greedy Deployment. It can be realized by first randomly picking a profile and assigning it to a node in the social network; then we assign the most similar unassigned profiles to their neighbors. Then iteratively for each unassigned node, the algorithm allocates it a profile that is the most similar to its neighbors. To measure the similarity, we exploit the commonly used Kendall τ coefficient:

$$\tau = \frac{|\text{concordant pairs}| - |\text{discordant pairs}|}{(1/2)\,n\,(n-1)}. \quad (16)$$

Community-Based Deployment. Our underlying idea is to match social network communities with clusters of preference profiles. We first conduct a community detection algorithm to identify communities in a social network (in the experiment, we apply [21]), which groups the community and determines the number of communities m automatically based on the maximization of the modularity. Next, we apply a clustering algorithm (in this experiment, we used k-means) to group preference profiles into m clusters based on Kendall τ similarity.

In the final step, we assign profiles to each node. The main idea is to assign each profile in the ith largest cluster to each node in the ith largest community. This, however, is not a straightforward task because the ith community and the corresponding cluster are likely to have different sizes. Here we propose a method to adjust the cluster sizes to match those of the communities. To accomplish this, we first sort both the sets of communities and their clusters by their respective sizes. Then from the largest community to the smallest, we compare the size of the ith community with that of the ith cluster. When the ith community is of the same size as the ith cluster, we randomly assign the profile in the cluster to a node in the community. When the cluster size is smaller, we add the unassigned points outside this cluster but closest to the cluster center into the cluster until its size matches the size of the community. When the cluster is larger, by contrast, we remove the points from the cluster that are farthest away from the center and join them to the closest unassigned neighbor cluster. Doing this iteratively will gradually assign profiles to nodes and guarantee that nodes in the same community have similar profiles. An example is illustrated in Figure 5, and pseudo code is listed in Algorithm 2.

Segregation Deployment. The idea is from the setting of Schelling's segregation model [22]: blacks and whites may not mind, even prefer, each other's presence, but people will move if they are the minority. At the beginning, we deploy profiles randomly. Then, in each iteration, the nodes which have fewer than 30% neighbors with positive Kendall τ similarity values will be selected, and their preference profiles will be shuffled. In this experiment, 1000 iterations are performed.

5.4. Preference Negotiation. Once profiles and communities have been assigned, the framework executes the core preference negotiation model to the network. There are two parameters that can have some impact on the process: the iterations of negotiations R and the susceptible ratio matrix E. R controls the negotiation iterations taken before the voters have to cast the votes. In the experiments below, we set $R = 20$.

We set ϵ as 0.5 for all voters by default for our case studies. As suggested in our proof, if E is a constant matrix c, then the resulting converged preference matrix S is indifferent no matter what c is (c only controls the speed of convergence).

6. Case Studies

In this section, we answer two questions based on the proposed simulation framework.

(1) To what extent does the negotiation process in a social network affect election results?

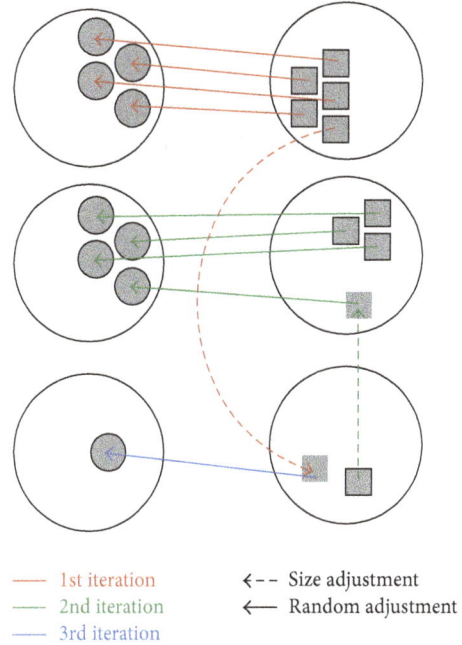

——— 1st iteration   ← - - Size adjustment
——— 2nd iteration   ←— Random adjustment
——— 3rd iteration

FIGURE 5: A diagram to demonstrate our deployment algorithm.

(2) Among the widely known voting schemas, Borda Count, k-approval, and plurality, which is the most vulnerable to vote-buying (i.e., easiest to be manipulated)?

In order to construct simulations, we use three collaboration networks (ca-GrQc containing 5,242 nodes and 2,890 edges, ca-HepPh containing 12,008 nodes and 237,010 edges, and ca-HepTh containing 9,877 nodes and 51,971 edges) as the underlying social network dataset. Once the negotiation process ends, the preference scoring vectors will be examined to determine the final rank of the candidates using different voting schemas: Borda Count, k-approval, and plurality [23]. We conduct experiments on all three plausible deployments proposed in Section 5.3. Ideally, we hope the simulation on all three deployment methods can produce similar conclusions, which would consequently offer users higher confidence about the results.

The Borda Count determines the final rank of the candidates by giving each candidate a certain number of points corresponding to the position in which it is ranked by each voter. Once all votes have been counted the candidate can be ranked by their total points. For each ballot, the top-ranked candidate will receive k points, the second $k - 1$ points, and so on.

The ranking in k-approval and plurality schemas is determined similar to Borda Count. The only difference among the three voting schemas is the definition of points to be given to each candidate. In the k-approval voting schema, the top k candidates in each ballot will each receive one point, while the rest will not receive any points. In the plurality voting schema, only the top candidate receives one point.

```
S: social-networks; P: preference-matrix
K: community-number
M[i]: the ith largest community
C[i]: the ith largest cluster
(M[], K) = CommunityDetection(S)
C[] = KmeansForPreference(P, K)
for i = 1 to K do
    if size(C[i]) = size(M[i]) then
        next
    else if size(C[i]) < size(M[i]) then
        add the unassigned preferences which is outside
        C[i] but closest to center(C([i]) into C[i] to
        match size(M[i])
    else
        remove the preferences most away from
        center(C[i]) to the closest unassigned neighbor
        cluster
    end if
    randomly assign the preferences in C[i] to nodes in
    M[i]
end for
return the pairs of (node, preference)
```

ALGORITHM 2: Preference deployment method.

6.1. Effectiveness of Negotiation.

The first question is whether the social-network-based negotiation process can significantly affect the election results. To quantify changes in a voters' preference profile through negotiations, we compute the average Kendall τ coefficient between the preference orders before and after each negotiation.

As shown in Figure 6, no matter which deployment method is used, the average Kendall *tau* coefficient generally decreases as we enter deeper rounds of negotiation. The slope is steepest in the beginning, revealing the fact that the effect of negotiation reaches the peak in the beginning and gradually declines, which matches the real world experience. Eventually the Kendall *tau* value decreases to below 0.5, implying that negotiation through social networks can significantly change the election results.

6.2. Vulnerability to Vote-Buying.

This section discusses a key question about elections: if an organization can boost the vote count of a candidate through manipulating certain seed nodes' preference profiles (pejoratively, we can call this "vote-buying"). We define one successful vote-buying to a voter as "raising the target candidate's preference score to slightly higher than the score required to obtain a vote from the voter." For example, if each voter is allowed to cast 3 votes, then the buyers would attempt raising its score to slightly above the 3rd place candidate. To quantify such manipulation cost, we defined it as the difference between the scores before and after manipulating. The scoring vector after vote-buying must be renormalized before further computation can proceed.

Discarding the effect of negotiation, it is intuitive to assume that the best promotion strategy is aiming at the voters whose costs are the smallest. Under such attack, which

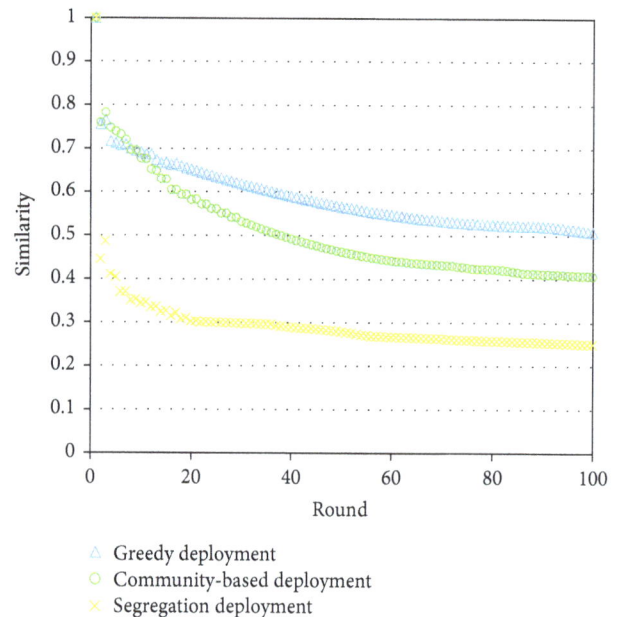

△ Greedy deployment
○ Community-based deployment
✕ Segregation deployment

FIGURE 6: The similarity (average Kendall τ coefficient) between the initial preference matrix and the preference matrix after negotiations.

voting schema is the most vulnerable to vote-buying? This answer can be deduced from Figure 7, where the x-axis stands for the budget spent while the y-axis stands for the rank of a given candidate for promotion. Note that the higher value of y stands for the less favorable, and a candidate has to move to lower position in order to be elected. The

(a) Greedy deployment in ca-GrQc

(b) Community-based deployment in ca-GrQc

(c) Segregation deployment in ca-GrQc

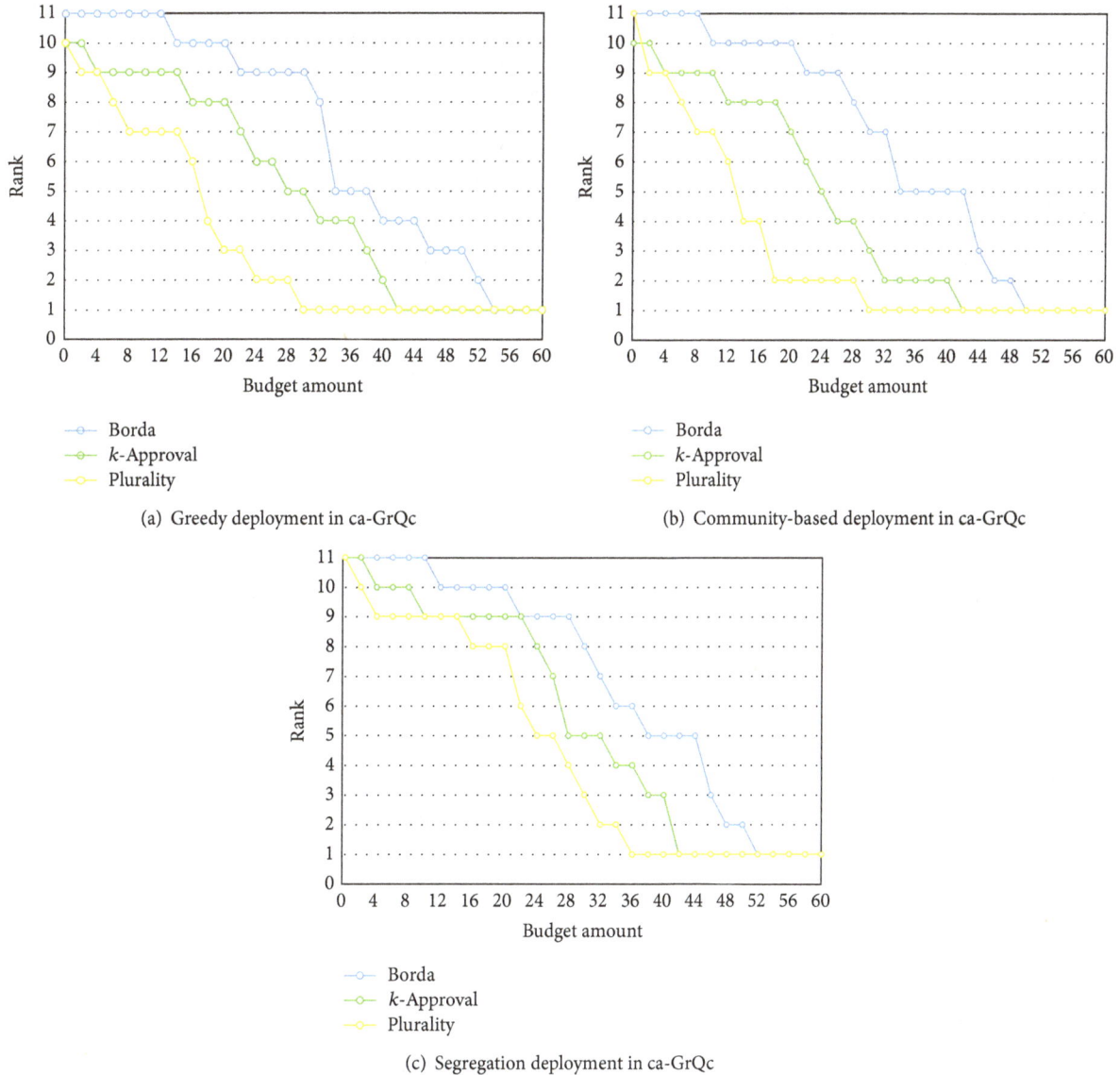

FIGURE 7: The figure shows the rank of promoted candidate in three voting schemas, where the x-axis stands for the budget spent while the y-axis stands for the rank of a given candidate for promotion. The figures in each row show the 3rd-round results with different deployments in networks ca-GrQc. The results in networks ca-HepTh and ca-HepPh are consistent. They are excluded due to the space.

results show that, regardless of the deployment, Borda Count schema consistently requires more budget to advance a target candidate, while the plurality schema is the most vulnerable to vote-buying since it generally requires less budget to advance a candidate to higher rank.

7. Conclusion

Analyzing the effect of social networks upon group decisions outcomes is a difficult problem because it is both costly and time consuming to perform user studies to collect people's private preferences. Indeed, it is the change of preferences through social propagation in particular that we care most about, and to our knowledge this is the first ever study

that provides not only theoretical analysis but also the empirical justification of this problem. This study provides an example of how to perform such research with limited data through exploiting algorithm and model design, theoretical justification, and computer simulation.

Our other significant contribution is that we provide an alternative evaluation plan and data to verify a preference propagation model. Acknowledging the lack of real world data to evaluate how the voter's preference can change through social diffusion, we have come up with a novel idea to identify a publicly available bibliography dataset to evaluate how researchers gradually change their research fields according to the influence of their collaborators. Our evaluation plan opens a new possibility that allows researchers working on preference diffusion problems to be able to evaluate their models without having to identify a highly private voter preference dataset.

Abbreviations

V: Individuals
C: Candidates
n: Number of individuals
k: Number of candidates
p_v: Preference profile vector of individual $v \in V$
s_v: Preference score vector for individual $v \in V$
S: Preference scoring matrix with size $|V| \times |C|$
G: Social network layer.

Conflict of Interests

The authors declare that there is no conflict of interests regarding the publication of this paper.

Acknowledgments

This work was supported by the National Science Council, National Taiwan University, and Intel Corporation under Grants NSC101-2911-I-002-001, NSC101-2628-E-002-028-MY2, and NTU102R7501.

References

[1] T. Liggett, *Interacting Particle Systems*, Grundlehren der Mathematischen Wissenschaften, Springer, 1985.

[2] M. Granovetter, "Threshold models of collective be- havior," *American Journal of Sociology*, vol. 83, no. 6, pp. 1420–1443, 1978.

[3] D. Kempe, J. Kleinberg, and É. Tardos, "Maximizing the spread of influence through a social network," in *Proceedings of the 9th ACM SIGKDD International Conference on Knowledge Discovery and Data Mining (KDD '03)*, pp. 137–146, August 2003.

[4] W. Chen, Y. Wang, and S. Yang, "Efficient influence maximization in social networks," in *Proceedings of the 15th ACM SIGKDD International Conference on Knowledge Discovery and Data Mining, KDD '09*, pp. 199–207, ACM, July 2009.

[5] D. Gruhl, D. Liben-Nowell, R. Guha, and A. Tomkins, "Information diffusion through blogspace," in *Proceedings of the 13th International World Wide Web Conference Proceedings (WWW '04)*, pp. 491–501, ACM, May 2004.

[6] J. Leskovec, M. Mcglohon, C. Faloutsos, N. Glance, and M. Hurst, "Cascading behavior in large blog graphs: patterns and a model," Tech. Rep., 2006.

[7] K. Saito, M. Kimura, K. Ohara, and H. Motoda, "Selecting information di usion models over social networks for behavioral analysis," *Machine Learning and Knowledge Discovery in Databases*, vol. 6323 of *Lecture Notes in Computer Science*, 2010.

[8] H. Ma, H. Yang, M. R. Lyu, and I. King, "Mining social networks using heat diffusion processes for marketing candidates selection," in *Proceedings of the 17th ACM Conference on Information and Knowledge Management (CIKM '08)*, pp. 233–242, October 2008.

[9] H. Yang, I. King, and M. R. Lyu, "DiffusionRank: a possible penicillin for web spamming," in *Proceedings of the 30th Annual International ACM SIGIR Conference on Research and Development in Information Retrieval (SIGIR '07)*, pp. 431–438, July 2007.

[10] J. J. Bartholdi III, C. A. Tovey, and M. A. Trick, "How hard is it to control an election?" *Mathematical and Computer Modelling*, vol. 16, no. 8-9, pp. 27–40, 1992.

[11] A. Gibbard, "Manipulation of voting schemes: a general result," *Econometrica*, vol. 41, no. 4, pp. 587–601, 1973.

[12] M. A. Satterthwaite, "Strategy-proofness and Arrow's conditions: existence and correspondence theorems for voting procedures and social welfare functions," *Journal of Economic Theory*, vol. 10, no. 2, pp. 187–217, 1975.

[13] F. C. Liu, "Modeling political individuals using the agent-based approach: a preliminary case study on po- litical experts and their limited in uence within com- munication networks," *Journal of Computers*, vol. 19, no. 4, pp. 8–19, 2009.

[14] S. Yoo, Y. Yang, F. Lin, and C. Moon II, "Mining social networks for personalized email prioritization," in *Proceedings of the 15th ACM SIGKDD International Conference on Knowledge Discovery and Data Mining (KDD '09)*, pp. 967–975, July 2009.

[15] C. D. Meyer, Ed., *Matrix Analysis and Applied Linear Algebra*, Society for Industrial and Applied Mathematics, Philadelphia, Pa, USA, 2000.

[16] M. H. DeGroot, "Reaching a consensus," *Journal of the American Statistical Association*, vol. 69, no. 345, pp. 118–121, 1974.

[17] R. L. Winkler, "The consensus of subjective probability distributions," *Management Science*, vol. 15, no. 2, pp. B61–B75, 1968.

[18] E. Mossel and G. Schoenebeck, "Reaching consensus on social networks," in *Proceedings of the International Conference on Supercomputing (ICS '10)*, pp. 214–229, 2010.

[19] J. Gehrke, P. Ginsparg, and J. Kleinberg, "Overview of the 2003 KDD Cup," *SIGKDD Explorations Newsletter*, vol. 5, no. 2, pp. 149–151, 2003.

[20] M. G. Kendall, "A new measure of rank correlation," *Biometrika*, vol. 30, no. 1-2, 1938.

[21] V. D. Blondel, J.-L. Guillaume, R. Lambiotte, and E. Lefebvre, "Fast unfolding of communities in large networks," *Journal of Statistical Mechanics: Theory and Experiment*, vol. 2008, no. 10, Article ID P10008, 2008.

[22] T. Schelling, *Models of Segregation*, The American Economic Review, 1969.

[23] P. E. Johnson, "Voting systems," http://pj.freefaculty.org/Ukraine/PJ3_VotingSystemsEssay.pdf.

13

Date Attachable Offline Electronic Cash Scheme

Chun-I Fan, Wei-Zhe Sun, and Hoi-Tung Hau

Department of Computer Science and Engineering, National Sun Yat-sen University, Kaohsiung 80424, Taiwan

Correspondence should be addressed to Chun-I Fan; cifan@faculty.nsysu.edu.tw

Academic Editors: T. Cao, M. Ivanovic, and F. Yu

Electronic cash (e-cash) is definitely one of the most popular research topics in the e-commerce field. It is very important that e-cash be able to hold the anonymity and accuracy in order to preserve the privacy and rights of customers. There are two types of e-cash in general, which are online e-cash and offline e-cash. Both systems have their own pros and cons and they can be used to construct various applications. In this paper, we pioneer to propose a provably secure and efficient offline e-cash scheme with date attachability based on the blind signature technique, where expiration date and deposit date can be embedded in an e-cash simultaneously. With the help of expiration date, the bank can manage the huge database much more easily against unlimited growth, and the deposit date cannot be forged so that users are able to calculate the amount of interests they can receive in the future correctly. Furthermore, we offer security analysis and formal proofs for all essential properties of offline e-cash, which are anonymity control, unforgeability, conditional-traceability, and no-swindling.

1. Introduction

Due to the rapid growth of the Internet and communication developments, electronic commerce has become much more popular and widely used than ever [1–8]. The mobile telecommunications have been developed from 2 G to 3.5 G. Furthermore, LTE Advanced, 4 G, and 5 G are being implemented to the market in recent years. With the convenience of mobile network, people can do shopping or electronic payments by using any devices with network capability instead of leaving home. As a result, electronic commerce has been emphasized nowadays. Electronic cash (e-cash) is definitely one of the most popular research topics among electronic commerce. E-cash and the traditional cash notes are very much alike except e-cash is digitized and used on Internet transactions; therefore, it is very important that e-cash be able to hold the accuracy, privacy, and all other security concerns.

A typical e-cash system usually consists of payers (customers), payees (shops), and a bank. There are two types of e-cash in general which are online e-cash [9–13] and offline e-cash [14–27]. Online e-cash system involves participation of the bank during transactions (the payment stage). Banks are able to check whether customers have double-spent the e-cash(s) or not, and if yes, banks can terminate the transactions at once. Thus, the bank has to be online during every

transaction and it may lead to a bottleneck of the system. On the other hand, while banks do not participate in the payment stage of offline e-cash systems, double-spending check is only held during the deposit stage. Yet, the bank is set to be offline, but the system design is usually much more complicated than the online type and it may lead to a longer transaction time. Since both systems have their own pros and cons, they are used under different circumstances.

Extending online and offline e-cash systems, many e-cash schemes with other different features have been proposed over the years. For instance, e-cash can be stored compactly such that the space to store these e-cash is much reduced [15, 16], e-cash is generated by multiauthorities instead of one bank only [25], exact payments e-cash [13], recoverable e-cash which can be recovered when an e-cash is lost [26], and so on.

Based on the majority of the existing approaches, we summarize that a secure e-cash system should satisfy the following requirements.

(i) *Anonymity*: no one, except the judge, can obtain any information of the e-cash owner's identity from the contents of e-cash.

(ii) *Unlinkability*: no one, except the judge, can link any e-cash payment contents.

(iii) *Unforgeability*: no one, except the bank, can generate a legal e-cash.

(iv) *Double-Spending Control*: banks should have the ability to check if the e-cash is double-spent or not. No e-cash is allowed to be spent twice or more in an e-cash system.

(v) *Conditional-Traceability*: the system should be able to trace and revoke the anonymity of users who violate any of the security rules so that they will receive penalties.

(vi) *No-swindling*: no one, except the real owner, can spend a valid offline e-cash successfully.

In order to perform double-spending checks, banks have to store information of e-cash(s) in their database. Thus, the database of banks grows in direct proportion to the number of e-cash(s) withdrawn. Embedding an expiration date into each e-cash has been considered since it helps the banks to manage the database more easily. On the other hand, customers have to exchange their expired e-cash(s) with banks for new ones so as to keep the validity of the e-cash. Furthermore, customers will receive interest from banks after cash is deposited. In order to guarantee customers will receive the right amount of interest, it is necessary for customers to attach the deposit date to their e-cash(s) and the date cannot be modified by anyone else [11]. So far, there are a number of online e-cash schemes with an expiration date attachment [9, 11, 28]. However, there are very few offline approaches [21].

In this paper, we are going to propose an efficient date attachable offline e-cash scheme and provide formal proofs on essential properties to it in the random oracle model. Considering the practical needs, we pioneer to embed two kinds of date, which are expiration data and deposit date, to the offline e-cash. Moreover, we will offer an *E-cash renewal protocol* in our scheme (Section 3.2.5). Users can exchange their unused expired e-cash for a new one with another valid expiration date more efficiently. Compared with other similar works, our scheme is efficient from the aspect of considering computation cost.

The rest of this paper is organized as follows. In Section 2, we briefly review techniques employed throughout our scheme. Our proposed scheme is described in Section 3 in detail. Security proofs and analysis are covered in Section 4. Features and performance comparisons are made in Section 5, and the conclusion is given in Section 6.

2. Preliminaries

In this section, we briefly review techniques used in our date attachable offline e-cash scheme.

2.1. Chaum's Blind Signature Scheme. Blind signature was first introduced by Chaum [29]. It has been widely used in e-cash protocols since it has been proposed. A signer will not be able to view the content of the message while she/he is signing the message. Afterwards, a user can get a message with the signature of the signer by unblinding the signed message. The protocol is described as follows.

(1) Initialization:

The signer randomly chooses two distinct large primes p and q, then computes $n = pq$ and $\phi(n) = (p-1)(q-1)$. Afterwards, the signer selects two integers e and d at random such that $ed \equiv 1 \pmod{\phi(n)}$. Finally, the signer publishes the public parameters (e, n) and a one-way hash function H.

(2) User → Signer: α

The user chooses a message m and a random integer r in $\mathbb{Z}_n^*$, then blinds the message by computing $\alpha = r^e H(m) \bmod n$ and sends it to the signer.

(3) Signer → User: t

After receiving α, the signer signs it with her/his private key d and sends it back to the user. The signed message will be $t = \alpha^d \bmod n$.

(4) Unblinding:

After receiving t from the signer, the user unblinds it by computing $s = r^{-1}t \bmod n$. The signature-message pair is (s, m).

(5) Verification:

The (s, m) can be verified by checking if $s^e \equiv H(m) \pmod{n}$ is true or not.

2.2. Chameleon Hashing Based on Discrete Logarithm. Chameleon hashing was proposed by Krawczyk and Rabin [30]. The chameleon hash function is associated with a one-time public-private key pair; it is a collision resistant function except for users who own a trapdoor for finding collision. Any user who knows the public key can compute the hashing, and for those who do not know the private key (trapdoor), it is impossible for them to find any two inputs which lead to the same hashing output. On the contrary, any user who knows the trapdoor can find the collision of given inputs. The construction of the chameleon hashing based on discrete logarithm is described as follows.

(1) *Setup*:

    (i) p, q: two large primes such that $p = kq + 1$,

    (ii) g: an element order q in $\mathbb{Z}_p^*$,

    (iii) x: private key in $\mathbb{Z}_q^*$,

    (iv) y: public key, where $y = g^x \bmod p$.

(2) *The function*: a message $m \in \mathbb{Z}_q^*$ is given and a random integer $r \in \mathbb{Z}_q^*$ is chosen. The hash is defined as CHAM-HASH$_y(m, r) = g^m y^r \bmod p$.

(3) *Collision*: for a user who knows x, she/he is able to find the collision of the hash for any given m, m' such that CHAM-HASH$_y(m, r) =$ CHAM-HASH$_y(m', r')$. The user derives r' in the equation $m + xr = m' + xr' \pmod{q}$.

3. The Proposed Date Attachable Offline Electronic Cash Scheme

In this section, we will introduce a new date attachable offline e-cash scheme. Considering the issues mentioned in Section 1, we propose a secure offline e-cash scheme with two specific kinds of date attached to the e-cash, which are expiration date and deposit date.

3.1. Outline of the Proposed Scheme. Here we are going to briefly describe the procedures of our scheme. The proposed scheme contains four protocols, *withdrawal protocol, payment protocol, deposit protocol,* and *e-cash renewal protocol.* A user withdraws an e-cash with an expiration date attached to it from the bank. A trusted computing platform (i.e., *judge device*) [31, 32], as stated in the proposed scheme, is installed in the bank to hold the identity information of all users and it will further help trace users when it is needed. It is impossible for anyone except the judge to obtain any information embedded in the device [33]. Nowadays, judge device can be implemented by the technique of Trusted Platform Module (TPM) [32, 34] in practice.

Before an e-cash is deposited, the depositor attaches the deposit date on the e-cash and sends it to the bank during the deposit stage. When the bank receives an e-cash, it will perform double-spending checking to verify whether the e-cash is doubly spent or not. The bank can derive secret parameters of the user who does double-spending and let the judge revoke the anonymity of the user. Besides, when an unused e-cash is expired, a user will be able to exchange it for a new one with a new expiration date. In our scheme, for the efficiency concerns, some of the unused parameters of users can remain unchanged while exchanging for a new valid e-cash. In the following sections, we will describe our scheme in detail.

3.2. The Proposed Scheme. Firstly, we define some notations as follows.

(1) H_1, H_2, H_3: three one-way hash functions, $H_1, H_2, H_3 : \{0,1\}^* \rightarrow \{0,1\}^n$.

(2) H_4, H_5: two one-way hash functions, $H_4, H_5 : \{0,1\}^* \rightarrow \{0,1\}^q$.

(3) $\widetilde{E}_x, \widetilde{D}_x$: a secure symmetric cryptosystem. Plaintext is both encrypted and decrypted with a symmetric key x.

(4) $\widehat{E}_{pk}, \widehat{D}_{sk}$: a secure asymmetric cryptosystem. Plaintext is encrypted with a public key pk and decrypted with the corresponding private key sk.

(5) (pk_j, sk_j): the public-private key pair of the judge.

(6) (e_b, d_b): the public-private key pair of bank.

(7) *Date*: expiration date. It represents an effective spending date of a withdrawn e-cash. Any e-cash withdrawn in the same period will have the same expiration date, and vice versa.

(8) ID_c: the identity of user C.

(9) l_k, l_r: the security parameters.

(10) *A judge device*: a tamper-resistant device which is issued by the judge. It is installed into the system of the bank. It is impossible to intercept or modify any information stored in the device.

3.2.1. Initialization. Initially, the bank randomly chooses two distinct large primes (p_b, q_b) and computes RSA parameters $n_b = p_b q_b$. It selects an integer e_b at random such that $\text{GCD}(\phi(n_b), e_b) = 1$, where $\phi(n_b) = (p_b - 1)(q_b - 1)$ and $1 < e_b < \phi(n_b)$. Then, it finds a d_b such that $e_b d_b \equiv 1 (\text{mod } \phi(n_b))$. Secondly, it also chooses two other large primes p and q and two generators g_1 and g_2 of order q in $\mathbb{Z}_p^*$. Then, the bank publishes $(n_b, e_b, p, q, g_1, g_2, pk_j, H_1, H_2, H_3, H_4, H_5, \widetilde{E}, \widetilde{D}, \widehat{E}, \widehat{D})$. Meanwhile, the judge embeds $(n_b, e_b, p, q, g_1, g_2, pk_j, sk_j, H_1, H_2, H_3, H_4, H_5, \widetilde{E}, \widetilde{D}, \widehat{E}, \widehat{D})$ into a judge device and issues it to the bank.

3.2.2. Withdrawal Protocol. Users run the withdrawal protocol with banks to get an e-cash, as shown in Figure 1, yet banks have to obtain information of users' identity, such as ID_c or account numbers, before the withdrawal protocol is proceeded. Therefore, users should perform an authentication with banks beforehand. Users can execute the withdrawal protocol by any devices that have the ability to compute and connect to the network. For instance, users can use mobile phones or computers to perform the withdrawal protocol and store the withdrawn e-cash. The detailed steps of the protocol are as follows.

(1) Bank $\rightarrow$ User: D

Firstly, the user prepares parameters for withdrawing an e-cash. The user chooses integers $a, x_1, x_2, r_1, r_2,$ and r_3 in random, where $a \in_R \mathbb{Z}_{n_b}^*$ and $x_1, x_2, r_1, r_2, r_3 \in_R \{0, 1, \ldots, q-1\}$ and selects a string $k \in_R \{0,1\}^{l_k}$ randomly. The user then computes (y_1, w_1, y_2, w_2), where $y_i = g_i^{x_i} \bmod p$ and $w_i = g_i^{r_i} \bmod p$ for $i = \{1, 2\}$. Secondly, the bank computes parameters for expiration date. It randomly chooses a r in $\mathbb{Z}_{n_b}^*$, prepares $D = \text{Date} \parallel r$ for some expiration date *Date*. The bank will send D to the user when she/he requests to withdraw an e-cash.

(2) User $\rightarrow$ Bank: (α, ϵ)

After receiving D, the user prepares $\epsilon = \widehat{E}_{pk_j}(k \parallel \text{ID}_c)$ and

$$\alpha = \left[a^{e_b} H_1^2(m \parallel D) \right]^{-1} \bmod n_b, \qquad (1)$$

where $m = (y_1 \parallel w_1 \parallel y_2 \parallel w_2 \parallel r_3)$. Finally, the user sends (α, ϵ) to the bank.

(3) Bank $\rightarrow$ Judge device: (ϵ, μ, D)

The bank sets $\mu = \text{ID}_c$, where ID_c is the identity of user C, and inputs it together with ϵ and D to the judge device.

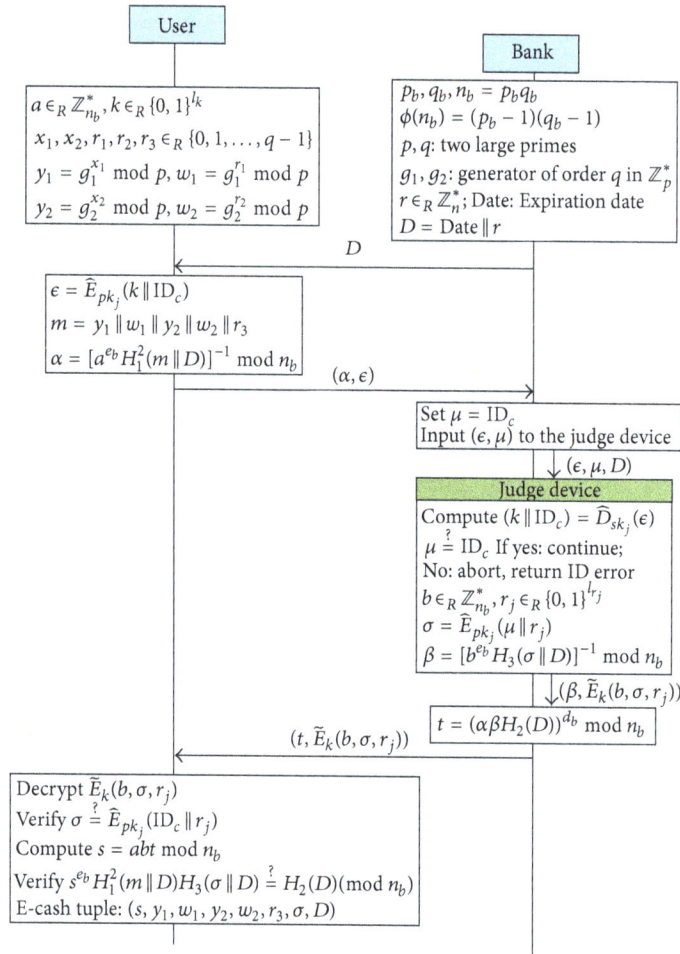

$$a \in_R \mathbb{Z}_{n_b}^*, k \in_R \{0,1\}^{l_k}$$
$$x_1, x_2, r_1, r_2, r_3 \in_R \{0,1,\ldots,q-1\}$$
$$y_1 = g_1^{x_1} \bmod p, w_1 = g_1^{r_1} \bmod p$$
$$y_2 = g_2^{x_2} \bmod p, w_2 = g_2^{r_2} \bmod p$$

$$p_b, q_b, n_b = p_b q_b$$
$$\phi(n_b) = (p_b - 1)(q_b - 1)$$
$$p, q: \text{ two large primes}$$
$$g_1, g_2: \text{ generator of order } q \text{ in } \mathbb{Z}_p^*$$
$$r \in_R \mathbb{Z}_n^*; \text{ Date: Expiration date}$$
$$D = \text{Date} \| r$$

D

$$\epsilon = \widehat{E}_{pk_j}(k \| \text{ID}_c)$$
$$m = y_1 \| w_1 \| y_2 \| w_2 \| r_3$$
$$\alpha = [a^{e_b} H_1^2(m \| D)]^{-1} \bmod n_b$$

(α, ϵ)

Set $\mu = \text{ID}_c$
Input (ϵ, μ) to the judge device

(ϵ, μ, D)

Judge device

Compute $(k \| \text{ID}_c) = \widehat{D}_{sk_j}(\epsilon)$
$\mu \stackrel{?}{=} \text{ID}_c$ If yes: continue;
No: abort, return ID error
$b \in_R \mathbb{Z}_{n_b}^*, r_j \in_R \{0,1\}^{l_{r_j}}$
$\sigma = \widehat{E}_{pk_j}(\mu \| r_j)$
$\beta = [b^{e_b} H_3(\sigma \| D)]^{-1} \bmod n_b$

$(\beta, \widetilde{E}_k(b, \sigma, r_j))$

$t = (\alpha \beta H_2(D))^{d_b} \bmod n_b$

$(t, \widetilde{E}_k(b, \sigma, r_j))$

Decrypt $\widetilde{E}_k(b, \sigma, r_j)$
Verify $\sigma \stackrel{?}{=} \widehat{E}_{pk_j}(\text{ID}_c \| r_j)$
Compute $s = abt \bmod n_b$
Verify $s^{e_b} H_1^2(m \| D) H_3(\sigma \| D) \stackrel{?}{=} H_2(D)(\bmod n_b)$
E-cash tuple: $(s, y_1, w_1, y_2, w_2, r_3, \sigma, D)$

FIGURE 1: Withdrawal protocol.

(4) Judge device $\rightarrow$ Bank: $(\beta, \widetilde{E}_k(b, \sigma, r_j))$

The judge device decrypts ϵ and checks if $\mu = \text{ID}_c$. If not, it returns "ID error" to the bank; or else, it picks a random integer $b \in_R \mathbb{Z}_{n_b}^*$ and a string $r_j \in_R \{0,1\}^{l_{r_j}}$ randomly. Then it computes $\sigma = \widehat{E}_{pk_j}(\mu \| r_j)$ and

$$\beta = [b^{e_b} H_3(\sigma \| D)]^{-1} \bmod n_b. \quad (2)$$

Finally, it encrypts (b, σ, r_j) by using the symmetric key k and outputs it together with β to the bank.

(5) Bank $\rightarrow$ User: $(t, \widetilde{E}_k(b, \sigma, r_j))$

After receiving $(\beta, \widetilde{E}_k(b, \sigma, r_j))$ from the judge device, it computes

$$t = (\alpha \beta H_2(D))^{d_b} \bmod n_b \quad (3)$$

and sends $(t, \widetilde{E}_k(b, \sigma, r_j))$ to the user.

(6) Verifications

After receiving $(t, \widetilde{E}_k(b, \sigma, r_j))$, the user firstly decrypts the ciphertext by using the symmetric key k

in order to obtain (b, σ, r_j). Secondly, she/he checks if his/her ID is embedded correctly by computing if $\sigma = \widehat{E}_{pk_j}(\text{ID}_c \| r_j)$ is true or not. Thirdly, she/he computes

$$s = abt \bmod n_b \quad (4)$$

and verifies s by checking if

$$s^{e_b} H_1^2(m \| D) H_3(\sigma \| D) = H_2(D)(\bmod n_b) \quad (5)$$

is true or not. Finally, when all verifications are done, the user gets the e-cash tuples (s, m, σ, D) and stores (x_1, x_2, r_1, r_2) for further payment usages.

3.2.3. Payment Protocol. When a user has to spend the e-cash, she/he performs the protocol as shown in Figure 2. The steps of the protocol are described as follows.

(1) User $\rightarrow$ Shop: $(s, m, \sigma, D, x_2, r_2)$

The user sends $(s, m, \sigma, D, x_2, r_2)$ to the shop, where D contains the expiration date of the e-cash.

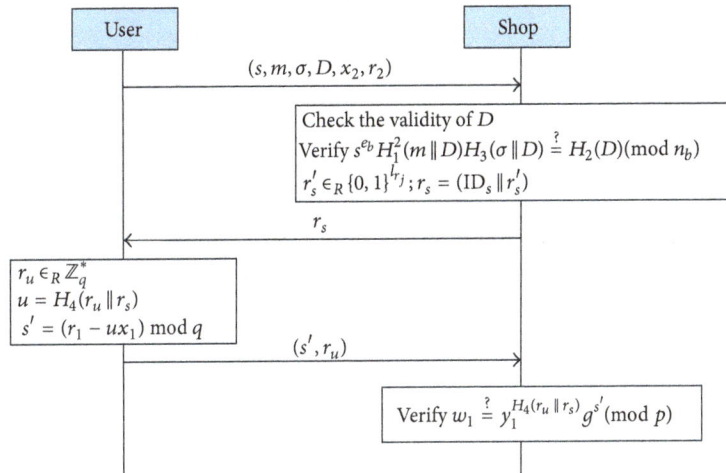

$$(s, m, \sigma, D, x_2, r_2)$$

Check the validity of D

Verify $s^{e_b} H_1^2(m \| D) H_3(\sigma \| D) \overset{?}{=} H_2(D)(\bmod n_b)$

$r_s' \in_R \{0, 1\}^{l_{r_j}}; r_s = (\text{ID}_s \| r_s')$

$$r_s$$

$r_u \in_R \mathbb{Z}_q^*$

$u = H_4(r_u \| r_s)$

$s' = (r_1 - ux_1) \bmod q$

$$(s', r_u)$$

Verify $w_1 \overset{?}{=} y_1^{H_4(r_u \| r_s)} g^{s'} (\bmod p)$

FIGURE 2: Payment protocol.

(2) Shop → User: r_s

The shop first checks D to verify if the e-cash is still within the expiration date or not. If not, it terminates the transaction. Otherwise, it continues to verify $s^{e_b} H_1^2(m \| D) H_3(\sigma \| D) = H_2(D)(\bmod n_b)$. If it is not valid, the protocol is aborted; or else, it selects a string $r_s' \in_R \{0, 1\}^{l_{r_j}}$ and sets a challenge $r_s = (\text{ID}_s \| r_s')$, where ID_s is the identity of the shop. Finally, it sends r_s to the user.

(3) User → Shop: (s', r_u)

After receiving r_s from the shop, the user randomly selects a $r_u \in_R \mathbb{Z}_q^*$ and computes a response to the challenge

$$s' = (r_1 - ux_1) \bmod q, \qquad (6)$$

where $u = H_4(r_u \| r_s)$. Then, the user sends (s', r_u) to the shop.

(4) Verifications

After receiving (s', r_u) from the user, the shop verifies if $w_1 = y_1^{H_4(r_u \| r_s)} g^{s'} (\bmod p)$ is true or not. If it is true, the shop will accept the e-cash. On the other hand, if it is not, the shop will reject it. Since it is an offline e-cash, the shop does not have to deposit it to the bank immediately. It can store the e-cash and deposit it later together with other received e-cash(s).

3.2.4. Deposit Protocol. As Figure 3 shows, shops attach the deposit date to their e-cash(s) and deposit them to banks in this protocol. Banks perform double-spending checks when they receive these e-cash(s). If any e-cash is double-spent, the bank will revoke the anonymity of the e-cash owner with the help of the judge. The steps are described in detail as follows.

(1) Shop → Bank: $(s, m, \sigma, D, d, r_4, s', r_u, r_s)$

The shop computes $r_4 = r_2 - x_2 H_5(d)$, where d is the deposit date, and sends $(s, m, \sigma, D, d, r_4, s', r_u, r_s)$ to the bank.

(2) Verifications

Firstly, the bank checks the correctness of expiration date D and deposit date d, respectively, and also checks if

$$w_2 = y_2^{H_5(d)} g_2^{r_4} \bmod p,$$
$$w_1 = y_1^{H_4(r_u \| r_s)} g_2^{s'} \bmod p \qquad (7)$$

are true or not. Secondly, the bank verifies if $s^{e_b} H_1^2(m \| D) H_3(\sigma \| D) = H_2(D)(\bmod n_b)$ and checks the uniqueness of (s, m, σ, D). Finally, if all of the above facts are verified successfully, the bank will accept and store the e-cash in its database and record $H_1(m \| D)$ in *exchange list*. Otherwise, it will reject this transaction and trace the owner of the e-cash.

3.2.5. E-Cash Renewal Protocol. In order to reduce the unlimited growth database problem of the bank, we have expiration date and renewal protocol in our scheme to achieve it, as shown in Figure 4. When an unused e-cash is expired, the user has to exchange it for another e-cash with a new expiration date from the bank.

(1) User → Bank: (s, ρ, σ, D)

The user recalls $m = (y_1, w_1, y_2, w_2, x_2, r_3)$ and prepares

$$\rho = H_1(m \| D) \qquad (8)$$

and sends it together with the unused (s, σ, D) to the bank.

(2) Verifications

Firstly, the bank checks the correctness of expiration date D and makes sure ρ does not exist in the *exchange list*. Secondly, the bank verifies if $s^{e_b} H_1(\rho) H_3(\sigma \| D) \equiv H_2(D)(\bmod n_b)$. Finally, if all of the above facts are verified successfully, the bank will accept to

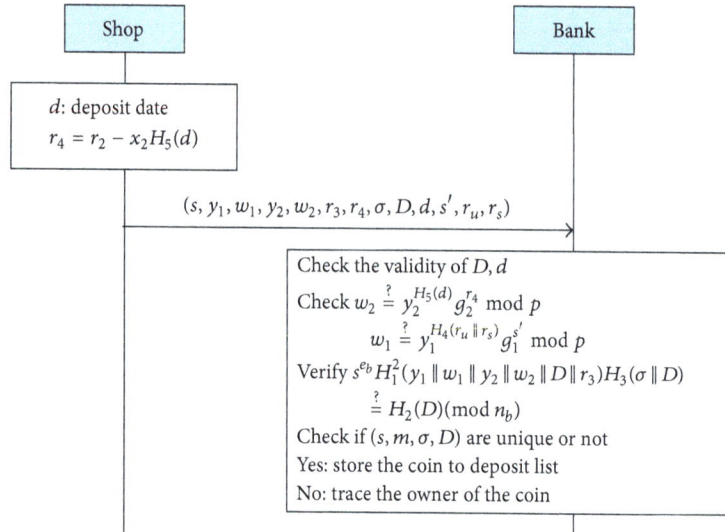

FIGURE 3: Deposit protocol.

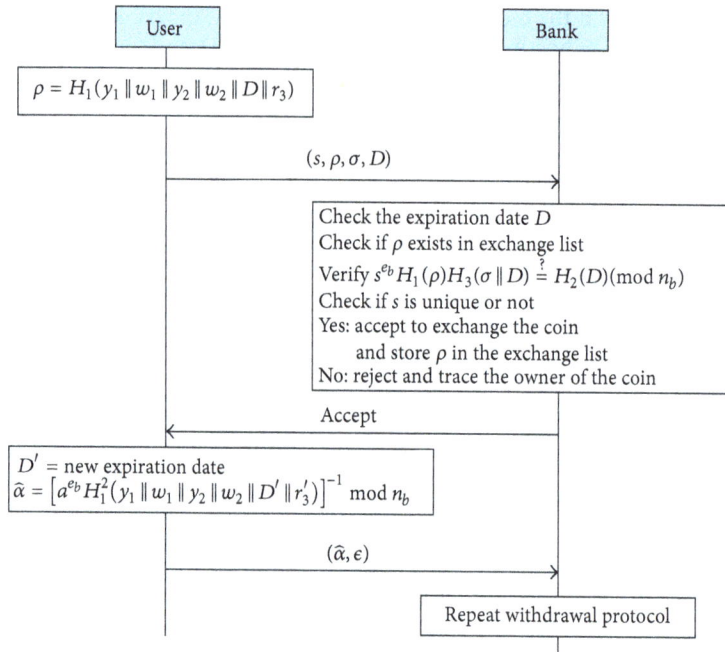

FIGURE 4: E-Cash renewal protocol.

exchange the e-cash. It will send a new expiration date D' and store ρ in the *exchange list*. Otherwise, it will reject the exchange request.

(3) User $\rightarrow$ Bank: $(\widehat{\alpha}, \epsilon)$

The user computes

$$\widehat{\alpha} = \left[a^{e_b} H_1^2(m' \parallel D') \right]^{-1} \bmod n_b, \qquad (9)$$

where $m' = (y_1, w_1, y_2, w_2, x_2, r_3')$, r_3' is a random, and D' is the new expiration date issued by the bank. The user sends $(\widehat{\alpha}, \epsilon, \mathrm{ID}_c)$ to the bank. Then the bank

repeats the withdrawal protocol in Section 3.2.2 from Step 2 with the user.

3.2.6. Double-Spending Checking and Anonymity Control. In our scheme, the identity of the users is anonymous in general except when the users violate any security rules and, therefore, their identities will be revealed.

(1) Double-Spending Checking

When an e-cash is being doubly spent, there must be two e-cash(s) with the same record prefixed by $(s, y_1, w_1, y_2, w_2, r_3, \sigma, D)$ stored in the database of the

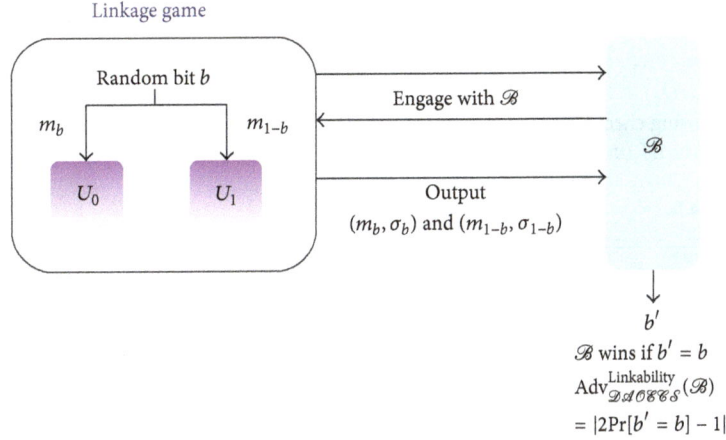

FIGURE 5: The game environment of linkage game.

bank. Therefore, the bank is able to detect any double-spent e-cash easily by checking the above parameters. For instance, the bank has received two e-cash(s),

$$\left(s, y_1, w_1, y_2, w_2, x_2, r_3, r_4, \sigma, D, d, s', r_u, r_s\right),$$
$$\left(s, y_1, w_1, y_2, w_2, x_2, r_3, \widehat{r_4}, \sigma, D, \widehat{d}, \widehat{s'}, \widehat{r_u}, \widehat{r_s}\right). \tag{10}$$

Thus, the bank can obtain two equations as follows:

$$s' \equiv r_1 - H_4\left(r_u \parallel r_s\right) x_1 \pmod{q},$$
$$\widehat{s'} \equiv r_1 - H_4\left(\widehat{r_u} \parallel \widehat{r_s}\right) x_1 \pmod{q}. \tag{11}$$

The bank can derive (x_1, r_1) from the above equations and send $(s, y_1, w_1, y_2, w_2, x_2, r_3, \sigma, D)$ and (x_1, r_1) to the judge to trace the owner of the e-cash.

(2) Revocation

The judge can trace any user who doubly spends e-cash(s) or violates any transaction regulations. When the judge receives $(s, y_1, w_1, y_2, w_2, x_2, r_3, \sigma, D)$ and (x_1, r_1) from the bank, it checks the following equations:

$$s^{e_b} H_1^2\left(m \parallel D\right) H_3\left(\sigma \parallel D\right) \overset{?}{\equiv} H_2\left(D\right)\left(\bmod n_b\right),$$

$$y_1 \overset{?}{\equiv} g_1^{x_1} \pmod{p}, \tag{12}$$

$$w_1 \overset{?}{\equiv} g_1^{r_1} \pmod{p}.$$

If all of the above equalities are true, the judge will decrypt σ and return the extracted ID_c to the bank.

4. Security Proofs

In this section, we provide security definitions and formal proofs of the following security features: unlinkability, unforgeability, traceability, and no-swindling for our proposed date attachable offline electronic cash scheme ($\mathscr{DAOECS}$).

4.1. E-Cash Unlinkability.
Based on the definition of unlinkability introduced by Abe and Okamoto [35] and Juels et al. [36], we formally define the unlinkability property of $\mathscr{DAOECS}$.

Definition 1 (The Linkage Game). Let U_0, U_1, and $\mathscr{J}$ be two honest users and the judge that follows $\mathscr{DAOECS}$, respectively. Let $\mathscr{B}$ be the bank that participates the following game with U_0, U_1, and $\mathscr{J}$. The game environment is shown in Figure 5.

Step 1. According to $\mathscr{DAOECS}$, $\mathscr{B}$ generates the bank's public key (e_b, n_b), the bank's private key (d_b, p_b, q_b), system parameters (p, q, g_1, g_2), the expiration date D, and the five public one-way hash functions H_1, H_2, H_3, H_4, and H_5. $\mathscr{J}$ generates the judge's public-private key pair (pk_j, sk_j).

Step 2. $\mathscr{B}$ generates $x_{1i}, x_{2i}, r_{1i}, r_{2i}, r_{3i}$ in random, where x_1, $x_2, r_1, r_2, r_3 \in_R \{0, 1, \dots, q-1\}$, and computes (y_{ki}, w_{ki}) for $k = \{1, 2\}$ and $i = \{0, 1\}$, where $y_{ki} = g_k^{x_k} \bmod p$ and $w_{ki} = g_k^{r_k} \bmod p$.

Step 3. We choose a bit $\widehat{b} \in \{0, 1\}$ randomly and place $(y_{1\widehat{b}}, w_{1\widehat{b}}, y_{2\widehat{b}}, w_{2\widehat{b}})$ and $(y_{11-\widehat{b}}, w_{11-\widehat{b}}, y_{21-\widehat{b}}, w_{21-\widehat{b}})$ on the private input tapes of U_0 and U_1, respectively, where $\widehat{b}$ is not disclosed to $\mathscr{B}$.

Step 4. $\mathscr{B}$ performs the withdrawal protocol of $\mathscr{DAOECS}$ with U_0 and U_1, respectively.

Step 5. If U_0 and U_1 output two e-cash(s) $(s_{\widehat{b}}, m_{\widehat{b}}, \sigma_{\widehat{b}}, D_{\widehat{b}})$ and $(s_{1-\widehat{b}}, m_{1-\widehat{b}}, \sigma_{1-\widehat{b}}, D_{1-\widehat{b}})$, where $m_i = (y_{1_i}, w_{1_i}, y_{2_i}, w_{2_i}, r_{3_i})$, on their private tapes, respectively, we give the two e-cash(s) in a random order to $\mathscr{B}$; otherwise, $\bot$ is given to $\mathscr{B}$.

Experiment Exp$_{\mathscr{A}}^{\text{FG-1}}(l_k)$

$\left(pk_j, sk_j, g_1, g_2, e_b, d_b, p_b, q_b, n_b, H_1, H_2, H_3, H_4, H_5\right) \leftarrow \text{Setup}(l_k)$

$\{(s_1, m_1, \sigma_1, D_1), \ldots, (s_{\ell+1}, m_{\ell+1}, \sigma_{\ell+1}, D_{\ell+1})\} \leftarrow \mathscr{A}^{\mathscr{O}_S}\left(pk_j, g_1, g_2, e_b, n_b, H_1, H_2, H_3, H_4, H_5\right)$

if the following checks are true, **return 1;**

    (i) $s_i^{e_b} H_1^2(m_i) H_3(\sigma_i \| D_i) \equiv H_2(D_i) \pmod{n_b}$, $\forall i \in \{1, \ldots, \ell+1\}$;

    (ii) $m_1, \ldots, m_{\ell+1}$ are all distinct

else **return 0;**

ALGORITHM 1: Experiment FG-1.

Step 6. $\mathscr{B}$ outputs $\widehat{b}' \in \{0, 1\}$ as the guess of $\widehat{b}$. The bank $\mathscr{B}$ wins the game if $\widehat{b}' = \widehat{b}$ and $\mathscr{J}$ has not revoked the anonymity of $(s_{\widehat{b}}, m_{\widehat{b}}, \sigma_{\widehat{b}}, D_{\widehat{b}})$ and $(s_{1-\widehat{b}}, m_{1-\widehat{b}}, \sigma_{1-\widehat{b}}, D_{1-\widehat{b}})$ to $\mathscr{B}$. We define the advantage of $\mathscr{B}$ as

$$\text{Adv}_{\mathscr{DAOECS}}^{\text{Linkability}}(\mathscr{B}) = \left|2\Pr\left[\widehat{b}' = \widehat{b}\right] - 1\right|, \quad (13)$$

where $\Pr[\widehat{b}' = \widehat{b}]$ denotes the probability of $\widehat{b}' = \widehat{b}$.

Definition 2 (Unlinkability). A $\mathscr{DAOECS}$ satisfies the unlinkability property if and only if the advantage $\text{Adv}_{\mathscr{DAOECS}}^{\text{Linkability}}(\mathscr{B})$ defined in Definition 1 is negligible.

Theorem 3. *A $\mathscr{DAOECS}$ satisfies the unlinkability property of Definition 2 if the adopted cryptosystems are semantically secure.*

Proof. If $\mathscr{B}$ is given $\perp$ in the Step 5 of the game, it will determine $\widehat{b}$ with probability $1/2$, which is exactly the same as a random guess of $\widehat{b}$.

Here, we assume that $\mathscr{B}$ gets two e-cash $(s_0, m_0, \sigma_0, D_0)$ and $(s_1, m_1, \sigma_1, D_1)$. Let $(\alpha_i, \beta_i, t_i, \epsilon_i, \widetilde{E}_{k_i}(b_i, \sigma_i, r_{j_i}))$, $i \in \{0, 1\}$, be the view of data exchanged between U_i and $\mathscr{B}$ in the withdrawal protocol (Section 3.2.2) and let $(x_{2_i}, r_{2_i}, r_{4_i}, r_{u_i}, r_{s_i}, s_i', d_i)$ be the view of data exchanged when $\mathscr{B}$ performs the payment protocol (Section 3.2.3) and the deposit protocol (Section 3.2.4) by using $(s_i, m_i, \sigma_i, D_i)$, where $i \in \{0, 1\}$.

For $(s, m, \sigma, D, x_2, r_2, r_4, r_u, r_s, s', d) \in$

$$\begin{Bmatrix} \left(s_0, m_0, \sigma_0, D_0, x_{2_0}, r_{2_0}, r_{4_0}, r_{u_0}, r_{s_0}, s_0', d_0\right), \\ \left(s_1, m_1, \sigma_1, D_1, x_{2_1}, r_{2_1}, r_{4_1}, r_{u_1}, r_{s_1}, s_1', d_1\right) \end{Bmatrix} \quad (14)$$

and $(\alpha_i, \beta_i, t_i, \epsilon_i, \widetilde{E}_{k_i}(b_i, \sigma_i, r_{j_i}))$, $i \in \{0, 1\}$, there always exists a pair (a_i', b_i') such that

$$a_i' = \left[\alpha_i H_1^2(m \| D)\right]^{-d_b} \bmod n_b \quad (\text{via }(1)),$$
$$b_i' = \left[\beta_i H_3(\sigma \| D)\right]^{-d_b} \bmod n_b \quad (\text{via }(2)). \quad (15)$$

And from (3), $t_i \equiv (\alpha_i \beta_i H_2(D))^{d_b} \pmod{n_b}$, (4) always holds as

$$s \equiv \left(a_i' b_i' t_i\right)$$
$$\equiv \left[\left(H_1^2(m \| D) H_3(\sigma \| D)\right)^{-1} H_2(D)\right]^{d_b} \pmod{n_b}. \quad (16)$$

Besides, $\widehat{E}_{pk_j}$ and $\widetilde{E}_{k_i}$ are semantically secure encryption functions. $\mathscr{B}$ cannot learn any information from ϵ_i and $\widetilde{E}_{k_i}(b_i, \sigma_i, r_{j_i})$.

From the above, given any $(s, m, \sigma, D) \in \{(s_0, m_0, \sigma_0, D_0), (s_1, m_1, \sigma_1, D_1)\}$ and (α_i, β_i, t_i), where $i \in \{0, 1\}$, there always exists a corresponding pair (a_i', b_i') such that (1), (2), (3), and (4) are satisfied.

Thus, go back to Step 6 of the game, the bank $\mathscr{B}$ succeeds in determining $\widehat{b}$ with probability $(1/2) + \varepsilon$, where ε is negligible since $\widehat{E}$ and $\widetilde{E}$ are semantically secure. Therefore, we have $\text{Adv}_{\mathscr{DAOECS}}^{\text{Linkability}}(\mathscr{B}) = 2\varepsilon$, which is negligible, so that $\mathscr{DAOECS}$ satisfies the unlinkability property. $\square$

4.2. E-Cash Unforgeability. In this section, we will formally prove that the proposed date attachable offline electronic cash scheme ($\mathscr{DAOECS}$) is secure against forgery attack. The forgery attack can be roughly divided into two types, one is the typical one-more forgery type (i.e., $(\ell, \ell+1)$-forgery) [37] and the other is the forgery on some specific expiration date D_i of an e-cash after sufficient communications with the signing oracle (i.e., bank). The details of definitions and our formal proofs will be described as follows.

Definition 4 (Forgery Game 1 in $\mathscr{DAOECS}$ (FG-1)). Let $l_k \in \mathbb{N}$ be a security parameter and $\mathscr{A}$ be an adversary in $\mathscr{DAOECS}$. $\mathscr{O}_S$ is an oracle which plays the role of the bank in $\mathscr{DAOECS}$ to be responsible for issuing e-cash(s) (i.e., (s, m, σ, D), where $m = (w_1, y_1, w_2, y_2, r_3, D)$) according to the queries from $\mathscr{A}$. $\mathscr{A}$ is allowed to query $\mathscr{O}_S$ for ℓ times; consider the experiment $\text{Exp}_{\mathscr{A}}^{\text{FG-1}}(l_k)$ shown in Algorithm 1. $\mathscr{A}$ wins the forgery game FG-1 if the probability $\Pr[\text{Exp}_{\mathscr{A}}^{\text{FG-1}}(l_k) = 1]$ of $\mathscr{A}$ is nonnegligible.

Definition 5 (Forgery Game 2 in $\mathscr{DAOECS}$ (FG-2)). Let $l_k \in \mathbb{N}$ be a security parameter and $\mathscr{A}$ be an adversary in $\mathscr{DAOECS}$. $\mathscr{O}_S$ is an oracle which plays the role of the bank in $\mathscr{DAOECS}$ to take charge of the following two events:

  (i) issue e-cash(s) (i.e., (s, m, σ, D), where $m = (w_1, y_1, w_2, y_2, r_3, D)$) according to the queries from $\mathscr{A}$,

  (ii) record the total number ℓ_{D_i} of each distinct expiration date D_i.

$\mathscr{A}$ is allowed to query $\mathscr{O}_S$ for ℓ times; consider the experiment $\text{Exp}_{\mathscr{A}}^{\text{FG-2}}(l_k)$ shown in Algorithm 2. $\mathscr{A}$ wins the forgery game

---

Experiment Exp$_{\mathscr{A}}^{\text{FG-2}}(l_k)$

$\left(pk_j, sk_j, g_1, g_2, e_b, d_b, p_b, q_b, n_b, H_1, H_2, H_3, H_4, H_5\right) \leftarrow \text{Setup}(l_k)$

$\{(s_i, m_i, \sigma_i, D^*), 1 \le i \le \ell_{D^*} + 1\} \leftarrow \mathscr{A}^{\mathscr{O}_S}\left(pk_j, g_1, g_2, e_b, n_b, H_1, H_2, H_3, H_4, H_5\right)$

if the following checks are true, **return 1;**

   (i) $s_i^{e_b} H_1^2(m_i) H_3(\sigma_i \parallel D^*) \equiv H_2(D^*) \pmod{n_b}, \forall i \in \{1, \ldots, \ell_{D^*} + 1\};$

   (ii) $m_1, \ldots, m_{\ell_{D^*}+1}$ are all distinct;

else **return 0;**

ALGORITHM 2: Experiment FG-2.

Experiment Exp$_{\mathscr{A}}^{\text{RSA-ACTI}}(k)$

$(N, e, d) \xleftarrow{R} KeyGen(k).$

$(y_1, \ldots, y_m) \leftarrow \mathscr{O}_t(N, e, k)$

$\{\pi, (x_1, y_1), \ldots, (x_n, y_n)\} \leftarrow \mathscr{A}^{\mathscr{O}_{\text{inv}}, \mathscr{O}_t}(N, e, k)$

if the following checks are true, **return 1;**

   (i) $\pi : \{1, \ldots, n\} \rightarrow \{1, \ldots, m\}$ is injective

   (ii) $x_i^e \equiv y_i \pmod{N}, \forall i \in \{1, \ldots, n\}$

   (iii) $n > q_h$

else **return 0;**

ALGORITHM 3

FG-2 if the probability $\Pr[\text{Exp}_{\mathscr{A}}^{\text{FG-2}}(k) = 1]$ of $\mathscr{A}$ is nonnegligible.

Here we introduce the hard problems used in our proof models.

Definition 6 (Alternative Formulation of RSA Chosen-Target Inversion Problem (RSA-ACTI)). Let $k \in \mathbb{N}$ be a security parameter and $\mathscr{A}$ be an adversary who is allowed to access the RSA-inversion oracle $\mathscr{O}_{\text{inv}}$ and the target oracle $\mathscr{O}_t$. $\mathscr{A}$ is allowed to query $\mathscr{O}_t$ and $\mathscr{O}_{\text{inv}}$ for m and q_h times, respectively. Consider Algorithm 3.

We say $\mathscr{A}$ breaks the RSA-ACTI problem if the probability $\Pr[\text{Exp}_{\mathscr{A}}^{\text{RSA-ACTI}}(k) = 1]$ of $\mathscr{A}$ is nonnegligible.

Definition 7 (The RSA Inversion Problem). Given (e, n), where n is the product of two distinct large primes p and q with roughly the same length and e is a positive integer relatively-prime to $(p - 1)(q - 1)$, and a randomly-chosen positive integer y less than n, find an integer x such that $x^e \equiv y \pmod{n}$.

Definition 8 (E-Cash Unforgeability). If there exists no probabilistic polynomial-time adversary who can win FG-1 or FG-2, then $\mathscr{DAOECS}$ is secure against forgery attacks.

Theorem 9. *For a polynomial-time adversary $\mathscr{A}$ who can win FG-1 or FG-2 with nonnegligible probability, there exists another adversary $\mathscr{S}$ who can break the RSA-ACTI problem or RSA inversion problem with nonnegligible probability.*

Proof. $\mathscr{S}$ simulates the environment and controls three hash oracles, $\mathscr{O}_{H_1}, \mathscr{O}_{H_2}, \mathscr{O}_{H_3}$ and an e-cash producing oracle $\mathscr{O}_S$ of $\mathscr{DAOECS}$ scheme to respond to different queries from $\mathscr{A}$ in the random oracle model and takes advantage of $\mathscr{A}$ to solve RSA-ACTI problem or RSA inversion problem, simultaneously. Then, for consistency, $\mathscr{S}$ maintains three lists $\mathscr{L}_{H_1}, \mathscr{L}_{H_2}$, and $\mathscr{L}_{H_3}$ to record every response of $\mathscr{O}_{H_1}, \mathscr{O}_{H_2}$, and $\mathscr{O}_{H_3}$, respectively.

Here we will start to do the simulation for the two games (i.e., FG-1 and FG-2) to prove $\mathscr{DAOECS}$ is secure against forgery attacks. The details of simulation are set below and illustrated in Figures 6 and 7, respectively.

Simulation in FG-1. In this proof model, $\mathscr{S}$ is allowed to query the oracles $\mathscr{O}_{\text{inv}}$ (i.e., $(\cdot)^d$) and $\mathscr{O}_t$ of RSA-ACTI problem defined in Definition 6 for helping $\mathscr{S}$ to produce e-cash(s) and the corresponding verifying key is (e, n).

(i) H_1 Query of $\mathscr{O}_{H_1}$

Initially, every blank record in $\mathscr{L}_{H_1}$ can be represented as $(\perp, \perp, \perp)$. When $\mathscr{A}$ sends m for querying the hash value $H_1(m)$, $\mathscr{S}$ will check the list $\mathscr{L}_{H_1}$:

   (a) if $m = m_i$ for some i, then $\mathscr{S}$ retrieves the corresponding $H_1(m_i)$ and returns it to $\mathscr{A}$;

   (b) else if $m = H_1(m_i)$ and $H_1^2(m_i) \ne \perp$ for some i, then $\mathscr{S}$ retrieves the corresponding $H_1^2(m_i)$ and returns it to $\mathscr{A}$;

   (c) else if $m = H_1(m_i)$ and $H_1^2(m_i) = \perp$ for some i, then $\mathscr{S}$ queries $\mathscr{O}_t$ to get an instance y and returns it to $\mathscr{A}$, then fills the record $(m_i, H_1(m_i), \perp)$ as $(m_i, H_1(m_i), y)$ in $\mathscr{L}_{H_1}$;

   (d) otherwise, $\mathscr{S}$ selects a random $\rho \in \mathbb{Z}_n$, records $(m, \rho, \perp)$ in $\mathscr{L}_{H_1}$, and returns ρ to $\mathscr{A}$.

(ii) H_2 Query of $\mathscr{O}_{H_2}$

When $\mathscr{A}$ asks for H_2 query by sending D to $\mathscr{S}$, $\mathscr{S}$ will look up the list $\mathscr{L}_{H_2}$:

   (a) if $D = D_i$ for some i, the corresponding τ will be retrieved and $\mathscr{S}$ will send $(\tau^e \bmod n)$ back to $\mathscr{A}$;

   (b) otherwise, $\mathscr{S}$ will select a random $\tau \in \mathbb{Z}_n$, record (D, τ) in $\mathscr{L}_{H_2}$, and return $(\tau^e \bmod n)$ back to $\mathscr{A}$.

(iii) H_3 Query of $\mathscr{O}_{H_3}$

While $\mathscr{A}$ sends (σ, D) to $\mathscr{S}$ for $H_3(\sigma \parallel D)$, $\mathscr{S}$ will look up the list $\mathscr{L}_{H_3}$:

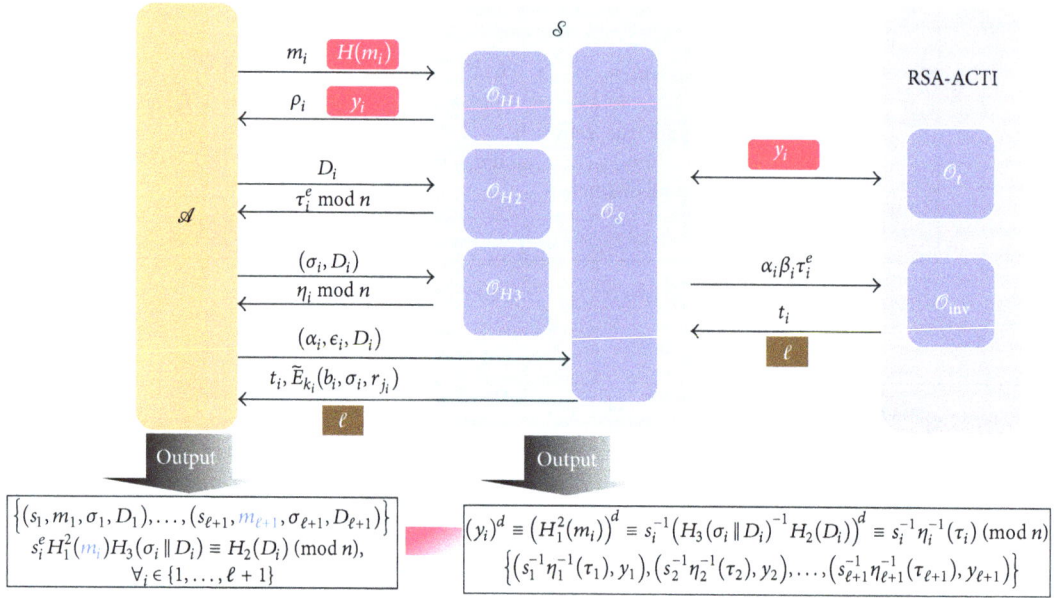

FIGURE 6: The proof model of FG-1.

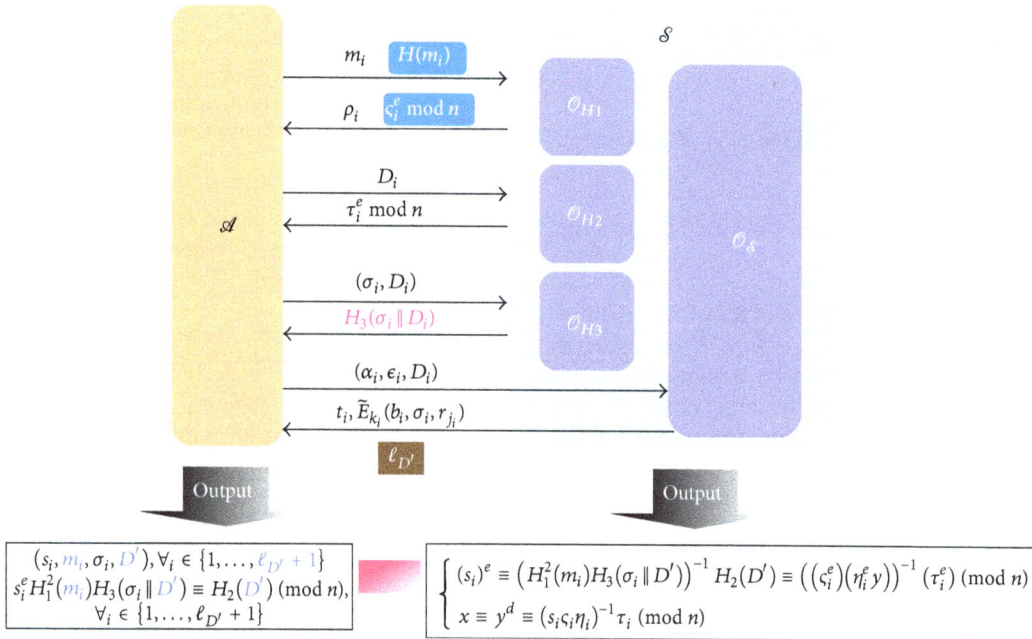

FIGURE 7: The proof model of FG-2.

(a) if $(\sigma, D) = (\sigma_i, D_i)$ for some i, the corresponding η will be retrieved and $(\eta^e \bmod n)$ will be returned to $\mathscr{A}$;

(b) otherwise, $\mathscr{S}$ will select a random $\eta \in \mathbb{Z}_n$, record $((\sigma, D), \eta)$ in $\mathscr{L}_{H_3}$, and return $(\eta^e \bmod n)$ back to $\mathscr{A}$.

(iv) E-Cash Producing Query of $\mathscr{O}_{\mathscr{S}}$

When $\mathscr{A}$ sends (α, ϵ, D) to $\mathscr{S}$, $\mathscr{S}$ will do the following steps:

(1) decrypt ϵ, obtain (k, ID);

(2) randomly select r_j and prepare $\sigma = \widehat{E}_{pk_j}(\text{ID} \parallel r_j)$;

(3) choose $\eta \in_R \mathbb{Z}_n$, set $H_3(\sigma \parallel D) = (\eta^e \bmod n)$, and store $((\sigma, D), \eta)$ in $\mathscr{L}_{H_3}$;

(4) select $b \in_R \mathbb{Z}_n^*$ and compute $\beta = (b^e \eta^e)^{-1} \bmod n$;

(5) retrieve or assign τ such that $H_2(D) = (\tau^e)$ as the $\mathscr{O}_{H_2}$ query described above;

(6) send $(\alpha \beta \tau^e)$ to oracle $\mathscr{O}_{\text{inv}}$ to get $t = (\alpha \beta \tau^e)^d \bmod n$;

(7) return $(t, \widetilde{E}_k(b, \sigma, r_j))$ back to $\mathscr{A}$.

Eventually, assume that $\mathscr{A}$ can successfully output $\ell + 1$ e-cash tuples

$$\{(s_1, m_1, \sigma_1, D_1) \cdots (s_{\ell+1}, m_{\ell+1}, \sigma_{\ell+1}, D_{\ell+1})\}, \quad (17)$$

where m_i are all distinct, $\forall i, 1 \leq i \leq \ell + 1$, such that $s_i^e H_1^2(m) H_3(\sigma_i \| D_i) = H_2(D_i) \pmod{n}$ after ℓ times to query $\mathcal{O}_S$ with nonnegligible probability $\epsilon_{\mathscr{A}}$.

According to $\mathscr{L}_{H_1}$, $\mathscr{L}_{H_2}$, and $\mathscr{L}_{H_3}$, $\mathscr{S}$ can compute and retrieve RSA-inversion instances ($\forall i, 1 \leq i \leq \ell + 1$)

$$(y_i)^d \equiv \left(H_1^2(m_i)\right)^d \equiv s_i^{-1}\left(H_3(\sigma_i \| D_i)^{-1} H_2(D_i)\right)^d \quad (18)$$

$$\equiv s_i^{-1} \eta_i^{-1}(\tau_i) \pmod{n}.$$

Via $\mathscr{A}$ querying the signing oracle $\mathcal{O}_S$ for ℓ times (i.e., query $\mathcal{O}_{\text{inv}}$ for ℓ times by $\mathscr{S}$), $\mathscr{S}$ can output $\ell + 1$ RSA-inversion instances

$$\{\left(s_1^{-1}\eta_1^{-1}(\tau_1), y_1\right), \left(s_2^{-1}\eta_2^{-1}(\tau_2), y_2\right), \ldots,$$
$$\left(s_{\ell+1}^{-1}\eta_{\ell+1}^{-1}(\tau_{\ell+1}), y_{\ell+1}\right)\} \quad (19)$$

and break the RSA-ACTI problem with nonnegligible probability at least $\epsilon_{\mathscr{A}}$.

Simulation in FG-2. Initially, $\mathscr{S}$ is given an instance (y, e, n) of RSA inversion problem defined in Definition 7 and simulates the environment as follows.

(i) H_1 Query of $\mathcal{O}_{H_1}$

Initially, every blank record in $\mathscr{L}_{H_1}$ can be represented as $(\perp, \perp, \perp)$. When $\mathscr{A}$ sends m for querying the hash value $H_1(m)$, $\mathscr{S}$ will check the list $\mathscr{L}_{H_1}$:

   (a) if $m = m_i$ for some i, then $\mathscr{S}$ retrieves the corresponding ρ_i and returns it to $\mathscr{A}$;

   (b) else if $m = H_1(m_i)$ and $H_1^2(m_i) \neq \perp$ for some i, then $\mathscr{S}$ retrieves the corresponding ς and returns $(\varsigma^e \bmod n)$ to $\mathscr{A}$;

   (c) else if $m = H_1(m_i)$ and $H_1^2(m_i) = \perp$ for some i, then $\mathscr{S}$ selects a random $\varsigma \in \mathbb{Z}_n$, returns $(\varsigma^e \bmod n)$ to $\mathscr{A}$, and then fills the record $(m_i, H_1(m_i), \perp)$ as $(m_i, H_1(m_i), \varsigma)$ in $\mathscr{L}_{H_1}$;

   (d) otherwise, $\mathscr{S}$ selects a random $\rho \in \mathbb{Z}_n$, records $(m, \rho, \perp)$ in $\mathscr{L}_{H_1}$, and returns ρ to $\mathscr{A}$.

(ii) H_2 Query of $\mathcal{O}_{H_2}$

When $\mathscr{A}$ asks for H_2 query by sending D to $\mathscr{S}$, $\mathscr{S}$ will look up the list $\mathscr{L}_{H_2}$:

   (a) if $D = D_i$ for some i, the corresponding τ will be retrieved and $\mathscr{S}$ will send $(\tau^e \bmod n)$ back to $\mathscr{A}$;

   (b) otherwise, $\mathscr{S}$ will select a random $\tau \in \mathbb{Z}_n$, record (D, τ) in $\mathscr{L}_{H_2}$, and return $(\tau^e \bmod n)$ back to $\mathscr{A}$.

(iii) H_3 Query of $\mathcal{O}_{H_3}$

While $\mathscr{A}$ sends (σ, D) to $\mathscr{S}$ for $H_3(\sigma \| D)$, $\mathscr{S}$ will look up the list $\mathscr{L}_{H_3}$:

   (a) if $(\sigma, D) = (\sigma_i, D_i)$ for some i, the corresponding $H_3(\sigma_i \| D_i)$ will be retrieved and returned to $\mathscr{A}$;

   (b) otherwise, $\mathscr{S}$ will select a random $\eta \in \mathbb{Z}_n$, set $H_3(\sigma \| D) = (\eta^e y \bmod n)$, record $((\sigma, D), \eta, H_3(\sigma \| D))$ in $\mathscr{L}_{H_3}$, and return $H_3(\sigma \| D)$ back to $\mathscr{A}$.

(iv) E-Cash Producing Query of $\mathcal{O}_S$

Let ℓ_{D_i} be a counter to record the number of queries on each expiration date D_i, which is initialized by 0. When $\mathscr{A}$ sends (α, ϵ, D) to $\mathscr{S}$, $\mathscr{S}$ will do the following steps:

   (1) decrypt ϵ, obtain (k, ID);

   (2) randomly select r_j and prepare $\sigma = \widehat{E}_{pk_j}(\text{ID} \| r_j)$;

   (3) choose $\eta \in_R \mathbb{Z}_n$, set $H_3(\sigma \| D) = (\alpha \eta^e \bmod n)$, and store $((\sigma, D), \perp, (\alpha \eta^e \bmod n))$ and (σ, D) in $\mathscr{L}_{H_3}$ and $\mathscr{L}_x$, respectively;

   (4) select $b \in_R \mathbb{Z}_n^*$ and compute $\beta = (b^e \alpha \eta^e)^{-1} \bmod n$;

   (5) retrieve or assign τ such that $H_2(D) = (\tau^e)$ as the $\mathcal{O}_{H_2}$ query described above;

   (6) compute $t \equiv (\alpha \beta \tau^e)^d \equiv ((b\eta)^{-1}\tau) \pmod{n}$;

   (7) set $\ell_D = \ell_D + 1$ and return $(t, \widetilde{E}_k(b, \sigma, r_j))$ back to $\mathscr{A}$.

Eventually, assume that $\mathscr{A}$ can successfully output $\ell_{D'} + 1$ e-cash tuples for some expiration date D'

$$\{(s_1, m_1, \sigma_1, D') \cdots (s_{\ell_{D'}+1}, m_{\ell_{D'}+1}, \sigma_{\ell_{D'}+1}, D')\} \quad (20)$$

such that $s_i^e H_1^2(m_i) H_3(\sigma_i \| D') = H_2(D') \pmod{n}$, $\forall i, 1 \leq i \leq \ell_{D'} + 1$, after $\ell_{D'}$ times to query $\mathcal{O}_S$ on D', with nonnegligible probability $\epsilon_{\mathscr{A}}$.

Assume some (σ_i, D'), $1 \leq i \leq \ell_{D'} + 1$, is not recorded in $\mathscr{L}_x$; then by the $\mathscr{L}_{H_1}$, $\mathscr{L}_{H_2}$, and $\mathscr{L}_{H_3}$, $\mathscr{S}$ can compute and retrieve

$$(s_i)^e \equiv \left(H_1^2(m_i) H_3\left(\sigma_i \| D'\right)\right)^{-1} H_2\left(D'\right)$$

$$\equiv \left((\varsigma_i^e)(\eta_i^e y)\right)^{-1}(\tau_i^e) \pmod{n}, \quad (21)$$

$$x \equiv y^d \equiv (s_i \varsigma_i \eta_i)^{-1} \tau_i \pmod{n}$$

and solve the RSA inversion problem with nonnegligible probability at least $\epsilon_{\mathscr{A}}$. $\square$

4.3. E-Cash Conditional-Traceability. In this section, we will prove that the ID information embedded in e-cash(s) cannot be replaced or moved out by any user against being traced after some misbehavior or criminals. The details of our proof model are illustrated in Figure 8.

Definition 10 (Tampering Game (TG)). Let $l_k \in \mathbb{N}$ be a security parameter and $\mathscr{A}$ be an adversary in $\mathscr{DAOECS}$. $\mathcal{O}_S$ is an oracle which plays the role of bank in $\mathscr{DAOECS}$

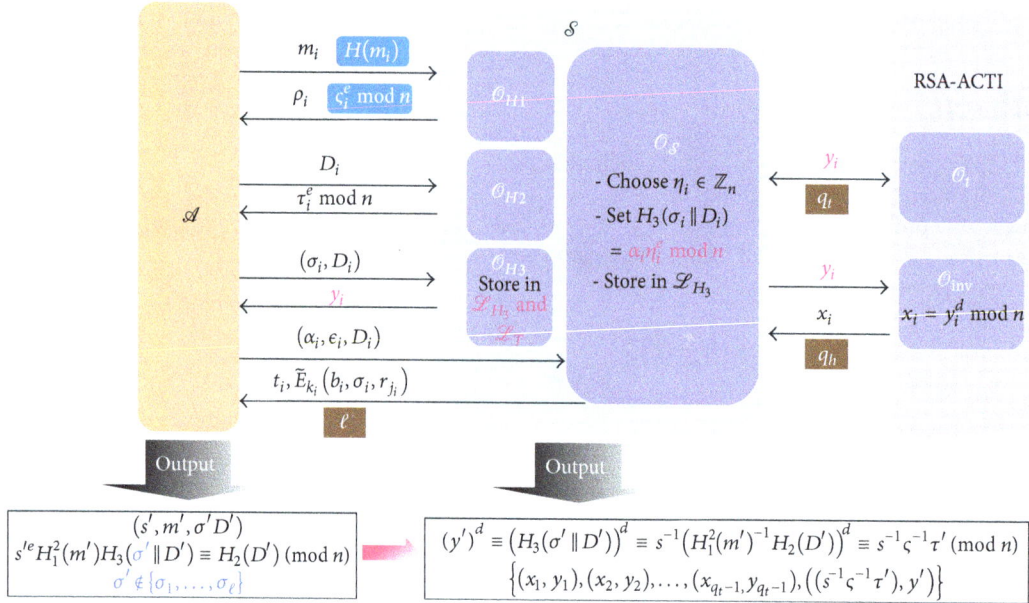

FIGURE 8: The proof model of TG.

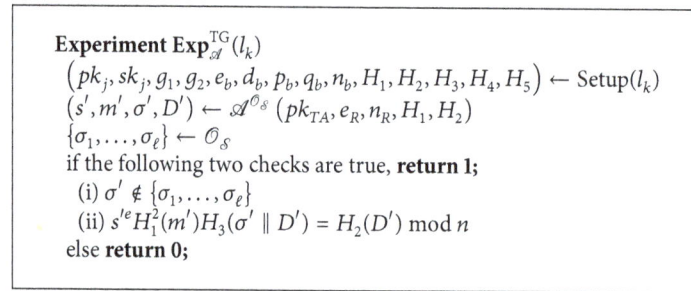

$$
\begin{aligned}
&\textbf{Experiment Exp}_{\mathscr{A}}^{\text{TG}}(l_k)\\
&\left(pk_j, sk_j, g_1, g_2, e_b, d_b, p_b, q_b, n_b, H_1, H_2, H_3, H_4, H_5\right) \leftarrow \text{Setup}(l_k)\\
&\left(s', m', \sigma', D'\right) \leftarrow \mathscr{A}^{\mathscr{O}_{\mathscr{S}}}\left(pk_{TA}, e_R, n_R, H_1, H_2\right)\\
&\{\sigma_1, \ldots, \sigma_\ell\} \leftarrow \mathscr{O}_{\mathscr{S}}\\
&\text{if the following two checks are true, } \textbf{return 1;}\\
&\quad\text{(i)}\ \sigma' \notin \{\sigma_1, \ldots, \sigma_\ell\}\\
&\quad\text{(ii)}\ s'^e H_1^2(m') H_3(\sigma' \parallel D') = H_2(D') \bmod n\\
&\text{else } \textbf{return 0;}
\end{aligned}
$$

ALGORITHM 4

to record parameters from the queries of $\mathscr{A}$ and issue e-cash(s) (i.e., (s, m, σ, D), where $m = (w_1, y_1, w_2, y_2, r_3, D)$) accordingly. $\mathscr{A}$ is allowed to query $\mathscr{O}_{\mathscr{S}}$ for ℓ times; consider Algorithm 4.

$\mathscr{A}$ wins the game if the probability $\Pr[\text{Exp}_{\mathscr{A}}^{\text{TG}}(k) = 1]$ of $\mathscr{A}$ is nonnegligible.

Definition 11 (E-Cash Traceability). If there exists no probabilistic polynomial-time adversary who can win the tracing game TG, then $\mathscr{DAOECS}$ satisfies the E-Cash Traceability.

Definition 12 (Alternative Formulation of RSA Known-Target Inversion Problem (RSA-AKTI)). Let $k \in \mathbb{N}$ be a security parameter and $\mathscr{A}$ be an adversary who is allowed to access the RSA-inversion oracle $\mathscr{O}_{\text{inv}}$ and the target oracle $\mathscr{O}_t$. $\mathscr{A}$ is allowed to query $\mathscr{O}_t$ and $\mathscr{O}_{\text{inv}}$ for q_t and q_h times ($q_h < q_t$), respectively. Consider Algorithm 5.

We say $\mathscr{A}$ breaks the RSA-AKTI problem if the probability $\Pr[\text{Exp}_{\mathscr{A}}^{\text{RSA-AKTI}}(k) = 1]$ of $\mathscr{A}$ is nonnegligible.

Theorem 13. *For a polynomial-time adversary $\mathscr{A}$ who can win the tracing game TG with nonnegligible probability, there exists*

$$
\begin{aligned}
&\textbf{Experiment Exp}_{\mathscr{A}}^{\text{RSA-AKTI}}(k)\\
&(N, e, d) \xleftarrow{R} KeyGen(k).\\
&(y_1, \ldots, y_{q_t}) \leftarrow \mathscr{O}_t(N, e, k)\\
&\{(x_1, y_1), \ldots, (x_{q_t}, y_{q_t})\} \leftarrow \mathscr{A}^{\mathscr{O}_{\text{inv}}, \mathscr{O}_t}(N, e, k)\\
&\text{if } x_i^e \equiv y_i \pmod{N}, \forall i \in \{1, \ldots, q_t\}, \textbf{return 1;}\\
&\text{else } \textbf{return 0;}
\end{aligned}
$$

ALGORITHM 5

another adversary $\mathscr{S}$ who can break the RSA-AKTI problem with nonnegligible probability.

Proof. $\mathscr{S}$ simulates the environment of $\mathscr{DAOECS}$ by controlling three hash oracles, $\mathscr{O}_{H_1}, \mathscr{O}_{H_2}, \mathscr{O}_{H_3}$, to respond hash queries and an e-cash producing oracle $\mathscr{O}_{\mathscr{S}}$ of $\mathscr{DAOECS}$ to respond e-cash producing queries from $\mathscr{A}$, respectively, in the random oracle model. Eventually, $\mathscr{S}$ will take advantage of $\mathscr{A}$'s capability to solve RSA-AKTI problem. Then, for consistency, $\mathscr{S}$ maintains three lists $\mathscr{L}_{H_1}, \mathscr{L}_{H_2}$, and $\mathscr{L}_{H_3}$ to record every response of $\mathscr{O}_{H_1}, \mathscr{O}_{H_2}$, and $\mathscr{O}_{H_3}$, respectively.

Besides, in the proof model, $\mathcal{S}$ is allowed to query the oracles $\mathcal{O}_{inv}$ (i.e., $(\cdot)^d$) and $\mathcal{O}_t$ of the RSA-AKTI problem defined in Definition 12 for helping $\mathcal{S}$ produce valid e-cash(s) and the corresponding verifying key is (e, n).

Here we will do the simulation for game TG to prove that $\mathcal{DAOECS}$ satisfies the e-cash traceability. Details are described as follows.

(i) H_1 Query of $\mathcal{O}_{H_1}$

Initially, every blank record in $\mathscr{L}_{H_1}$ can be represented as $(\bot, \bot, \bot)$. When $\mathcal{A}$ sends m for querying the hash value $H_1(m)$, $\mathcal{S}$ will check the list $\mathscr{L}_{H_1}$:

(a) if $m = m_i$ for some i, then $\mathcal{S}$ retrieves the corresponding $H_1(m_i)$ and return it to $\mathcal{A}$;

(b) else if $m = H_1(m_i)$ and $H_1^2(m_i) \neq \bot$ for some i, then $\mathcal{S}$ retrieves the corresponding ς_i and returns $(\varsigma_i^e \bmod n)$ to $\mathcal{A}$;

(c) else if $m = H_1(m_i)$ and $H_1^2(m_i) = \bot$ for some i, then $\mathcal{S}$ chooses $\varsigma \in_R \mathbb{Z}_n$, sets $H_1^2(m_i) = (\varsigma^e \bmod n)$, and returns $H_1^2(m_i)$ to $\mathcal{A}$ then fills the original record $(m_i, H_1(m_i), \bot)$ as $(m_i, H_1(m_i), \varsigma)$ in $\mathscr{L}_{H_1}$;

(d) otherwise, $\mathcal{S}$ selects a random $\rho \in \mathbb{Z}_n$, sets $H_1(m_i) = \rho$, records $(m, H_1(m_i), \bot)$ in $\mathscr{L}_{H_1}$, and returns ρ to $\mathcal{A}$.

(ii) H_2 Query of $\mathcal{O}_{H_2}$

When $\mathcal{A}$ asks for H_2 query by sending D to $\mathcal{S}$, $\mathcal{S}$ will look up the list $\mathscr{L}_{H_2}$:

(a) if $D = D_i$ for some i, the corresponding τ will be retrieved and $\mathcal{S}$ will send $(\tau^e \bmod n)$ back to $\mathcal{A}$;

(b) otherwise, $\mathcal{S}$ will select a random $\tau \in \mathbb{Z}_n$, record (D, τ) in $\mathscr{L}_{H_2}$, and return $(\tau^e \bmod n)$ back to $\mathcal{A}$.

(iii) H_3 Query of $\mathcal{O}_{H_3}$

While $\mathcal{A}$ sends (σ, D) to $\mathcal{S}$ for $H_3(\sigma)$, $\mathcal{S}$ will look up the list $\mathscr{L}_{H_3}$:

(a) if $(\sigma, D) = (\sigma_i, D_i)$ for some i, the corresponding y_i will be retrieved and returned to $\mathcal{A}$;

(b) otherwise, $\mathcal{S}$ will query $\mathcal{O}_t$ to get an instance y; record y and $((\sigma, D), y)$ in $\mathscr{L}_T$ and $\mathscr{L}_{H_3}$, respectively;

(c) return y back to $\mathcal{A}$.

(iv) E-Cash Producing Query of $\mathcal{O}_{\mathcal{S}}$

While $\mathcal{A}$ sends (α, ϵ, D) to $\mathcal{S}$, $\mathcal{S}$ will do the following steps:

(1) decrypt ϵ, obtain (k, ID);

(2) randomly select r_j and prepare $\sigma = \widehat{E}_{pk_j}(\mathrm{ID} \parallel r_j)$;

(3) choose $\eta \in_R \mathbb{Z}_n$, set $H_3(\sigma \parallel D) = (\alpha\eta^e \bmod n)$, and store $((\sigma, D), H_3(\sigma \parallel D))$ in $\mathscr{L}_{H_3}$;

(4) select $b \in_R \mathbb{Z}_n^*$ and compute $\beta = (b^e \alpha \eta^e)^{-1} \bmod n$;

(5) retrieve or assign τ such that $H_2(D) = (\tau^e)$ as the $\mathcal{O}_{H_2}$ query described above;

(6) compute $t \equiv (\alpha\beta\tau^e)^d \equiv ((b\eta)^{-1}\tau) \pmod{n}$;

(7) return $(t, \widetilde{E}_k(b, \sigma, r_j))$ back to $\mathcal{A}$.

Assume that $\mathcal{A}$ can successfully output an e-cash tuples (s', m', σ', D'), where σ' never appeals as a part for some $\mathcal{O}_{\mathcal{S}}$ query such that $s'^e H_1^2(m')H_3(\sigma' \parallel D') \equiv H_2(D') \pmod{n}$; then by $\mathscr{L}_{H_1}$, $\mathscr{L}_{H_2}$, and $\mathscr{L}_{H_3}$, $\mathcal{S}$ can derive

$$\left(y'\right)^d \equiv \left(H_3\left(\sigma' \parallel D'\right)\right)^d \equiv s'^{-1}\left(H_1^2\left(m'\right)^{-1}H_2\left(D'\right)\right)^d \tag{22}$$
$$\equiv s'^{-1}\varsigma'^{-1}\tau' \pmod{n}.$$

Let $|\mathscr{L}_T| = q_t$ and $\mathscr{L}_T = \{y_1, \ldots, y_{q_t}\}$. $\mathcal{S}$ sends $y_i \in (\mathscr{L}_T - \{y'\})$, $1 \le i \le (q_t - 1)$, to $\mathcal{O}_{inv}$ and obtains $q_t - 1$ x_i such that $x_i = y_i^d \bmod n$.

Eventually $\mathcal{S}$ can output q_t RSA-inversion instances

$$\left\{(x_1, y_1), (x_2, y_2), \ldots, \left(x_{q_t-1}, y_{q_t-1}\right), \left(\left(s'^{-1}\varsigma'^{-1}\tau'\right), y'\right)\right\} \tag{23}$$

after querying $\mathcal{O}_{inv}$ for q_h times, where $q_h = q_t - 1 < q_t$ and thus, it breaks the RSA-AKTI problem with nonnegligible probability at least $\epsilon_{\mathcal{A}}$. $\square$

4.4. E-Cash No-Swindling. In typical online e-cash transactions, when an e-cash has been spent in previous transactions, another spending will be detected immediately owing to the double-spending check procedure. However, in an offline e-cash model, the merchant may accept a transaction involving a double-spent e-cash first and then do the double-spending check later. In this case, the original owner of the e-cash may suffer from loss. Therefore, a secure offline e-cash scheme should guarantee the following two events.

(i) No one, except the real owner, can spend a fresh and valid offline e-cash successfully.

(ii) No one can double spend an e-cash successfully.

Roughly, it can be referred to as *e-cash no-swindling* property. In this section, we will define the no-swindling property and formally prove that our scheme is secure against swindling attacks.

Definition 14 (Swindling Game in $\mathcal{DAOECS}$). Let $l_k \in \mathbb{N}$ be a security parameter and $\mathcal{A}$ be an adversary in $\mathcal{DAOECS}$. $\mathcal{O}_B$ is an oracle issuing generic e-cash(s) (i.e., $(s, y_1, w_1, x_2, r_2, r_3, \sigma, D)$) of $\mathcal{DAOECS}$ to $\mathcal{A}$. $\mathcal{O}_{off}$ is an oracle to show the expanding form $(s, y_1, w_1, x_2, r_2, r_3, \sigma, D, r_s, s')$ for the payment according to the input (s, m, σ, D). Consider the two experiments SWG-1 and SWG-2 shown in Algorithms 6 and 7, respectively.

$\mathcal{A}$ wins the game if the probability $\Pr[\mathrm{Exp}_{\mathcal{A}}^{\mathrm{SWG-1}}(l_k) = 1]$ or $\Pr[\mathrm{Exp}_{\mathcal{A}}^{\mathrm{SWG-2}}(l_k) = 1]$ of $\mathcal{A}$ is nonnegligible.

Experiment Exp$_{\mathscr{A}}^{\text{SWG-1}}(l_k)$

$\left(pk_j, sk_j, g_1, g_2, e_b, d_b, p_b, q_b, n_b, p, q, H_1, H_2, H_3, H_4, H_5\right) \leftarrow \text{Setup}\,(l_k)$

$\{(s, w_1, y_1, w_2, y_2, r_3, \sigma, D, r_u, r_s, s')\} \leftarrow \mathscr{A}^{\mathcal{O}_B, \mathcal{O}_{\text{off}}}\left(pk_j, g_1, g_2, e_b, n_b, p, q, H_1, H_2, H_3, H_4, H_5\right)$

if the following checks are true, **return 1;**

(i) $s^{e_b} H_1^2 (y^{H_4(r_u \| r_s)} g^{s'} \bmod p \| y_1 \| w_2 \| y_2 \| D \| r_3) H_3(\sigma \| D) = H_2(D) \bmod n_b$;

(ii) $(s, w_1, y_1, w_2, y_2, r_3, \sigma, D)$ never be a query to $\mathcal{O}_{\text{off}}$

else **return 0;**

ALGORITHM 6: Experiment SWG-1.

Experiment Exp$_{\mathscr{A}}^{\text{SWG-2}}(l_k)$

$(pk_j, sk_j, g_1, g_2, e_b, d_b, p_b, q_b, n_b, p, q, H_1, H_2, H_3, H_4, H_5) \leftarrow \text{Setup}(l_k)$

$\{(s, w_1, y_1, w_2, y_2, r_3, \sigma, D, r_u, r_s, s')\} \leftarrow \mathscr{A}^{\mathcal{O}_B, \mathcal{O}_{\text{off}}}\left(pk_j, g_1, g_2, e_b, n_b, p, q, H_1, H_2, H_3, H_4, H_5\right)$

if the following checks are true, **return 1;**

(i) $s^{e_b} H_1^2 (y^{H_4(r_u \| r_s)} g^{s'} \bmod p \| y_1 \| w_2 \| y_2 \| D \| r_3) H_3(\sigma \| D) = H_2(D) \bmod n_b$;

(ii) $(s, w_1, y_1, w_2, y_2, r_3, \sigma, D)$ is allowed to be queried to $\mathcal{O}_{\text{off}}$ for once;

(iii) $(s, w_1, y_1, w_2, y_2, r_3, \sigma, D, r_s, s')$ is not obtained from $\mathcal{O}_{\text{off}}$

else **return 0;**

ALGORITHM 7: Experiment SWG-2.

Definition 15 (E-Cash No-Swindling). If there exists no probabilistic polynomial-time adversary who can win the swindling game defined in Definition 14, then $\mathscr{DAOECS}$ satisfies e-cash no-swindling.

Theorem 16. *For a polynomial-time adversary $\mathscr{A}$ who can win the swindling game SWG with nonnegligible probability, there exists another adversary $\mathscr{S}$ who can solve the discrete logarithm problem with nonnegligible probability.*

Proof. Consider the swindling game defined in Definition 14. $\mathscr{S}$ simulates the environment by controlling the hash oracles, $\mathcal{O}_{H_4}$, to respond hash queries on H_4 of $\mathscr{DAOECS}$ in the random oracle model. Eventually, $\mathscr{S}$ will take advantage of $\mathscr{A}$'s capability to solve the discrete logarithm problem. Then, for consistency, $\mathscr{S}$ maintains a list $\mathscr{L}_{H_4}$ to record every response of $\mathcal{O}_{H_4}$. $\mathscr{S}$ is given all parameters $(pk_j, sk_j, g_1, g_2, e_b, d_b, p_b, q_b, n_b, p, q, H_1, H_2, H_3, H_4, H_5)$ of $\mathscr{DAOECS}$ and an instance y^* of discrete logarithm problem (i.e., $y^* = g^{x^*} \bmod p$). Here we will describe the simulations for the two experiments $\text{Exp}_{\mathscr{A}}^{\text{SWG-1}}$ and $\text{Exp}_{\mathscr{A}}^{\text{SWG-2}}$, individually.

The simulation for $\text{Exp}_{\mathscr{A}}^{\text{SWG-1}}$ is illustrated in Figure 9 and each oracle is constructed as follows.

(i) Oracle $\mathcal{O}_B$

Initially, $\mathscr{S}$ guesses that the generic e-cash produced from νth query will be the attack target. When $\mathscr{A}$ sends ith query to $\mathcal{O}_B$ for an e-cash, $\mathcal{O}_B$ will do the following:

(a) select $r_1, x_1, r_3 \in_R \mathbb{Z}_q$ and $y_2, w_2 \in_R \mathbb{Z}_p$;

(b) if $i = \nu$,

(1) compute $(w_1 = (y^*)^{r_1} \bmod p)$ and $(y_1 = g^{x_1} \bmod p)$;

(c) if $i \neq \nu$,

(1) compute $(w_1 = g^{r_1} \bmod p)$ and $(y_1 = g^{x_1} \bmod p)$;

(d) prepare $s = ((H_1^2(m)H_3(\sigma \| D))^{-1}H_2(D))^{d_b} \bmod n_b$, where $m = (w_1, y_1, w_2, y_2, r_3, D)$;

(e) record $(i, (s, m, \sigma, D), (r_1, x_1))$ in list $\mathscr{L}_B$ and return (s, m, σ, D) to $\mathscr{A}$.

(ii) Oracle $\mathcal{O}_{\text{off}}$

When $\mathscr{A}$ sends a valid e-cash tuple $(s, w_1, y_1, w_2, y_2, r_3, \sigma, D, r_s)$ to $\mathcal{O}_{\text{off}}$, it will look up the list $\mathscr{L}_B$:

(a) if $(s, w_1, y_1, w_2, y_2, r_3, \sigma, D)$ exists with prefix index ν, then abort;

(b) otherwise, $\mathcal{O}_{\text{off}}$ will retrieve the corresponding (r_1, x_1); choose a random r_u, compute $u = H_4(r_u \| r_s)$ and $(s' = r_1 - ux_1 \bmod q)$, and send $(s, w_1, y_1, w_2, y_2, r_3, \sigma, D, r_u, r_s, s')$ back to $\mathscr{A}$.

Assume that $\mathscr{A}$ can successfully output a valid offline e-cash expansion tuple $(s^*, w_1^*, y_1^*, w_2^*, y_2^*, r_3^*, \sigma^*, D^*, r_u^*, r_s^*, s'^*)$, where $(s^*, w_1^*, y_1^*, w_2^*, y_2^*, r_3^*, \sigma^*, D^*)$ is prefixed with ν and postfixed with (r_1^*, x_1^*) in $\mathscr{L}_B$. Then, since $w_1^* = y_1^{*H_4(r_u^* \| r_s^*)} g^{s'^*} \bmod p$ and $w_1^* = (y^*)^{r_1^*}$, $\mathscr{S}$ can derive

$$x^* = (r_1^*)^{-1}\left(x_1^* H_4\left(r_u^* \| r_s^*\right) + s'^*\right) \bmod q \quad (24)$$

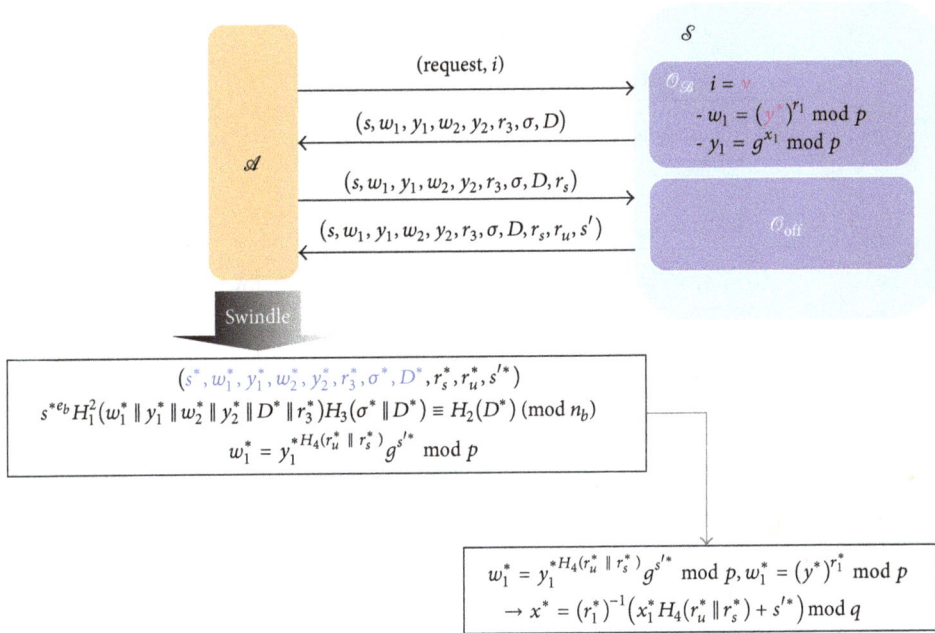

FIGURE 9: The proof model of SWG-1.

and solve the discrete logarithm problem with nonnegligible probability at least $(1/q_{\mathcal{O}_B})\epsilon_{\mathcal{A}}$, where $q_{\mathcal{O}_B}$ is the total number of $\mathcal{O}_B$ query.

The simulation for $\text{Exp}_{\mathcal{A}}^{\text{SWG-2}}$ is illustrated in Figure 10 and each oracle is constructed as follows.

(i) Oracle $\mathcal{O}_B$

Initially, $\mathcal{S}$ guesses that the generic e-cash produced from νth query will be the attack target. When $\mathcal{A}$ sends ith query to $\mathcal{O}_B$ for an e-cash, $\mathcal{O}_B$ will do the followings.

(a) if $i = \nu$:

(1) select $s', u, x_1, r_3 \in_R \mathbb{Z}_q$ and $y_2, w_2 \in_R \mathbb{Z}_p$;
(2) compute $(y_1 = (y^*)^{x_1} \bmod p)$ and $(w_1 = y_1^u g^{s'} \bmod p)$;
(3) prepare $s = ((H_1^2(m)H_3(\sigma \parallel D))^{-1} H_2(D))^{d_b} \bmod n_b$, where $m = (w_1, y_1, w_2, y_2, r_3, D)$;
(4) record $(i, (s, m, \sigma, D), (u, s'))$ in list $\mathcal{L}_{\mathcal{B}}$.

(b) if $i \neq \nu$:

(1) select $r_1, x_1, r_3 \in_R \mathbb{Z}_q$ and $y_2, w_2 \in_R \mathbb{Z}_p$;
(2) compute $(w_1 = g^{r_1} \bmod p)$ and $(y_1 = g^{x_1} \bmod p)$;
(3) prepare $s = ((H_1^2(m)H_3(\sigma \parallel D))^{-1} H_2(D))^{d_b} \bmod n_b$, where $m = (w_1, y_1, w_2, y_2, r_3, D)$;
(4) record $(i, (s, m, \sigma, D), (r_1, x_1))$ in list $\mathcal{L}_{\mathcal{B}}$.

(c) return (s, m, σ, D) to $\mathcal{A}$.

(ii) Oracle $\mathcal{O}_{\text{off}}$

A status parameter sta is initialized by 0. When $\mathcal{A}$ sends a valid e-cash tuple $(s, w_1, y_1, w_2, y_2, r_3, \sigma, D, r_s)$ to $\mathcal{O}_{\text{off}}$, it will look up the list $\mathcal{L}_{\mathcal{B}}$:

(a) if $(s, w_1, y_1, w_2, y_2, r_3, \sigma, D)$ exists with prefix index ν and sta $= 0$, $\mathcal{O}_{\text{off}}$ will perform the following procedures:

(1) set sta $= 1$
(2) retrieve the corresponding (u, s') from $\mathcal{L}_{\mathcal{B}}$ and choose a random r_u;
(3) set $H_4(r_u \parallel r_s) = u$ and record $((r_u \parallel r_s), u)$ in $\mathcal{L}_H$;
(4) record $(s, w_1, y_1, w_2, y_2, r_3, \sigma, D, r_u, r_s, s')$ in list $\mathcal{L}_{\text{off}}$;
(5) send $(s, w_1, y_1, w_2, y_2, r_3, \sigma, D, r_u, r_s, s')$ back to $\mathcal{A}$;

(b) if $(s, w_1, y_1, w_2, y_2, r_3, \sigma, D)$ exists with prefix index $\neq \nu$, $\mathcal{O}_{\text{off}}$ will retrieve the corresponding (r_1, x_1), choose random r_u and u, set $H_4(r_u \parallel r_s) = u$, record $((r_u \parallel r_s), u)$ in $\mathcal{L}_H$, compute $(s' = r_1 - ux_1 \bmod q)$, and send $(s, w_1, y_1, w_2, y_2, r_3, \sigma, D, r_u, r_s, s')$ back to $\mathcal{A}$.

(c) Otherwise, abort.

(iii) Oracle $\mathcal{O}_{H_4}$

While $\mathcal{A}$ sends $(r_u \parallel r_s)$ to query for $H_4(r_u \parallel r_s)$, $\mathcal{O}_{H_4}$ will check the list $\mathcal{L}_H$:

(a) if $(r_u \parallel r_s)$ exists as the prefix of some record, $\mathcal{O}_{H_4}$ will retrieve the corresponding u and return it to $\mathcal{A}$;

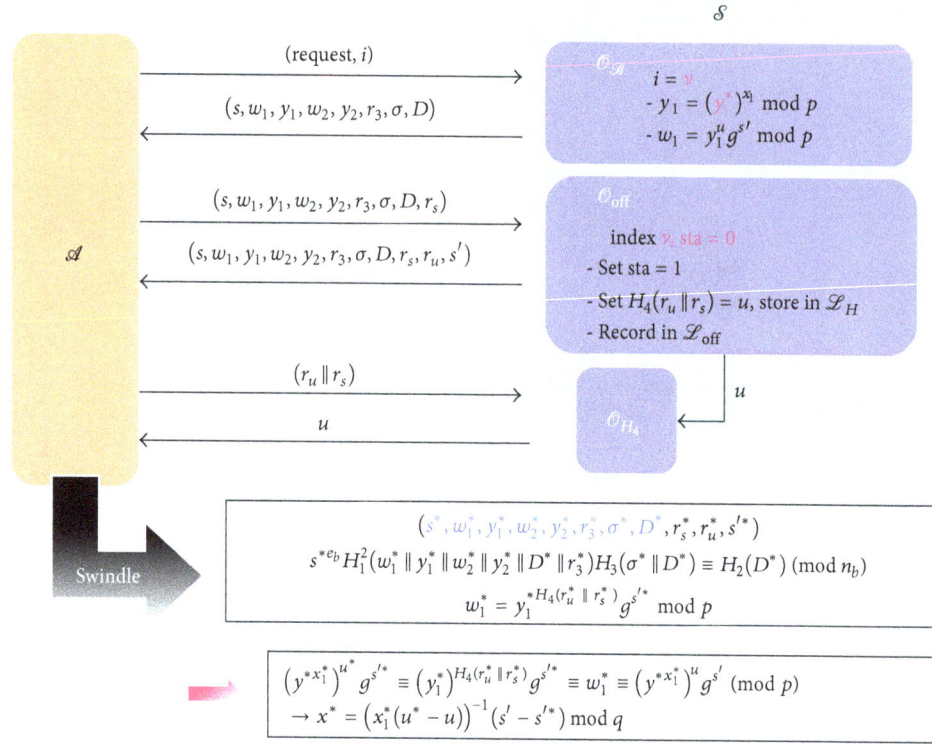

FIGURE 10: The proof model of SWG-2.

(b) otherwise, $\mathcal{O}_{H_4}$ will choose a random u, record $((r_u \parallel r_s), u)$ in $\mathcal{L}_H$, and return u to $\mathcal{A}$.

Assume that $\mathcal{A}$ can successfully output a valid offline e-cash expansion tuple $(s^*, w_1^*, y_1^*, w_2^*, y_2^*, r_3^*, \sigma^*, D^*, r_u^*, r_s^*, s'^*)$, where $(s^*, w_1^*, y_1^*, w_2^*, y_2^*, r_3^*, \sigma^*, D^*)$ is prefixed with v and postfixed with (u, s') in $\mathcal{L}_{\mathcal{B}}$ and $H_4(r_u^* \parallel r_s^*) \neq u$.

Then, via $\mathcal{L}_{\mathcal{H}}$, since

$$\left(y^{*x_1^*}\right)^{u^*} g^{s'^*} \equiv \left(y_1^*\right)^{H_4(r_u^* \| r_s^*)} g^{s'^*} \equiv w_1^*$$
$$\equiv \left(y^{*x_1^*}\right)^{u} g^{s'} \pmod{p}, \tag{25}$$

$\mathcal{S}$ can derive

$$x^* = \left(x_1^* (u^* - u)\right)^{-1} \left(s' - s'^*\right) \bmod q \tag{26}$$

and solve the discrete logarithm problem with nonnegligible probability at least $(1/q_{\mathcal{O}_B})\epsilon_{\mathcal{A}}$, where $q_{\mathcal{O}_B}$ is the total number of $\mathcal{O}_B$ query. □

Summarize the proof models for the two experiments shown above, if there exists a polynomial-time adversary who can win the swindling game with nonnegligible probability, then there exists another one who can solve the discrete logarithm problem with nonnegligible probability. It implies that there exists no p.p.t. adversary who can win the swindling game, and our proposed offline e-cash scheme $\mathcal{DAOECS}$ satisfies no-swindling property.

5. E-Cash Advanced Features and Performance Comparisons

In this section, we compare the e-cash features and performance of our proposed scheme with other schemes given in [9, 13–15, 21, 22, 27, 38–40]. We analyze the features and performance of the aforementioned schemes and form a table (Table 1) for the summary.

5.1. Features Comparisons. All the schemes mentioned above fulfill the basic security requirements stated in Section 1, which are anonymity, unlinkability, unforgeability, and no double-spending. Besides these features, there can be other advanced features on an e-cash system discussed in the literatures. We focus on three other advanced features, which are traceability, date attachability, and no-swindling, and we compare the proposed scheme with the aforementioned schemes.

We also propose an e-cash renewal protocol for users to exchange a new valid e-cash with their unused but expired e-cash(s); therefore, users do not have to deposit the e-cash before it expires and withdraw a new e-cash again. Our proposed e-cash renewal protocol reduces the computation cost by 49.5% as compared to withdrawal and deposit protocols, which is almost half of the effort of getting a new e-cash, at the user side. It does a great help to the users since their devices usually have a weaker computation capability, such as smart phones.

TABLE 1: Advanced features and performance comparisons.

	Ours	[38]	[14]	[15]	[9]	[21]	[22]	[39]	[40]	[13]	[27]
Advanced features											
On/off-line	Off	Off	Off	Off	On	Off	Off	Off	Off	On	Off
Conditional-traceability	Yes	Yes	No	Yes	No	Yes	Yes	Yes	Yes	Yes	No
Date attachability	Yes	No	No	No	Yes	Yes	No	No	No	No	Yes
No-swindling	Yes	No	No	No	—	No	Yes	No	No	—	No
Renewal protocol	Yes	—	Yes	—	No	Yes	Yes	—	—	—	Yes
Formal proof	Yes	Yes	No	Yes	No	No	Yes	Yes	Yes	Yes	No
Performance											
Transaction cost*	$5E+7M$ $+7H+1inv$ $+1A$ $\approx 1454M$	$14E+14M$ $+1H+5A$ $\approx 3375M$	$6E+8M$ $\approx 1448M$	$23E+14M$ $+1A$ $\approx 5534M$	$2E+2M$ $+2H$ $\approx 966M$	$5E+9M$ $+1H+1inv$ $+2A$ $\approx 1450M$	$2E \approx 480M$	$18E+15M$ $+2H+8A$ $\approx 4337M$	$31E+22M$ $+6H+10A$ $\approx 7468M$	$22E+11M$ $+4A$ $\approx 5291M$	$6E+8M$ $+1H$ $\approx 1449M$
Communication cost◇	1092	576	1288	939	769	644	300	828	968	1536	728

According to [41], $H \approx M$, $E \approx M$, $E \approx inv \approx 240M$.

E: a modular exponentiation; M: a modular multiplication; H: a hash operation; zkp: a zero-knowledge proof.

A: a modular addition; inv: a modular inversion.

*The computation cost of withdrawal and payment protocols at user side.

◇The communication cost of each transaction at user side in bytes.

5.2. Performance Comparisons. According to [41], we can summarize and induce the computation cost of all operations as follows. The computation cost of a modular exponentiation computation is about 240 times of the computation cost of a modular multiplication computation, while the computation cost of a modular inversion almost equals to that of a modular exponentiation. Also, the computation cost of a hash operation almost equals to that of a modular multiplication.

With the above assumptions, the total computation cost of users during withdrawal and payment phases of our proposed scheme can be induced as 1452 times of a modular multiplication computation, while other works [9, 13–15, 21, 22, 27, 38–40] need 3375, 1448, 5534, 966, 1450, 480, 4337, 7468, 5291, and 1449 times of a modular multiplication computation to finish withdrawal and payment phases at the user ends.

According to [15], we assume the RSA parameters n, p, q are 1024, 512, and 512 bits, respectively. We adopt AES and SHA-1 as the symmetric cryotsystem and one-way hash function used in all protocols, respectively; therefore, the signed message and hash massage are in 128 and 160 bits, respectively. We assume the expiration date is in 32 bits.

With the above assumptions, we compute the communication cost of each offline transaction, withdrawal, and payment, at the user side. Our scheme needs 2048 bits for withdrawing an e-cash and 6688 bits for spending an e-cash, which is 1092 bytes for each transaction.

The details of the comparisons are summarized in Table 1.

6. Conclusion

In this paper, we have presented earlier a provably secure offline electronic cash scheme with an expiration date and a deposit date attached to it. Besides, we have also designed an e-cash renewal protocol, where users can exchange their unused and expired e-cash(s) for new ones more efficiently. Compared with other similar works, our scheme is efficient from the aspect of considering computation cost of the user side and satisfying all security properties, simultaneously. Except for anonymity, unlinkability, unforgeability, and no double-spending, we also formally prove that our scheme achieves conditional-traceability and no-swindling. Not only does our scheme help the bank to manage their huge databases against unlimited growth, but also it strengthens the preservation of users' privacy and rights as well.

Conflict of Interests

The authors declare that there is no conflict of interests regarding the publication of this paper.

Acknowledgments

This work was partially supported by the National Science Council of Taiwan under Grants NSC 102-2219-E-110-002, NSYSU-KMU Joint Research Project (NSYSUKMU 2013-I001), and Aim for the Top University Plan of the National Sun Yat-sen University and Ministry of Education, Taiwan.

References

[1] H. Chen, P. P. Y. Lam, H. C. B. Chan, T. S. Dillon, J. Cao, and R. S. T. Lee, "Business-to-consumer mobile agent-based internet commerce system (MAGICS)," *IEEE Transactions on Systems, Man and Cybernetics C: Applications and Reviews*, vol. 37, no. 6, pp. 1174–1189, 2007.

[2] S. C. Fan and Y. L. Lai, "A study on e-commerce applying in Taiwan's restaurant franchise," in *Proceedings of the IET International Conference on Frontier Computing. Theory, Technologies and Applications*, pp. 324–329, August 2010.

[3] D. R. W. Holton, I. Nafea, M. Younas, and I. Awan, "A class-based scheme for E-commerce web servers: formal specification and performance evaluation," *Journal of Network and Computer Applications*, vol. 32, no. 2, pp. 455–460, 2009.

[4] Z. Jie and X. Hong, "E-commerce security policy analysis," in *Proceedings of the International Conference on Electrical and Control Engineering (ICECE '10)*, pp. 2764–2766, June 2010.

[5] D. R. Liuy and T. F. Hwang, "An agent-based approach to flexible commerce in intermediary-Centric electronic markets," *Journal of Network and Computer Applications*, vol. 27, no. 1, pp. 33–48, 2004.

[6] S. J. Lin and D. C. Liu, "An incentive-based electronic payment scheme for digital content transactions over the Internet," *Journal of Network and Computer Applications*, vol. 32, no. 3, pp. 589–598, 2009.

[7] H. Wang, Y. Zhang, J. Cao, and V. Varadharajan, "Achieving Secure and Flexible M-Services through Tickets," *IEEE Transactions on Systems, Man, and Cybernetics A:Systems and Humans*, vol. 33, no. 6, pp. 697–708, 2003.

[8] C. Yue and H. Wang, "Profit-aware overload protection in E-commerce Web sites," *Journal of Network and Computer Applications*, vol. 32, no. 2, pp. 347–356, 2009.

[9] C. C. Chang and Y. P. Lai, "A flexible date-attachment scheme on e-cash," *Computers and Security*, vol. 22, no. 2, pp. 160–166, 2003.

[10] C. L. Chen and J. J. Liao, "A fair online payment system for digital content via subliminal channel," *Electronic Commerce Research and Applications*, vol. 10, no. 3, pp. 279–287, 2011.

[11] C. I. Fan, W. K. Chen, and Y. S. Yeh, "Date attachable electronic cash," *Computer Communications*, vol. 23, no. 4, pp. 425–428, 2000.

[12] C. I. Fan and W. Z. Sun, "Efficient encoding scheme for date attachable electronic cash," in *Proceedings of the 24th Workshop on Combinatorial Mathematics and Computation Theory*, pp. 405–410, 2007.

[13] T. Nakanishi, M. Shiota, and Y. Sugiyama, "An efficient online electronic cash with unlinkable exact payments," *Information Security*, vol. 3225, pp. 367–378, 2004.

[14] Y. Baseri, B. Takhtaei, and J. Mohajeri, "Secure untraceable offline electronic cash system," *Scientia Iranica*, vol. 20, pp. 637–646, 2012.

[15] J. Camenisch, S. Hohenberger, and A. Lysyanskaya, "Compact e-cash," in *Proceedings of the 24th Annual International Conference on the Theory and Applications of Cryptographic Techniques: Advances in Cryptology (EUROCRYPT '05)*, pp. 302–321, May 2005.

[16] J. Camenisch, S. Hohenberger, and A. Lysyanskaya, "Balancing accountability and privacy using E-cash," in *Security and Cryptography for Networks*, vol. 4116 of *Lecture Notes in Computer Science*, pp. 141–155, 2006.

[17] J. Camenisch, A. Lysyanskaya, and M. Meyerovich, "Endorsed e-cash," in *Proceedings of the IEEE Symposium on Security and Privacy*, pp. 101–115, May 2007.

[18] S. Canard, A. Gouget, and J. Traoré, "Improvement of efficiency in (unconditional) anonymous transferable E-cash," in *Financial Cryptography and Data Security*, vol. 5143 of *Lecture Notes in Computer Science*, pp. 202–214, 2008.

[19] D. Chaum, A. Fiat, and M. Naor, "Untraceable electronic cash," in *Advances in Cryptology-CRYPTO '88*, vol. 403 of *Lecture Notes in Computer Science*, pp. 319–327, Springer, Berlin, Germany, 1990.

[20] G. Davida, Y. Frankel, Y. Tsiounis, and M. Yung, "Anonymity control in E-cash systems," in *Proceedings of the First International Conference on Financial Cryptography*, pp. 1–16, 1997.

[21] Z. Eslami and M. Talebi, "A new untraceable off-line electronic cash system," *Electronic Commerce Research and Applications*, vol. 10, no. 1, pp. 59–66, 2011.

[22] C. I. Fan, V. S. M. Huang, and Y. C. Yu, "User efficient recoverable off-line e-cash scheme with fast anonymity revoking," *Mathematical and Computer Modelling*, vol. 58, pp. 227–237, 2013.

[23] X. Hou and C. H. Tan, "Fair traceable off-line electronic cash in wallets with observers," in *Proceedings of the 6th International Conference on Advanced Communication Technology*, pp. 595–599, February 2004.

[24] X. Hou and C. H. Tan, "A new electronic cash model," in *Proceedings of the International Conference on Information Technology: Coding and Computing*, pp. 374–379, April 2005.

[25] W. S. Juang, "A practical anonymous off-line multi-authority payment scheme," *Electronic Commerce Research and Applications*, vol. 4, no. 3, pp. 240–249, 2005.

[26] J. K. Liu, V. K. Wei, and S. H. Wong, "Recoverable and untraceable e-cash," in *International Conference on Trends in Communications (EUROCON '01)*, vol. 1, pp. 132–135, 2001.

[27] C. Wang, H. Sun, H. Zhang, and Z. Jin, "An improved off-line electronic cash scheme," in *Proceedings of the 5th International Conference on Computational and Information Sciences (ICCIS '13)*, pp. 438–441, 2013.

[28] W. S. Juang, "D-cash: a flexible pre-paid e-cash scheme for date-attachment," *Electronic Commerce Research and Applications*, vol. 6, no. 1, pp. 74–80, 2007.

[29] D. Chaum, "Blind signatures for untraceable payments," in *Advances in Cryptology-CRYPTO '82*, Lecture Notes in Computer Science, pp. 199–203, Springer, Berlin, Germany, 1983.

[30] H. Krawczyk and T. Rabin, "Chameleon signatures," in *Proceedings of the Network and Distributed System Security Symposium (NDSS '00)*, pp. 143–154, 2000.

[31] S. Pearson, *Trusted Computing Platforms: TCPA Technology in Context*, Prentice Hall, New York, NY, USA, 2002.

[32] S. Pearson, "Trusted computing platforms: the next security solution," Tech. Rep. HPL-2002-221, Hewllet-Packard Laboratorie, 2002.

[33] C. I. Fan and V. S. M. Huang, "Provably secure integrated on/off-line electronic cash for flexible and efficient payment," *IEEE Transactions on Systems, Man and Cybernetics C: Applications and Reviews*, vol. 40, no. 5, pp. 567–579, 2010.

[34] S. Bajikar, Trusted platform module (TPM) based security on notebook pcs—white paper, Mobile Platform Group, Intel Corporation, 2002.

[35] M. Abe and T. Okamoto, "Provably secure partially blind signatures," in *Proceedings of the 20th Annual International Cryptology Conference on Advances in Cryptology (CRYPTO '00)*, pp. 271–286, Springer, 2000.

[36] A. Juels, M. Luby, and R. Ostrovsky, "Security of blind digital signatures," in *Proceedings of the 17th Annual International Cryptology Conference on Advances in Cryptology (CRYPTO '97)*, pp. 150–164, Springer, 1997.

[37] M. Bellare, C. Namprempre, D. Pointcheval, and M. Semanko, "The one-more-RSA-inversion problems and the security of chaum's blind signature scheme," *Journal of Cryptology*, vol. 16, no. 3, pp. 185–215, 2003.

[38] S. Brands, "Untraceable off-line cash in wallets with observers (extended abstract)," *CRYPTO*, pp. 302–318, 1993.

[39] Y. Hanatani, Y. Komano, K. Ohta, and N. Kunihiro, "Provably secure electronic cash based on blind multisignature schemes," *Financial Cryptography*, vol. 4107, pp. 236–250, 2006.

[40] C. Popescu, "An off-line electronic cash system with revokable anonymity," in *Proceedings of the 12th IEEE Mediterranean Electrotechnical Conference*, pp. 763–767, May 2004.

[41] A. Menezes, P. van Oorschot, and S. Vanstone, *Handbook of Applied Cryptography*, CRC Press, New York, NY, USA, 1997.

14

Average Gait Differential Image Based Human Recognition

Jinyan Chen[1] and Jiansheng Liu[2]

[1] *School of Computer Software, Tianjin University, Tianjin 300072, China*
[2] *College of Science, Jiangxi University of Science and Technology, Ganzhou 330200, China*

Correspondence should be addressed to Jinyan Chen; chenjinyan@tju.edu.cn

Academic Editor: Fei Yu

The difference between adjacent frames of human walking contains useful information for human gait identification. Based on the previous idea a silhouettes difference based human gait recognition method named as average gait differential image (AGDI) is proposed in this paper. The AGDI is generated by the accumulation of the silhouettes difference between adjacent frames. The advantage of this method lies in that as a feature image it can preserve both the kinetic and static information of walking. Comparing to gait energy image (GEI), AGDI is more fit to representation the variation of silhouettes during walking. Two-dimensional principal component analysis (2DPCA) is used to extract features from the AGDI. Experiments on CASIA dataset show that AGDI has better identification and verification performance than GEI. Comparing to PCA, 2DPCA is a more efficient and less memory storage consumption feature extraction method in gait based recognition.

1. Introduction

With the development of information and Internet technology, it is very necessary to authenticate and authorize human securely. The rapid growth of e-commerce also needs a reliable identification method to ensure safety transaction. As a promising authentication method, biometrics is attracting more and more attention. Biometrics overcomes the inherent flaws and limitations of conventional identification technology and brings a highly secure identification and authentication method. Traditional biometrical resources include fingerprint, face, and iris, which have been widely used for authentication. However, these biometrical features have the following disadvantage. (1) These features cannot be taken in a relative long distance. (2) User's cooperation is required to get good results. As a new biometrics method, gait based human identification overcomes the above limitation and is attracting more and more researchers.

Human gait is the manner of one walking, which was firstly studied in medical field. Doctors analyzed the human gait to find out whether patients had health problems [1, 2]. Later researchers found that just like fingerprint and iris, almost everyone has his distinctive walking style [3, 4]. So it was believed that gait could also be used as a biological feature to recognize the person. Although suffering from clothing, shoes, view angel, or environmental context, human gait is still a promising identification method.

Human walking can be considered as an images sequence; however, most of the current model-free gait based identification methods extract features from image sequence without considering its contained spatiotemporal information. The method proposed in this paper focuses on the difference among the images sequence while constructing the feature image. The procedure can be described as follows. The silhouettes were normalized to the same height and aligned by the centroid. Then the difference between two adjacent silhouettes was accumulated to get the average gait differential image (AGDI) which is used as the feature image of one walking. Two-dimensional principal component analysis is used to extract feature from AGDI.

2. Related Work and Our Contribution

Usually recognition based on human gait includes several different approaches like walking, running, and jumping. In this paper we would like to restrict the recognition to walking. Currently human gait recognition can be divided into two categories: model-based methods and motion-based ones.

Model-based approaches aim to describe human movement using a mathematical model. Cunado et al. [5] used Hough transform to extract the positions of arms, legs, and torso and then use articulated pendulum to match those moving body parts. Yoo et al. [6] divided the body into head, neck, waist, leg, and arm by image segmentation and then got the moving curves of these body parts, respectively. Lee and Grimson [7] used 7 ellipses to model the human body and applied the ellipses' movement features to identify human. Yam et al. [8, 9] used dynamically coupled oscillator to describe and analyze the walking and running style of a person. Tafazzoli and Safabakhsh [10] constructed movements model based on anatomical proportions; then, Fourier transform was used to analyze human walking.

Model-free methods focused on the statistics information derived from the human gait. Cheng et al. [11] took HMM and manifold to analyze the relationship between the human and their gait images. Chen et al. [12] used parallel HMM to describe the features of human gait. Kale et al. [13] used "frieze" patterns to get features from image sequence and use them to identify a human. Murase and Sakai [14] speeded up the comparison of human gait by parametric eigenspace representation. Little and Boyd [15] derived scale-independent scalar features from optical flow information of walking figures to recognize individuals. Wang et al. [16] extracted feature by unwrapping the outer contour of silhouette and use PCA to reduce the dimension of the feature. Lee et al. [17] adopted product of Fourier coefficients as a distance measure between contours to recognize gait. Hu [18] combined the enhanced Gabor (EG) representation of the gait energy image and the regularized local tensor discriminate analysis (RLTDA) method in human identification. Hong et al. [19] proposed probabilistic framework to identify a human. Wang et al. [20] proposed spatiotemporal information analysis to get the features of human walking. Collins et al. [21] extract key frames from the image sequence and compare the key frames similarity by normalized correlation. Sarkar et al. [22] estimated the similarity between the gallery image sequence and the probe image sequence by directly computing the correlation between the frame pairs. Chen [23] proposes image correlation based human identification method.

Our method is similar to gait energy image (GEI) proposed by Yu et al. [24], Han and Bhanu [25], and frame difference energy image (FDEI) proposed by Chen et al. [26]. The major difference lies in the approach to generate the feature image. GEI is obtained by directly adding up every normalized silhouette. FDEI is obtained by taking the difference from every adjacent two frames and then combined with the "denoised" GEI. In this paper the difference between every two frames will be accumulated to generate average gait differential image. We also enhanced the feature extraction method by using 2DPCA, which has been used in the application of face recognition [27].

In comparison with the works of state of the art, the contributions of this paper are as follows.

Gait Representation Method. We propose a new gait feature representation which is called average gait differential image.

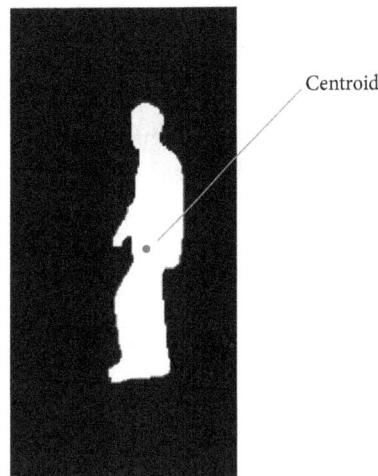

FIGURE 1: The centroid of a silhouette.

Comparing to GEI, our method has the advantage of better performance.

Feature Extraction Method. Two-dimensional principal component analysis (2DPCA) is used to extract features from AGDI, which can be more efficient and save more storage comparing to the widely used one-dimensional principal component analysis (PCA).

3. Average Gait Differential Image (AGDI) Representation

3.1. The Construction of AGDI. The construction of average differential image can be expressed in the following steps.

Silhouette Segmentation. Gauss model is used to get the background model from the original images sequence. To eliminate the effect of noise, every image is blurred by Gauss filter. The method proposed by Wang et al. [16] is used to extract walking object from the original images.

Normalization. To exclude the distance effect, every silhouette is normalized to the same height using bicubic interpolation.

Alignment and Subtraction. We define the centroid (x_c, y_c) of a silhouette as follows:

$$x_c = \frac{1}{n}\sum_{i=1}^{n} x_i,$$

$$y_c = \frac{1}{n}\sum_{i=1}^{n} y_i,$$

(1)

where n is the number of pixels in the silhouette. Figure 1 shows the centroid of a silhouette.

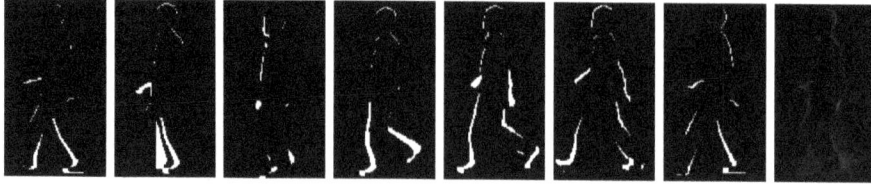

FIGURE 2: Differential images and average gait differential image.

Suppose that I_j and I_{j+1} are two adjacent images aligned by the centroid; the gait differential image D_j can then be defined as follows:

$$D_j(x, y) = \begin{cases} 0 \text{ if } I_j(x, y) = I_{j+1}(x, y) \\ 1 \text{ if } I_j(x, y) \neq I_{j+1}(x, y), \end{cases} \quad (2)$$

where j is the frame number in the image sequence and x and y are values in the 2D image coordinate.

Get the Average Gait Differential Image. By overlapping all the differential images of one human gait cycle, we can get the following average gait differential image:

$$G(x, y) = \frac{1}{N-1} \sum_{j=1}^{N-1} D_j(x, y), \quad (3)$$

where N is the number of frames in the complete gait cycle(s) of a silhouette sequence. Figure 2 show the differential images in a gait cycle and the average gait differential image, respectively.

3.2. Feature Extraction.

Although intheprevious section we have compressed the human gait features into one image, the dimensionality of the average gait differential image is still very large. The most commonly used dimensional reduction method is principal component analysis (PCA). In traditional PCA method, every two-dimensional image must be transformed into one-dimensional vector, leading to a covariance matrix with large size. This large matrix will use massive memory storage and is difficult to be evaluated accurately.

To reduce memory storage and speed up the calculation, this paper adopts the two-dimensional principal component analysis (2DPCA) to reduce the dimensionality, which was first proposed by Yang et al. [27] in the recombination of human face. Our final target is to project average gait differential image G, a $m \times n$ random matrix, onto a m-dimension projected vector Y which is called the projected feature vector of image G by the following linear transformation [27]:

$$Y = GW, \quad (4)$$

where W denotes a n-dimensional unitary column vector. To preserve the features of G, W should make Y have the maximum scatter. We define S_y as the scatter of Y [27] as follows:

$$\begin{aligned} S_y &= E(Y - E(Y))(Y - E(Y))^T \\ &= E(GW - E(GW))(GW - E(GW))^T \quad (5) \\ &= E(G - E(G))WW^T(G - E(G))^T. \end{aligned}$$

The trace of S_y can be expressed as

$$\text{tr}(S_y) = W^T \left(E(G - E(G))^T (G - E(G)) \right) W. \quad (6)$$

Here, we can define the image covariance matrix as

$$C_t = E(G - E(G))^T (G - E(G)). \quad (7)$$

In this paper, average gait differential image for each individual $(1, 2, \ldots, M)$ is expressed as $G_1, G_2 \cdots G_M$, and then C_t can be calculated by

$$C_t = \sum_{i=1}^{M} \left(G_i - \overline{G} \right)^T \left(G_i - \overline{G} \right). \quad (8)$$

Our target is to find a series of W_{opt} in formula (6) to make $\text{tr}(S_y)$ have the maximum value. According to [13], the optimal projection axis W_{opt} is the unitary orthogonal eigenvector of C_t corresponding to the largest eigenvalue. We define the first d unitary orthogonal eigenvector as $W_1, W_2, \ldots, W_d$; that is,

$$\{W_1, \ldots W_d\} = \arg \max \left(W^T C_t W \right)$$

$$W_i W_j = 0, \quad i \neq j, \, i, j = 1, \ldots d, \quad (9)$$

$$W_i W_j = 1, \quad i = j, \, j = 1, \ldots d.$$

The first d optimal projection vectors, $W_1, \ldots, W_d$, are used to extract features from the average different images. That is to say, given an average gait differential image X, let

$$Y_k = GW_k, \quad k = 1, 2 \ldots d. \quad (10)$$

Then we get a series of projected feature vectors, $Y_1 \cdots Y_d$, which are different from those scalar counterparts obtained from PCA. By using 2DPCA, the original $m \times n$ image is projected to a $m \times d$ $(d \leq n)$ feature matrix Y as

$$Y = \begin{bmatrix} Y_1 \\ \vdots \\ Y_d \end{bmatrix}. \quad (11)$$

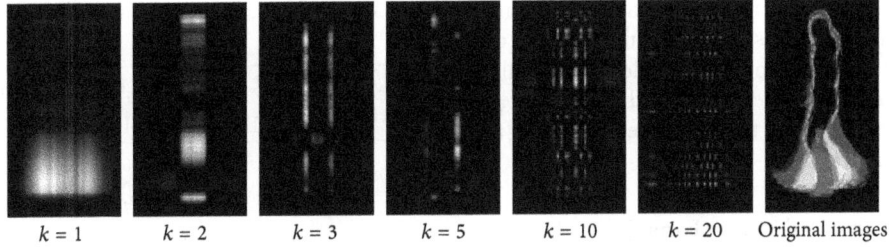

$$k = 1 \qquad k = 2 \qquad k = 3 \qquad k = 5 \qquad k = 10 \qquad k = 20 \qquad \text{Original images}$$

FIGURE 3: The reconstructed subimages ($k = 1, 2, 3, 5, 10, 20$) and the original images.

The distance between two feature matrixes is defined as

$$d\left(Y\left(i\right), Y\left(j\right)\right) = \sum_{k=1}^{d} \left\| Y_k^{(i)} - Y_k^{(j)} \right\|, \qquad (12)$$

where $\left\| Y_k^{(i)} - Y_k^{(j)} \right\|$ means the Euclidean distance between two vectors.

3.3. Identification and Verification.

Following the pattern proposed by Sarkar et al. [22], we evaluate performance for both identification and verification scenarios.

In the scenario of identification, every images sequence in the gallery (training set) is transformed to a $m \times d$ feature matrix $Y(i)$ by the method described in Section 3.2. Given a probe silhouette sequence, its transformed feature matrix is defined as P. This probe P is assigned to person k by using the nearest neighbor method:

$$d\left(Y\left(k\right), P\right) = \min_i d\left(Y\left(i\right) - P\right). \qquad (13)$$

In the scenario of verification, the similarity between two feature matrixes is defined as the negative of distance; that is,

$$\text{Sim}\left(Y\left(i\right), Y\left(j\right)\right) = -\sum_{k=1}^{d} \left\| Y_k^{(i)} - Y_k^{(j)} \right\|. \qquad (14)$$

In this paper the similarity between a probe, P_j, and $Y(i)$ in the gallery is defined as z-normed similarity [28]:

$$\text{Sim}\left(P_j, Y\left(i\right)\right) = \frac{\text{Sim}\left(P_j, Y\left(i\right)\right) - \text{Mean}_i \text{Sim}\left(P_j, Y\left(i\right)\right)}{\text{s.d.}_i \text{Sim}\left(P_j, Y\left(i\right)\right)}, \qquad (15)$$

where s.d. is standard deviation.

FAR (false acceptance rate), FRR (false rejection rate), and EER (equal error rate) are used to evaluate the performance of verification [22].

3.4. DPCA-Based Average Gait Differential Image Reconstruction.

In PCA the principal components and eigenvectors can be combined to reconstruct the original matrix. Similarly, 2DPCA can also be used to reconstruct an average gait differential silhouette.

Suppose that the eigenvectors corresponding to the largest d eigenvectors of C_t are $W_1, \dots, W_d$; that is,

$$Y_k = GW_k, \quad k = 1, \dots, d,$$

$$[Y_1 \cdots Y_d] = X [W_1 \cdots W_d]. \qquad (16)$$

According to formula (9), $W_1 \cdots W_d$ are normal orthogonal vectors so the new reconstructed image $\widetilde{G}$ can be expressed as

$$\widetilde{G} = G [W_1 \cdots W_d][W_1 \cdots W_d]^T = [Y_1 \cdots Y_d][W_1 \cdots W_d]^T$$

$$= \sum_{k=1}^{d} Y_k W_k^T. \qquad (17)$$

4. Experiments and Analysis

4.1. Data and Parameters.

CASIA gait database (Dataset B) [29], one of the largest gait databases in gait-research field, is used in the following experiment. Dataset B consists of 124 subjects (93 males and 31 females) captured from 11 view angles (ranging from 0 to 180° degree with view angle interval of 18). For every person there are six normal walking sequences (named normal-01 $\cdots$ normal-06) conducted from every view angle. Every walking sequence contains 3–8 gait cycles (about 40–100 frames). The video frame size is 320×240 pixels, and the frame rate is 25 fps. We use all the 124 objects in Dataset B to carry out our experiments.

In all the following experiments, 2DPCA method was used to get features from the images and 20 eigenvectors corresponding to the first 20 eigenvalues are used to produce features ($d = 20$). The size of original image is 240×320 except for special declaration.

For each person, from every view angle, we select the 39 frames from sequence normal-01 as the training data (gallery) and 13 frames (except for special declaration) from sequence normal-02 as the test data (probe).

For every view angle, each time we leave one training image sequence out and use the remainder as the training set. In the scenario of identification we calculate the distance between the probe corresponding to the leave out training image sequence and the 124 classes (including the leave out image sequence). In the scenario of verification, we calculate the similarity between the probe corresponding to the leave out training image sequence and the 124 classes (including the leave out image sequence).

4.2. The Reconstruction of Subimage.

Formula (17) indicates that we can reconstruct the subimage from the W_k and Y_k. Figure 3 shows the result of the reconstruction. For the consideration of illustration we normalize the brightness of every image into the range of 0–255.

As showed in Figure 3, the first and the second ($k = 1$, $k = 2$ in formula (17)) subimages corresponding to large eigenvectors of C_t contain the most energy of the original images. With the increase of k, the subimage contains more detailed information.

We also demonstrate the eigenvalue calculated by 2DPCA. Figure 4 shows the magnitude of the eigenvalues that quickly converges to zero.

4.3. Performance Evaluation

4.3.1. Comparison of AGDI and GEI.

We compare the performance of our AGDI base method with that of gait energy image based method (In this paper, we use real template for GEI method [25].). Table 1 shows the rank 1 and rank 5 identification rates comparing with GEI.

To compare the performance of verification, we also evaluate the FAR (false acceptance rate) and FRR (false rejection rate) for AGDI and GEI. The ROC (receiver operating characteristic) curves under view angles 0°, 90°, and 180° are shown in Figures 5(a)–5(c). The comparison of EERs (equal error rate) is shown in Figure 5(d).

From Table 1 and Figure 5, we can see that almost under every view angle AGDI has better performance comparing to GEI (except that it is comparable under 0 view angel).

As also can be seen in Table 2, the best performance was obtained from the walking sequence taken from 0°, 90°, and 180°, while the worst was obtained from the walking sequence taken from 36° and 54°. This is probably due to the least visual deformation in the former degrees but more in the latter ones.

4.3.2. The Effect of Images Amount.

From the definition of AGDI (formula (3)) we can see that the AGDI image is the average value of differential images. It should be expected that the use of more images as sample would contribute to a more precise result. To demonstrate this effect, a test was conducted by selectively choosing 13, 26, and 39 (approximately corresponding to 1, 2, and 3 gait cycles) images from 90 degree in sequences normal 01-02 as test dataset probe. Figure 5 shows the experimental result.

Indeed in Figure 5 the performance of 26 and 39 images is much better than that of 13.

4.3.3. Comparison of 2DPCA and PCA.

We also design an experiment to compare the performance of 2DPCA and PCA, which were applied in the step of feature extraction,

TABLE 1: Comparison of identification performance of AGDI and GEI.

View angle	Rank 1 performance		Rank 5 performance	
	AGDI	GEI	AGDI	GEI
0°	72%	68%	88%	89%
18°	54%	37%	73%	60%
36°	35%	22%	51%	44%
54°	44%	26%	55%	45%
72°	66%	44%	86%	66%
90°	81%	77%	93%	90%
108°	78%	62%	92%	85%
126°	46%	36%	73%	53%
144°	49%	34%	72%	52%
162°	59%	35%	75%	48%
180°	88%	84%	94%	93%

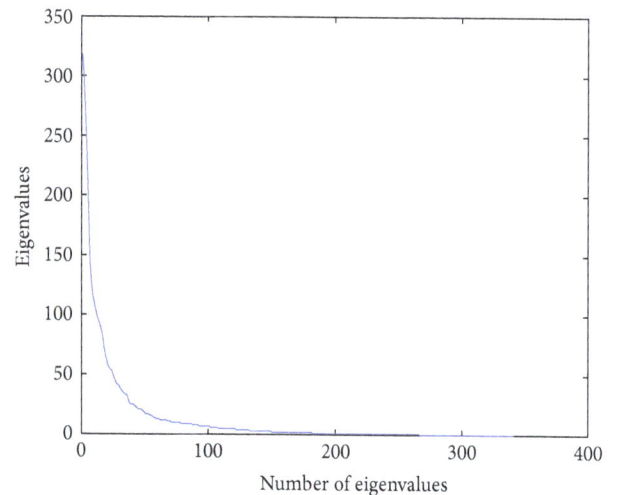

FIGURE 4: The magnitude of eigenvalue.

respectively. The data set view angel is 90° and every frame is resized to 120×160.

As illustrated in Figure 7, the performance of 2DPCA, achieving the maximum at about 25 dimensions, is much better than PCA.

The key step for both PCA and 2DPCA is to get the eigenvalue and eigenvector from the covariance matrix Ct. For PCA method, every line of the covariance matrix corresponds to an image, as does the whole covariance matrix for the 2DPCA. That is, if the image size is $m \times n$, for PCA, the covariance matrix will be an $(m \times n) \times (m \times n)$ matrix, while for 2DPCA it is just an $m \times m$ matrix. We resize the silhouette to different sizes and compare the CPU time of PCA and 2DPCA for the step of feature extraction.

From Table 2 we can see that 2DPCA is more efficient than PCA, especially when the image is large.

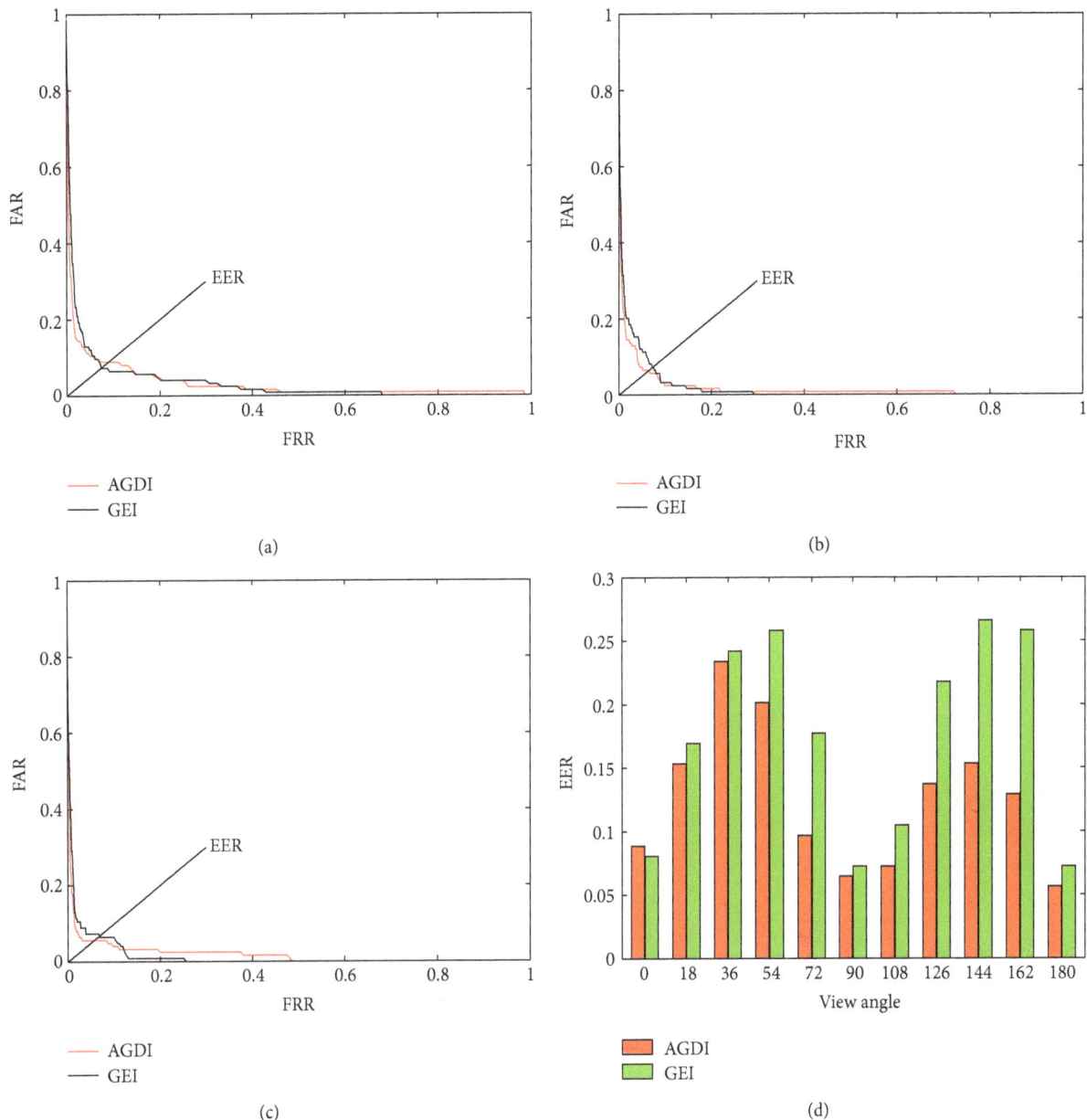

FIGURE 5: (a)–(c) The comparison of ROC curves of AGDI and GEI with view angles 0°, 90°, and 180°, respectively. (d) The comparison of EERs of AGDI and GEI with view angles 0°–180°.

TABLE 2: Comparison of CPU time (ms) for PCA and 2DPCA feature extraction (CPU: Intel Core i3 2.30 GHz; RAM: 4 GB).

Feature extraction method	Image size					
	32×24	64×48	96×72	128×96	160×120	192×144
2DPCA	17 ms	47 ms	105 ms	167 ms	257 ms	431 ms
PCA	117 ms	318 ms	1273 ms	4288 ms	8896 ms	17876 ms

5. Conclusions

An average gait differential image based human recognition method is proposed in this paper (Figure 6). The Kernel idea of AGDI is to apply the average of differential image as the feature image and use the two-dimensional principal component analysis to extract features. Experiments on CASIA dataset show the following. (1) Comparing to GEI, AGDI method achieves better identification and verification performance. (2) Comparing to PCA, 2DPCA is more efficient and needs lower memory storage. (3) The 0, 90, and 180 degrees silhouettes are more fit to AGDI base recognition.

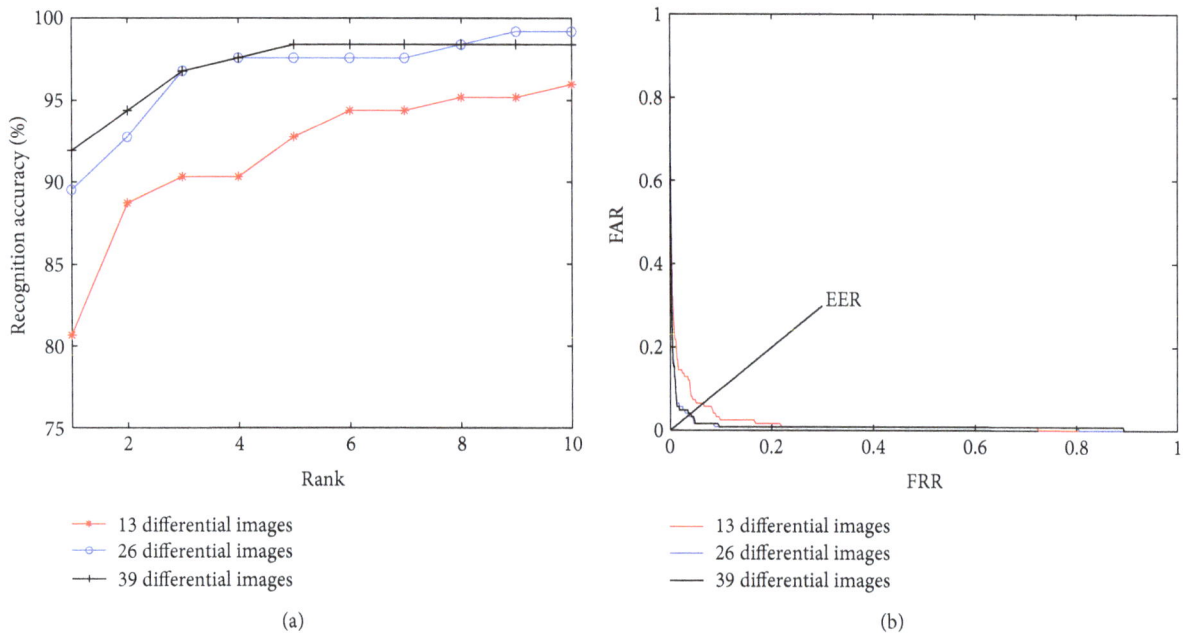

FIGURE 6: (a) Recognition accuracy of different probe sizes. (b) The comparison of ROC curves of different probe sizes.

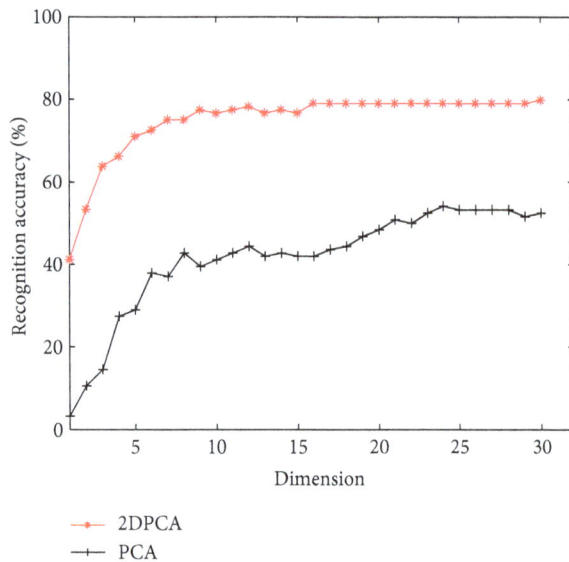

FIGURE 7: The rank 1 recognition accuracy comparison of 2DPCA and PCA.

Conflict of Interests

The authors declare that they have no financial or personal relationships with other people or organizations that can inappropriately influence their work. There are no professional or other personal interests of any nature or kind in any product, service, and/or company that could be construed as influencing the position presented in, or the review of, the paper.

Acknowledgments

This work is funded by Ph.D. Programs Foundation of Ministry of Education of China (no. 20100032120011). Dataset used in this paper is provided by Institute of Automation, Chinese Academy of Sciences [29].

References

[1] M. W. Whittle, "Clinical gait analysis: a review," *Human Movement Science*, vol. 15, no. 3, pp. 369–387, 1996.

[2] S. Lobet, C. Detrembleur, F. Massaad, and C. Hermans, "Three-dimensional gait analysis can shed new light on walking in patients with haemophilia," *The Scientific World Journal*, vol. 2013, Article ID 284358, 7 pages, 2013.

[3] S. V. Stevenage, M. S. Nixon, and K. Vince, "Visual analysis of gait as a cue to identity," *Applied Cognitive Psychology*, vol. 13, no. 6, pp. 513–526, 1999.

[4] C. Ben Abdelkader, R. Cutler, and L. Davis, "Stride and cadence as a biometric in automatic person identification and verification," in *Proceedings of the 5th IEEE International Conference on Automatic Face and Gesture Recognition (FGR '02)*, 2002.

[5] D. Cunado, M. S. Nixon, and J. N. Carter, "Automatic extraction and description of human gait models for recognition purposes," *Computer Vision and Image Understanding*, vol. 90, no. 1, pp. 1–41, 2003.

[6] J. H. Yoo, D. Hwang, and M. S. Nixon, "Gender classification in human gait using support vector machine," in *Advanced Concepts for Intelligent Vision Systems: 7th International Conference, ACIVS 2005, Antwerp, Belgium, September 20-23, 2005. Proceedings*, vol. 3708 of *Lecture Notes in Computer Science*, pp. 138–145, Springer, 2005.

[7] L. Lee and W. E. L. Grimson, "Gait analysis for recognition and classification," in *Proceedings of 5th IEEE International*

Conference on Automatic Face Gesture Recognition, pp. 155–162, 2002.

[8] C. Y. C. Yam, M. S. Nixon, and J. N. Carter, "Automated person recognition by walking and running via model-based approaches," *Pattern Recognition*, vol. 37, no. 5, pp. 1057–1072, 2004.

[9] C. Yam, M. S. Nixon, and J. N. Carter, "Gait recognition by walking and running: a model-based approach," in *Proceedings of the 5th Asian Conference on Computer Vision*, Melbourne, Australia, 2002.

[10] F. Tafazzoli and R. Safabakhsh, "Model-based human gait recognition using leg and arm movements," *Engineering Applications of Artificial Intelligence*, vol. 23, no. 8, pp. 1237–1246, 2010.

[11] M.-H. Cheng, M.-F. Ho, and C.-L. Huang, "Gait analysis for human identification through manifold learning and HMM," *Pattern Recognition*, vol. 41, no. 8, pp. 2541–2553, 2008.

[12] C. Chen, J. Liang, H. Zhao, H. Hu, and J. Tian, "Factorial HMM and parallel HMM for gait recognition," *IEEE Transactions on Systems, Man and Cybernetics C: Applications and Reviews*, vol. 39, no. 1, pp. 114–123, 2009.

[13] A. Kale, A. Roy-Chowdhury, and R. Chellappa, "Gait-based human identification from a monocular video sequence," in *Handbook on Pattern Recognition and Computer Vision*, C. H. Cheng and P. S. P. Wang, Eds., World Scientific Publishing Company, 3rd edition, 2005.

[14] H. Murase and R. Sakai, "Moving object recognition in eigenspace representation: gait analysis and lip reading," *Pattern Recognition Letters*, vol. 17, no. 2, pp. 155–162, 1996.

[15] J. J. Little and J. E. Boyd, "Recognizing people by their gait: the shape of motion," *Videre: Journal of Computer Vision Research*, vol. 1, no. 2, 1998.

[16] L. Wang, T. Tan, H. Ning, and W. Hu, "Silhouette analysis-based gait recognition for human identification," *IEEE Transactions on Pattern Analysis and Machine Intelligence*, vol. 25, no. 12, pp. 1505–1518, 2003.

[17] C. P. Lee, A. W. C. Tan, and S. C. Tan, "Gait recognition via optimally interpolated deformable contours," *Pattern Recognition Letters*, vol. 34, no. 6, pp. 663–669, 2013.

[18] H. Hu, "Enhanced gabor feature based classification using a regularized locally tensor discriminant model for multiview gait recognition," *IEEE Transactions on Circuits and Systems for Video Technology*, vol. 23, no. 7, pp. 1274–1286, 2013.

[19] S. Hong, H. Lee, and E. Kim, "Probabilistic gait modelling and recognition," *Computer Vision*, vol. 7, no. 1, pp. 56–70, 2013.

[20] C. Wang, J. Zhang, L. Wang, J. Pu, and X. Yuan, "Human identification using temporal information preserving gait template," *IEEE Transactions on Pattern Analysis and Machine Intelligence*, vol. 34, no. 11, pp. 2164–2176, 2012.

[21] R. T. Collins, R. Gross, and S. Jianbo, "Silhouette-based human identification from body shape and gait," in *Proceedings of 5th IEEE International Conference on Automatic Face Gesture Recognition*, 2002.

[22] S. Sarkar, P. J. Phillips, Z. Liu, I. R. Vega, P. Grother, and K. W. Bowyer, "The humanID gait challenge problem: data sets, performance, and analysis," *IEEE Transactions on Pattern Analysis and Machine Intelligence*, vol. 27, no. 2, pp. 162–177, 2005.

[23] J. Chen, "Gait correlation analysis based human identification," *The Scientific World Journal*, vol. 2014, Article ID 168275, 8 pages, 2014.

[24] S. Yu, T. Tan, K. Huang, K. Jia, and X. Wu, "A study on gait-based gender classification," *IEEE Transactions on Image Processing*, vol. 18, no. 8, pp. 1905–1910, 2009.

[25] J. Han and B. Bhanu, "Individual recognition using gait energy image," *IEEE Transactions on Pattern Analysis and Machine Intelligence*, vol. 28, no. 2, pp. 316–322, 2006.

[26] C. Chen, J. Liang, H. Zhao, H. Hu, and J. Tian, "Frame difference energy image for gait recognition with incomplete silhouettes," *Pattern Recognition Letters*, vol. 30, no. 11, pp. 977–984, 2009.

[27] J. Yang, D. Zhang, A. F. Frangi, and J.-Y. Yang, "Two-dimensional PCA: a new approach to appearance-based face representation and recognition," *IEEE Transactions on Pattern Analysis and Machine Intelligence*, vol. 26, no. 1, pp. 131–137, 2004.

[28] P. J. Phillips, P. Grother, R. Micheals, D. M. Blackburn, E. Tabassi, and M. Bone, "Face recognition vendor test 2002," in *Proceedings of the IEEE International Workshop on Face Recognition*, 2003.

[29] "CASIA Gait Database," 2009, http://www.sinobiometrics.com/.

15

Efficient Certificate-Based Signcryption Secure against Public Key Replacement Attacks and Insider Attacks

Yang Lu and Jiguo Li

College of Computer and Information Engineering, Hohai University, No. 8, Focheng Xi Road, Jiangning District, Nanjing, Jiangsu 211100, China

Correspondence should be addressed to Yang Lu; luyangnsd@163.com

Academic Editor: Tianjie Cao

Signcryption is a useful cryptographic primitive that achieves confidentiality and authentication in an efficient manner. As an extension of signcryption in certificate-based cryptography, certificate-based signcryption preserves the merits of certificate-based cryptography and signcryption simultaneously. In this paper, we present an improved security model of certificate-based signcryption that covers both public key replacement attack and insider security. We show that an existing certificate-based signcryption scheme is insecure in our model. We also propose a new certificate-based signcryption scheme that achieves security against both public key replacement attacks and insider attacks. We prove in the random oracle model that the proposed scheme is chosen-ciphertext secure and existentially unforgeable. Performance analysis shows that the proposed scheme outperforms all the previous certificate-based signcryption schemes in the literature.

1. Introduction

Public key cryptography (PKC) is an important technique to realize network and information security. In traditional PKC, a public key infrastructure (PKI) is used to provide an assurance to the users about the relationship between a public key and the holder of the corresponding private key by certificates. However, the need for PKI-supported certificates is considered the main difficulty in the deployment and management of traditional PKC. To simplify the management of the certificates, Shamir [1] introduced the concept of identity-based cryptography (IBC) in which the public key of each user is derived directly from his identity, such as an IP address or an e-mail address, and the corresponding private key is generated by a trusted third party called private key generator (PKG). The main practical benefit of IBC lies in the reduction of need for public key certificates. However, if the KGC becomes dishonest, it can impersonate any user using its knowledge of the user's private key. This is due to the key escrow problem inherent in IBC. In addition, private keys must be sent to the users over secure channels, so private key distribution in IBC becomes a very daunting task.

To fill the gap between traditional PKC and IBC, Al-Riyami and Paterson [2] proposed a new paradigm called certificateless public key cryptography (CL-PKC) in Asiacrypt 2003. CL-PKC eliminates the key escrow problem inherent in IBC. At the same time, it preserves the advantage of IBC which is the absence of certificates and their heavy management overhead. In CL-PKC, a trusted third party called key generating center (KGC) is involved in the process of issuing a partial secret key for each user. The user independently generates its public/private key pair and combines the partial secret key from the KGC with its private key to generate the actual decryption key. By way of contrast to the PKG in IBC, the KGC does not have access to the user's decryption key. Therefore, CL-PKC solves the key escrow problem. However, as partial secret keys must be sent to the users over secure channels, CL-PKC suffers from the distribution problem.

In Eurocrypt 2003, Gentry [3] introduced the notion of certificate-based cryptography (CBC). CBC provides an implicit certification mechanism for a traditional PKI and allows for a periodical update of certificate status. As in traditional PKC, each user in CBC generates his own public/private key pair and requests a certificate from a trusted

TABLE 1: Properties of the related public key cryptosystems.

	Do not require trusted third party	Implicit certificates	Key escrow free	Key distribution free
Traditional PKC	×	×	√	√
IBC	×	√	×	×
CL-PKC	×	√	√	×
CBC	×	√	√	√

third party called certifier. The certificate will be pushed only to the owner of the public/private key pair and act as a partial decryption key or a partial signing key. This additional functionality provides an efficient implicit certificate mechanism. For example, in the encryption scenario, a receiver needs both his private key and certificate to decrypt a ciphertext sent to him, while the message sender need not be concerned about the certificate revocation problem. The feature of implicit certification allows us to eliminate third-party queries for the certificate status and simply the public key revocation problem so that CBC does not need infrastructures like CRL and OCSP. Therefore, CBC can be used to construct an efficient PKI requiring fewer infrastructures than the traditional one. Although CBC may be inefficient when a certifier has a large number of users, this problem can be overcome by using *subset covers* [3]. Furthermore, there are no key escrow problem (since the certifier does not know the private keys of users) and key distribution problem (since the certificates need not be kept secret) in CBC.

Table 1 summarizes the comparison of the above cryptosystems.

Since its advent, CBC has attracted great interest in the research community and many schemes have been proposed, including many encryption schemes (e.g., [4–10]) and signature schemes (e.g., [11–16]). As an extension of the signcryption [17] in CBC, Li et al. [18] introduced the concept of certificate-based signcryption (CBSC) that provides the functionalities of encryption and signature simultaneously. As far as we know, there exist three CBSC schemes in the literature so far. In [18], Li et al. proposed the first CBSC scheme based on Chen and Malone-Lee's identity-based signcryption scheme [19]. However, they did not give a formal proof of their security claim. A subsequent paper by Luo et al. [20] proposed the second CBSC scheme alone with a security model of CBSC. Recently, Li et al. [21] proposed a publicly verifiable CBSC scheme that is provably secure in the random oracle model.

Our Motivations and Contributions. In this paper, we focus on the construction of a CBSC scheme that resists both the public key replacement attacks and the insider attacks.

Public key replacement attack was first introduced into CL-PKC by Al-Riyami and Paterson [2]. In this attack, an adversary who can replace a user's public key with a value of its choice dupes any other third parties to encrypt messages or verify signatures using a false public key. It seems that this attack does not have effect on CBC since a certifier is employed for issuing a certificate for each user. Unfortunately, some previous research works [13, 16, 22] have demonstrated that it does. In CBC, the certifier does issue the certificates.

However, as introduced above, CBC adopts the implicit certificate mechanism so that only the owner of a certificate needs to check the validity of his certificate and others need not be concerned about the status of his certificate. Thus, a malicious user is able to launch the public key replacement attack against an ill-designed certificate-based cryptographic scheme. We observe that Luo et al.'s CBSC scheme [20] is insecure under this attack. The concrete attack can be found in Section 4 of this paper.

Insider security [23] refers to the security against the attacks made by the insider (i.e., the sender or the receiver). It requires that, even if a sender's private key is compromised, an attacker should not be able to designcrypt the message generated by the sender and, even with a receiver's private key, an attacker should not be able to forge a valid signcryption as if generated by the same sender. In contrast to outsider security [23] that refers to the security against the attacks made by the outsider (i.e., any third party except the sender and the receiver), insider security can provide the stronger security for signcryption schemes [24, 25]. Therefore, it has been accepted as a necessary security requirement for a signcryption scheme to achieve. However, none of the previous constructions of CBSC [18, 20, 21] has considered insider security. The previous security models of CBSC [20, 21] only cover the case where the CBSC scheme is attacked by the outsiders. Actually, the public key replacement attack presented in Section 4 also shows that Luo et al.'s CBSC scheme [20] fails in providing insider security.

The main contributions of this paper are as follows.

(1) We extend previous works by proposing an improved security model for CBSC that accurately models both the public key replacement attacks and the insider attacks. We show that Luo et al.'s CBSC scheme [20] is insecure in our security model.

(2) We develop a new CBSC scheme and formally prove its security in our improved security model. In the random oracle, we prove that the proposed scheme is chosen-ciphertext secure and existentially unforgeable. To the best of our knowledge, it is the first signcryption scheme that achieves security under both the public key replacement attacks and the insider attacks in the certificate-based cryptographic setting. Furthermore, compared with the previous CBSC schemes, our scheme enjoys better performance, especially in the computation efficiency.

Paper Organization. The rest of this paper is organized as follows. In the next section, we briefly review some preliminaries required in this paper. In Section 3, we present an improved security model of CBSC. In Section 4, we show that Luo et al.'s CBSC scheme is insecure in our security model. The proposed CBSC scheme is described and analyzed in Section 5. Finally, we draw our conclusions in Section 6.

2. Preliminaries

Let k be a security parameter and p a k-bit prime number. Let G be an additive cyclic group of prime order p and G_T a multiplicative cyclic group of the same order, and let P be a generator of G. A bilinear pairing is a map $e : G \times G \to G_T$ satisfying the following properties.

(i) Bilinearity: for all P_1, $P_2 \in G$, and all a, $b \in Z_p^*$, we have $e(aP_1, bP_2) = e(P_1, P_2)^{ab}$.

(ii) Nondegeneracy: $e(P, P) \neq 1$.

(iii) Computability: for all P_1, $P_2 \in G$, $e(P_1, P_2)$ can be efficiently computed.

The security of our CBSC scheme is based on the following hard problems.

Definition 1. The computational Diffie-Hellman (CDH) problem in G is, given a tuple $(P, aP, bP) \in G^3$ for unknown $a, b \in Z_p^*$, to compute $abP \in G$.

Definition 2 (see [26]). The bilinear Diffie-Hellman (BDH) problem in (G, G_T) is, given a tuple $(P, aP, bP, cP) \in G^4$ for unknown $a, b, c \in Z_p^*$, to compute $e(P, P)^{abc} \in G_T$.

Definition 3 (see [27]). The collusion attack algorithm with q-traitors (q-CAA) problem in G is given a tuple $(P, \alpha P, (\omega_1 + \alpha)^{-1}P, \ldots, (\omega_q + \alpha)^{-1}P, \omega_1, \ldots, \omega_q) \in G^{q+2} \times (Z_p^*)^q$ for unknown $\alpha \in Z_p^*$, to compute $(\omega^* + \alpha)^{-1}P$ for some value $\omega^* \notin \{\omega_1, \ldots, \omega_q\}$.

Definition 4 (see [28]). The modified bilinear Diffie-Hellman inversion for q-values (q-mBDHI) problem in G is given a tuple $(P, \alpha P, (\omega_1 + \alpha)^{-1}P, \ldots, (\omega_q + \alpha)^{-1}P, \omega_1, \ldots, \omega_q) \in G^{q+2} \times (Z_p^*)^q$ for unknown $\alpha \in Z_p^*$, to compute $e(P, P)^{(\omega^* + \alpha)^{-1}}$ for some value $\omega^* \in Z_p^* - \{\omega_1, \ldots, \omega_q\}$.

3. Improved Security Model for CBSC Schemes

In this section, we present an improved security model for CBSC that covers both public key replacement attack and insider security. Below, we first briefly review the definition of CBSC.

Formally, a CBSC scheme is specified by the following five algorithms.

(i) *Setup(k)*: on input a security parameter $k \in Z^+$, this algorithm generates a master key *msk* and a list of public parameters *params*. This algorithm is performed by a certifier. After the algorithm is performed, the certifier publishes *params* and keeps *msk* secret.

(ii) *UserKeyGen(params)*: on input the public parameters *params*, this algorithm generates a private key and public key pair (SK_U, PK_U) for a user with identity id_U.

(iii) *CertGen(params, msk, id_U, PK_U)*: on input the public parameters *params*, the master key *msk*, a user's identity id_U, and public key PK_U, this algorithm generates a certificate $Cert_U$. This algorithm is performed by a certifier. After this algorithm is performed, the certifier sends the certificate $Cert_U$ to the user id_U via an open channel.

(iv) *Signcrypt(params, M, id_S, PK_S, SK_S, Cert_S, id_R, PK_R)*: on input the public parameters *params*, a sender's identity id_S, public key PK_S, private key SK_S and certificate $Cert_S$, a receiver's identity id_R, and public key PK_R, this algorithm generates a ciphertext σ.

(v) *Designcrypt(params, σ, id_R, PK_R, SK_R, Cert_R, id_S, PK_S)*: on input the public parameters *params*, a ciphertext σ, the receiver's identity id_R, public key PK_R, private key SK_R and certificate $Cert_R$, the sender's identity id_S, and public key PK_S, this algorithm outputs either a plaintext M or a special symbol $\perp$ indicating a designcryption failure.

As introduced in [3], the adversaries against a certificate-based cryptographic scheme should be divided into two types: Type I and Type II. Type I adversary (denoted by A_I) models an uncertified user while Type II adversary (denoted by A_{II}) models an honest-but-curious certifier who is equipped with the master key. In order to capture public key replacement attack, the Type I adversary A_I in our CBSC security model is allowed to replace any user's public key. Note that the Type II adversary A_{II} should not be allowed to make public key replacement attacks; otherwise, it may trivially break the security of a CBSC scheme using a man-in-the-middle attack.

A CBSC scheme should satisfy both confidentiality (i.e., indistinguishability against adaptive chosen-ciphertext attacks (IND-CBSC-CCA2)) and unforgeability (i.e., existential unforgeability against adaptive chosen-messages attacks (EUF-CBSC-CMA)).

The confidentiality security of a CBSC scheme is defined via the following two games: "IND-CBSC-CCA2 Game-I" and "IND-CBSC-CCA2 Game-II," in which a Type I adversary A_I and a Type II adversary A_{II} interact with a challenger, respectively.

IND-CBSC-CCA2 Game-I. This game is played between A_I and a challenger.

Setup. The challenger runs the algorithm *Setup(k)* to generate *msk* and *params*. It then returns *params* to A_I and keeps *msk* to itself.

Phase 1. In this phase, A_I makes requests to the following oracles adaptively.

(i) $O^{CreateUser}$: on input an identity id_U, if id_U has already been created; the challenger outputs the current public key PK_U associated with id_U. Otherwise, it performs the algorithm $UserKeyGen(params)$ to generate a private/public key pair (SK_U, PK_U), inserts (id_U, PK_U, SK_U) into a list, and outputs PK_U. In this case, id_U is said to be created. We assume that other oracles defined below only respond to an identity which has been created.

(ii) $O^{ReplacePublicKey}$: on input an identity id_U and a value PK'_U, the challenger replaces the current public key of the identity id_U with PK'_U. Note that the current value of a user's public key is used by the challenger in any computations or responses to A_I's requests. This oracle models the ability of a Type I adversary to convince a legitimate user to use a false public key and thus enables our security model to capture the public key replacement attacks attempted by the Type I adversary A_I.

(iii) $O^{GenerateCertificate}$: on input an identity id_U, the challenger responds with a certificate $Cert_U$ by running the algorithm $CertGen(params, msk, id_U, PK_U)$.

(iv) $O^{ExtractPrivateKey}$: on input an identity id_U, the challenger responds with a private key SK_U. Here, A_I is disallowed to query this oracle on any identity for which the public key has been replaced. This restriction is imposed due to the fact that it is unreasonable to expect the challenger to be able to provide a private key of a user for which it does not know the private key.

(v) $O^{Signcryption}$: on input a message M, a sender's identity id_S, and a receiver's identity id_R, the challenger responds with $\sigma = Signcrypt(params, M, id_S, PK_S, SK_S, Cert_S, id_R, PK_R)$. Note that it is possible that the challenger is not aware of the sender's private key if the associated public key has been replaced. In this case, we require A_I to provide it. In addition, we do not consider attacks targeting ciphertexts where the identities of the sender and receiver are the same. So, we disallow queries where $id_S = id_R$.

(vi) $O^{Designcryption}$: on input a ciphertext σ, a sender's identity id_S, and a receiver's identity id_R, the challenger responds with the result of $Designcrypt(params, \sigma, id_R, PK_R, SK_R, Cert_R, id_S, PK_S)$. Note that it is possible that the challenger is not aware of the receiver's private key if the associated public key has been replaced. In this case, we require A_I to provide it. Again, we disallow queries where $id_S = id_R$.

Challenge. Once A_I decides that Phase 1 is over, it outputs two equal-length messages (M_0, M_1) and two distinct identities (id_S^*, id_R^*). The challenger picks a random bit b, computes

$\sigma^* = Signcrypt(params, M_b, id_S^*, PK_S^*, SK_S^*, Cert_S^*, id_R^*, PK_R^*)$, and returns σ^* as the challenge ciphertext to A_I.

Phase 2. In this phase, A_I continues to issues queries as in Phase 1.

Guess. Finally, A_I outputs a guess $b' \in \{0, 1\}$. We say that A_I wins the game if $b = b'$ and the following conditions are simultaneously satisfied: (1) A_I cannot query $O^{GenerateCertificate}$ on the identity id_R^* at any point; (2) A_I cannot query $O^{ExtractPrivateKey}$ on an identity if the corresponding public key has been replaced; (3) in Phase 2, A_I cannot query $O^{Designcryption}$ on $(\sigma^*, id_S^*, id_R^*)$ unless the public key of the sender id_S^* or that of the receiver id_R^* has been replaced after the challenge was issued. We define A_I's advantage in this game to be $2|\Pr\{b = b'\} - 1/2|$.

IND-CBSC-CCA2 Game-II. This game is played between A_{II} and a challenger.

Setup. The challenger runs the algorithm $Setup(k)$ to generate msk and $params$. It then returns $params$ and msk to A_{II}.

Phase 1. In this phase, A_{II} adaptively asks a polynomial bounded number of queries as in *IND-CBSC-CCA2 Game-I.* The only restriction is that A_{II} cannot replace public keys of any users. In addition, A_{II} need not make any queries to $O^{GenerateCertificate}$ since it can compute the certificates for any identities by itself with the master key msk.

Challenge. Once A_{II} decides that Phase 1 is over, it outputs two equal-length messages (M_0, M_1) and two distinct identities (id_S^*, id_R^*). The challenger picks a random bit b, computes $\sigma^* = Signcrypt(params, M_b, id_S^*, PK_S^*, SK_S^*, Cert_S^*, id_R^*, PK_R^*)$, and returns σ^* as the challenge ciphertext to A_{II}.

Phase 2. In this phase, A_{II} continues to issue queries as in Phase 1.

Guess. Finally, A_{II} outputs a guess $b' \in \{0, 1\}$. We say that A_{II} wins the game if $b = b'$ and the following two conditions are both satisfied: (1) A_{II} cannot query $O^{ExtractPrivateKey}$ on the identity id_R^* at any point; (2) A_{II} cannot query $O^{Designcryption}$ on $(\sigma^*, id_S^*, id_R^*)$ in Phase 2. We define A_{II}'s advantage in this game to be $2|\Pr\{b = b'\} - 1/2|$.

Definition 5. A CBSC scheme is said to be IND-CBSC-CCA2 secure if no probabilistic polynomial time (PPT) adversary has nonnegligible advantage in the above two games.

Remark 6. The oracle $O^{ReplacePublicKey}$ defined in the game *IND-CBSC-CCA2 Game-I* models the ability of a Type I adversary to convince a legitimate user to use a false public key. It enables our security model to capture the public key replacement attacks attempted by the Type I adversary A_I.

Remark 7. The adversary in the above definition of message confidentiality is allowed to be challenged on a ciphertext generated using a corrupted sender's private key and certificate. This condition corresponds to the stringent requirement of insider security for confidentiality of signcryption [23]. This means that our security model ensures that the confidentiality of signcryption is preserved even if a sender's private key is corrupted.

The unforgeability security of a CBSC scheme is defined via the following two games: "*EUF-CBSC-CMA Game-I*" and "*EUF-CBSC-CMA Game-II*," in which a Type I adversary A_I and a Type II adversary A_{II} interact with a challenger, respectively.

EUF-CBSC-CMA Game-I. This game is played between A_I and a challenger.

Setup. The challenger runs the algorithm $Setup(k)$ to generate msk and $params$. It then returns $params$ to A_I and keeps msk to itself.

Query. In this phase, A_I can adaptively ask a polynomial bounded number of queries as in the game *IND-CBSC-CCA2 Game-I*.

Forge. Finally, A_I outputs a forgery $(\sigma^*, id_S^*, id_R^*)$. We say that A_I wins the game if the result of $Designcrypt(params, \sigma^*, id_R^*, PK_R^*, SK_R^*, Cert_R^*, id_S^*, PK_S^*)$ is not the $\perp$ symbol and the following conditions are simultaneously satisfied: (1) A_I cannot query $O^{GenerateCertificate}$ on the identity id_S^* at any point; (2) A_I cannot query $O^{ExtractPrivateKey}$ on an identity if the corresponding public key has been replaced; (3) σ^* is not the output of any $O^{Signcryption}$ query on (M^*, id_S^*, id_R^*), where M^* is a message. We define A_I's advantage in this game to be the probability that it wins the game.

EUF-CBSC-CMA Game-II. This game is played between A_{II} and a challenger.

Setup. The challenger runs the algorithm $Setup(k)$ to generate msk and $params$. It then returns $params$ and msk to A_{II}.

Query. In this phase, A_{II} can adaptively ask a polynomial bounded number of queries as in the game *IND-CBSC-CCA2 Game-II*.

Forge. Finally, A_{II} outputs a forgery $(\sigma^*, id_S^*, id_R^*)$. We say that A_{II} wins the game if the result of $Designcrypt(params, \sigma^*, id_R^*, PK_R^*, SK_R^*, Cert_R^*, id_S^*, PK_S^*)$ is not the $\perp$ symbol and the following conditions are simultaneously satisfied: (1) A_{II} cannot query $O^{ExtractPrivateKey}$ on the identity id_S^*; (2) σ^* is not the output of any $O^{Signcryption}$ query on (M^*, id_S^*, id_R^*), where M^* is a message. We define A_{II}'s advantage in this game to be the probability that it wins the game.

Definition 8. A CBSC scheme is said to be EUF-CBSC-CMA secure if no PPT adversary has nonnegligible advantage in the above two games.

Remark 9. The adversary in the above definition of signature unforgeability may output a ciphertext generated using a

corrupted receiver's private key and certificate. Again, this condition corresponds to the stringent requirement of insider security for unforgeability of signcryption [23]. Hence, our security model also ensures that the unforgeability of signcryption is preserved even if a receiver's private key is corrupted.

4. Cryptanalysis of Luo et al.'s CBSC Scheme

In this section, we give the review and attack of Luo et al.'s CBSC scheme [20].

4.1. Review of Luo et al.'s CBSC Scheme. Luo et al.'s CBSC scheme consists of the following six algorithms.

(i) *Setup:* given a security parameter k, the certifier performs as follows: generate two cyclic groups G and G_T of prime order p such that there exists a bilinear pairing map $e : G \times G \rightarrow G_T$; select a random element $s \in Z_p^*$ and a random generator $P \in G$, and compute $P_{pub} = sP$; select four hash functions $H_1 : \{0,1\}^n \times G \rightarrow G$, $H_2 : \{0,1\}^n \times G \times G \rightarrow G$, $H_3 : G \times G \times \{0,1\}^n \rightarrow Z_p^*$, and $H_4 : G_T \rightarrow \{0,1\}^n$; set the public parameters $params = \{p, G, G_T, e, n, P, P_{pub}, H_1, H_2, H_3, H_4\}$ and the master key $msk = s$.

(ii) *UserKeyGen:* given $params$, a user with identity $id_U \in \{0,1\}^n$ chooses a random $x_U \in Z_p^*$ as his private key SK_U and then computes his public key $PK_U = x_U P$.

(iii) *CertGen:* to generate a certificate for the user with identity id_U and public key PK_U, the certifier computes $Q_U = H_1(id_U, PK_U)$ and outputs the certificate $Cert_U = sQ_U$.

(iv) *Sender Signcrypt:* to send a message $M \in \{0,1\}^n$ to the receiver id_R, the sender id_S does the following: randomly choose $r \in Z_p^*$ and compute $R = rP$ and $T = H_2(id_S, PK_S, R)$; compute $h = H_3(R, S, M)$ and $V = r^{-1}(Cert_S + SK_S \cdot T + h \cdot P_{pub})$; compute $W = e(PK_S, PK_R)^r$ and then $C = M \oplus H_4(W)$; set the ciphertext $\sigma = (C, R, V)$.

(v) *Receiver Decrypt:* when receiving a ciphertext $\sigma = (C, R, V)$ from the sender id_S, the receiver id_R does the following: compute $M = C \oplus H_4(W)$ where $W = e(R, SK_R \cdot PK_S)$; forward the message M and signature (R, V) to the algorithm *Receiver Verify*.

(vi) *Receiver Verify:* to verify the sender id_S's signature (R, V) on the message M, the receiver id_R does the following: compute $S = H_2(id_S, PK_S, R)$ and $h = H_3(R, S, M)$; check whether $e(R, V) = e(P_{pub}, Q_S)e(PK_S, S)e(P, P_{pub})^h$. If the check holds, output M; otherwise, output $\perp$.

4.2. Attack on Luo et al.'s CBSC Scheme. A Type I adversary who is capable of replacing any user's public key can forge a valid signcryption on any message M from id_S to id_R by performing the following steps.

(1) Replace the sender id_S's public key with $PK'_S = x'_S P_{pub}$, where x'_S is a random value chosen from Z^*_p.

(2) Choose a random value $r' \in Z^*_p$ and compute $R' = r' P_{pub}$ and $T' = H_2(id_S, PK'_S, R')$.

(3) Choose a random message $M \in \{0,1\}^n$ and compute $V' = r'^{-1}(Q_S + x'_S T' + h' P)$, where $Q_S = H_1(id_S, PK'_S)$ and $h' = H_3(R', T', M)$.

(4) Randomly choose $C' \in \{0,1\}^n$ and set $\sigma' = (C', R', V')$ as the signcryption of the message M. Note that if the adversary has corrupted the receiver id_R's private key SK_R, it can compute $C' = M \oplus H_4(W')$, where $W' = e(R', SK_R, PK'_S)$.

The ciphertext $\sigma' = (C', R', V')$ passes the verification test as shown below:

$$e\left(P_{pub}, Q_S\right) e\left(PK'_S, T'\right) e\left(P, P_{pub}\right)^{h'}$$
$$= e\left(P_{pub}, Q_S + x'_S T' + h' P\right)$$
$$= e\left(r' P_{pub}, r'^{-1}\left(Q_S + x'_S T' + h' P\right)\right) \quad (1)$$
$$= e\left(R', V'\right).$$

This proves that the forged signcryption is valid.

Note that Luo et al.'s scheme also doses not resist insider attacks since the adversary can forge a valid signcryption using the corrupted receiver id_R's private key in the step (4).

5. Our Proposed CBSC Scheme

5.1. Description of the Scheme. Our CBSC scheme is constructed from the certificate-based encryption scheme proposed by Lu et al. [8]. It consists of the following five algorithms.

(i) *Setup(k):* given a security parameter k, the certifier performs the following: generate two cyclic groups G and G_T of a k-bit prime order p such that there exists a bilinear pairing map $e : G \times G \rightarrow G_T$; choose two random generators $P, Q \in G$ and compute $g = e(P, Q)$; choose a random element $\alpha \in Z^*_p$ and set $P_{pub} = \alpha P$; select three hash functions $H_1 : \{0,1\}^* \times G_T \rightarrow Z^*_p$, $H_2 : G_T \times G_T \rightarrow \{0,1\}^n$ and $H_3 : \{0,1\}^* \rightarrow Z^*_p$, where n is the bit-length of the message to be signcrypted; set the public parameters $params = \{p, G, G_T, e, n, P, Q, P_{pub}, g, H_1, H_2, H_3\}$ and the master key $msk = \alpha$.

(ii) *UserKeyGen(params):* given $params$, a user with identity $id_U \in \{0,1\}^*$ chooses a random $x_U \in Z^*_p$ as his private key SK_U and then computes his public key $PK_U = g^{x_U}$.

(iii) *CertGen(params, msk, id_U, PK_U):* to generate a certificate for a user with identity id_U and public key PK_U, the certifier computes $Cert_U = (H_1(id_U, PK_U) + \alpha)^{-1} Q$. The user id_U can check the

validness of the certificate $Cert_U$ by verifying whether $e(H_1(id_U, PK_U)P + P_{pub}, Cert_U) = g$.

(iv) *Signcrypt(params, M, id_S, PK_S, SK_S, Cert_S, id_R, PK_R):* to send a message $M \in \{0,1\}^n$ to the receiver id_R, the sender id_S does the following: randomly choose $r \in Z^*_p$ and compute $R_1 = g^r$ and $R_2 = (PK_R)^r$; compute $U = r(H_1(id_R, PK_R)P + P_{pub})$ and $C = M \oplus H_2(R_1, R_2)$; compute $V = (h \cdot SK_S + r) \cdot Cert_S$, where $h = H_3(M, U, R_1, R_2, id_S, PK_S, id_R, PK_R)$; set the ciphertext $\sigma = (C, U, V)$.

(v) *Designcrypt(params, σ, id_R, PK_R, SK_R, Cert_R, id_S, PK_S):* to designcrypt a ciphertext $\sigma = (C, U, V)$ from the sender id_S, the receiver id_R does the following: compute $R_1 = e(U, Cert_R)$ and $R_2 = e(U, Cert_R)^{SK_R}$; compute $M = C \oplus H_2(R_1, R_2)$ and then check whether $e(H_1(id_S, PK_S)P + P_{pub}, V)(PK_S)^{-h} = R_1$, where $h = H_3(M, U, R_1, R_2, id_S, PK_S, id_R, PK_R)$. If the check holds, output M; otherwise, output $\perp$.

The consistency of our scheme can be easily verified by the following equalities:

(1) $e(U, Cert_R) = e(r(H_1(id_R, PK_R)P + P_{pub}), (H_1(id_R, PK_R) + \alpha)^{-1}Q) = e(P, Q)^r = g^r$;

(2) $e(U, Cert_R)^{SK_R} = (g^r)^{SK_R} = (PK_R)^r$;

(3)

$$e\left(H_1\left(id_S, PK_S\right)P + P_{pub}, V\right)\left(PK_S\right)^{-h}$$
$$= e\left(\left(H_1\left(id_S, PK_S\right) + \alpha\right)^{-1}P, \left(h \cdot SK_S + r\right) \cdot Cert_S\right)$$
$$\cdot \left(PK_S\right)^{-h} \quad (2)$$
$$= e\left(P, \left(h \cdot SK_S + r\right)Q\right) \cdot \left(PK_S\right)^{-h}$$
$$= e(P, Q)^r = R_1.$$

5.2. Security Proof

Theorem 10. *The CBSC scheme above is IND-CBSC-CCA2 secure under the hardness of the q-mBDHI and BDH problems in the random oracle model.*

This theorem can be proved by combining the following two lemmas.

Lemma 11. *If a Type I adversary A_I has advantage ε against our CBSC scheme when asking at most q_{cu} to $O^{CreateUser}$ queries, q_{sc} queries to $O^{Signcryption}$, q_{dsc} queries to $O^{Designcryption}$, and q_i queries to random oracles $H_1 \sim H_3$, then there exists an algorithm B to solve the $(q_1 - 1)$-mBDHI problem with advantage*

$$\varepsilon' \geq \frac{\varepsilon}{q_1(q_2 + 2q_3 + 2q_{sc})}\left(1 - q_{sc}\frac{q_2 + 2q_3 + 2q_{sc}}{2^k}\right)$$
$$\times \left(1 - \frac{q_{dsc}}{2^k}\right). \quad (3)$$

Proof. Assume that B is given a random q-mBDHI instance $(P, \alpha P, (\omega_1 + \alpha)^{-1}P, \ldots, (\omega_q + \alpha)^{-1}P, \omega_1, \ldots, \omega_q)$, where $q = q_1 - 1$. B interacts with A_I as follows.

In the setup phase, B randomly chooses $t \in Z_p^*$ and sets $P_{pub} = \alpha P$, $Q = tP$, and $g = e(P, Q)$. Furthermore, it randomly chooses a value $\omega^* \in Z_p^*$ such that $\omega^* \notin \{\omega_1, \ldots, \omega_q\}$ and an index $\theta \in [1, q_1]$. Then, B starts *IND-CBSC-CCA2 Game-I* by supplying A_I with *params* $= \{p, G, G_T, e, n, P, Q, P_{pub}, g, H_1, H_2, H_3\}$, where $H_1 \sim H_3$ are random oracles controlled by B. A_I can make queries on these random oracles at any time during the game. Note that the corresponding master key is $msk = \alpha$ which is unknown to B.

Now, B starts to respond to various queries as follows:

H_1 *Queries.* We assume that q_1 queries to H_1 are distinct. B maintains a list H_1List of tuples $\langle id_i, PK_i, h_{1,i}, Cert_i \rangle$. On input (id_i, PK_i), B does the following.

(1) If (id_i, PK_i) already appears on H_1List in a tuple $\langle id_i, PK_i, h_{1,i}, Cert_i \rangle$, then B returns $h_{1,i}$ to A_I.

(2) Else if the query is on the θth distinct (id_θ, PK_θ), then B inserts $\langle id_\theta, PK_\theta, \omega^*, \bot \rangle$ into H_1List and returns $h_{1,\theta} = \omega^*$ to A_I. Note that the certificate for the identity id_θ is $Cert_\theta = t(\omega^* + \alpha)^{-1}P$ which is unknown to B.

(3) Else B sets $h_{1,i}$ to be ω_j ($j \in [1, q]$) which has not been used and computes $Cert_i = t(\omega_j + \alpha)^{-1}P$. It then inserts $\langle id_i, PK_i, h_{1,i}, Cert_i \rangle$ into H_1List and returns $h_{1,i}$.

H_2 *Queries.* B maintains a list H_2List of tuples $\langle R_1, R_2, h_2 \rangle$. On input (R_1, R_2), B does the following.

(1) If (R_1, R_2) already appears on H_2List in a tuple $\langle R_1, R_2, h_2 \rangle$, B returns h_2 to A_I.

(2) Otherwise, it returns a random $h_2 \in \{0, 1\}^n$ and inserts $\langle R_1, R_2, h_2 \rangle$ into H_2List.

H_3 *Queries.* B maintains a list H_3List of tuples $\langle M, U, R_1, R_2, id_S, PK_S, id_R, PK_R, h_3, C \rangle$. On input $(M, U, R_1, R_2, id_S, PK_S, id_R, PK_R)$, B does the following.

(1) If $(M, U, R_1, R_2, id_S, PK_S, id_R, PK_R)$ already appears on H_3List in a tuple $\langle M, U, R_1, R_2, id_S, PK_S, id_R, PK_R, h_3, C \rangle$, B returns h_3 to A_I.

(2) Otherwise, it returns a random $h_3 \in Z_p^*$ to A_I. To anticipate possible subsequent queries to $O^{Designcryption}$, it additionally simulates the random oracle H_2 on its own to obtain $h_2 = H_2(R_1, R_2)$ and then inserts $\langle M, U, R_1, R_2, id_S, PK_S, id_R, PK_R, h_3, C = M \oplus h_2 \rangle$ into H_3List.

$O^{CreateUser}$ *Queries.* B maintains a list KeyList of tuples $\langle id_i, PK_i, SK_i, \text{flag}_i \rangle$ which is initially empty. On input (id_i), B does the following.

(1) If id_i already appears on KeyList in a tuple $\langle id_i, PK_i, SK_i, \text{flag}_i \rangle$, B returns PK_i to A_I directly.

(2) Otherwise, B randomly chooses $x_i \in Z_p^*$ as the private key SK_i for the identity id_i and computes the corresponding public key as $PK_i = g^{x_i}$. It then inserts $\langle id_i, PK_i, SK_i, 0 \rangle$ into KeyList and returns PK_i to A_I.

$O^{ReplacePublicKeyr}$ *Queries.* On input (id_i, PK_i'), B searches id_i in KeyList to find a tuple $\langle id_i, PK_i, SK_i, \text{flag}_i \rangle$ and updates the tuple with $\langle id_i, PK_i', SK_i, 1 \rangle$.

$O^{ExtractPrivateKey}$ *Queries.* On input (id_i), B searches id_i in KeyList to find a tuple $\langle id_i, PK_i, SK_i, \text{flag}_i \rangle$. If $\text{flag}_i = 0$, it returns SK_i to A_I; otherwise, it rejects this query.

$O^{GenerateCertificate}$ *Queries.* On input (id_i), B does the following.

(1) If $(id_i, PK_i) = (id_\theta, PK_\theta)$, then B aborts.

(2) Otherwise, B searches id_i in H_1List to find a tuple $\langle id_i, PK_i, h_{1,i}, Cert_i \rangle$ and then returns $Cert_i$ to A_I. If H_1List does not contain such a tuple, B queries H_1 on (id_i, PK_i) first.

$O^{Signcryption}$ *Queries.* On input (M, id_S, id_R), B performs as follows.

(1) If $(id_S, PK_S) \neq (id_\theta, PK_\theta)$, B can answer the query according to the specification of the algorithm *Signcrypt* since it knows the sender id_S's private key and certificate.

(2) Otherwise, B randomly chooses $r, h_3 \in Z_p^*$, $h_2 \in \{0, 1\}^n$ and sets $U = r(H_1(id_\theta, PK_\theta)P + P_{pub}) - h_3 SK_\theta (H_1(id_R, PK_R)P + P_{pub})$, $V = rCert_R$, $C = M \oplus h_2$, $R_1 = e(U, Cert_R)$, and $R_2 = e(U, Cert_R)^{SK_R}$, respectively. It is easy to verify that $e(H_1(id_\theta, PK_\theta)P + P_{pub}, V) \cdot (PK_\theta)^{-h_3} = e(U, Cert_R)$. Then, B inserts $\langle R_1, R_2, h_2 \rangle$ and $\langle M, U, R_1, R_2, id_\theta, PK_\theta, id_R, PK_R, h_3, C \rangle$ into H_2List and H_3List respectively, and returns the ciphertext $\sigma = (C, U, V)$ to A_I. Note that B fails if H_2List or H_3List is already defined in the corresponding value, but this only happens with probability smaller than $(q_2 + 2q_3 + 2q_{sc})/2^k$.

$O^{Designcryption}$ *Queries.* On input $(\sigma = (C, U, V), id_S, id_R)$, B does the following.

(1) If $(id_R, PK_R) \neq (id_\theta, PK_\theta)$, B can answer the query according to the specification of the algorithm *Designcrypt* since it knows the receiver id_R's private key and certificate.

(2) Otherwise, B searches in H_3List for all tuples of the form $\langle M, U, R_1, R_2, id_S, PK_S, id_\theta, PK_\theta, h_3, C \rangle$. If no such tuple is found, then σ is rejected. Otherwise,

each one of them is further examined. For a tuple $\langle M, U, R_1, R_2, id_S, PK_S, id_\theta, PK_\theta, h_3, C \rangle$, B first checks whether $e(H_1(id_S, PK_S)P + P_{\text{pub}}, V) \cdot (PK_S)^{-h_3} = R_1$. If the tuple passes the verification, then B returns M in this tuple to A_I. If no such tuple is found, σ is rejected. Note that a valid ciphertext is rejected with probability smaller than $q_{dsc}/2^k$ across the whole game.

In the challenge phase, A_I outputs $(M_0, M_1, id_S^*, id_R^*)$, on which it wants to be challenged. If $(id_R^*, PK_R^*) \neq (id_\theta, PK_\theta)$, then B aborts. Otherwise, B randomly chooses $C^* \in \{0,1\}^n$, $r^* \in Z_p^*$, and $V^* \in G$, computes $U^* = r^*P$, and returns $\sigma^* = (C^*, U^*, V^*)$ to A_I as the challenge ciphertext. Observe that the decryption of C^* is $C^* \oplus H_2(e(U^*, Cert_\theta), e(U^*, Cert_\theta)^{SK_\theta})$.

In the guess phase, A_I outputs a bit which is ignored by B. Note that A_I cannot recognize that σ^* is not a valid ciphertext unless it queries H_2 on $(e(U^*, Cert_\theta), e(U^*, Cert_\theta)^{SK_\theta})$ or H_3 on $(M_b, U^*, (e(U^*, Cert_\theta), e(U^*, Cert_\theta)^{SK_\theta}), id_S^*, PK_S^*, id_\theta, PK_\theta)$, where $b \in \{0, 1\}$. Standard arguments can show that a successful A_I is very likely to query H_2 on $(e(U^*, Cert_\theta), e(U^*, Cert_\theta)^{SK_\theta})$ or H_3 on $(M_b, U^*, (e(U^*, Cert_\theta), e(U^*, Cert_\theta)^{SK_\theta}), id_S^*, PK_S^*, id_\theta, PK_\theta)$ if the simulation is indistinguishable from a real attack environment. To produce a result, B picks a random tuple $\langle R_1, R_2, h_2 \rangle$ or $\langle M, U, R_1, R_2, id_S, PK_S, id_R, PK_R, h_3, C \rangle$ from H_2List or H_3List. With probability $1/(q_2 + 2q_3 + 2q_{sc})$ (as H_2List, H_3List contain at most $q_2 + q_3 + q_{sc}, q_3 + q_{sc}$ tuples, resp.), the chosen tuple will contain the value $R_1 = e(U^*, Cert_\theta)$. Because $e(U^*, Cert_\theta) = e(r^*P, t(\omega^* + \alpha)^{-1}P) = e(P, P)^{tr^*(\omega^* + \alpha)^{-1}}$, B returns $T = R_1^{(tr^*)^{-1}}$ as the solution to the given q-mBDHI problem.

We now derive B's advantage in solving the q-mBDHI problem. From the above construction, the simulation fails if any of the following events occurs: (1) E_1: in the challenge phase, B aborts because $(id_R^*, PK_R^*) \neq (id_\theta, PK_\theta)$; (2) E_2: A_I makes an $O^{GenerateCertificate}$ query on (id_θ, PK_θ); (3) E_3: B aborts in answer one of A_I's $O^{Signcryption}$ queries because of a collision on H_2 or H_3; (4) E_4: B rejects a valid ciphertext at some point of the game.

We clearly have that $\Pr[\neg E_1] = 1/q_1$ and $\neg E_1$ implies $\neg E_2$. We also already observed that $\Pr[E_3] \leq (q_2 + 2q_3 + 2q_{sc})/2^k$ and $\Pr[E_4] \leq q_{dsc}/2^k$. Thus, we have that

$$\Pr\left[\neg E_1 \wedge \neg E_2 \wedge \neg E_3 \wedge \neg E_4\right] \geq \frac{1}{q_1}\left(1 - q_{sc}\frac{q_2 + 2q_3 + 2q_{sc}}{2^k}\right)$$
$$\times \left(1 - \frac{q_{dsc}}{2^k}\right). \tag{4}$$

Since B selects the correct tuple from H_2List or H_3List with probability $1/(q_2 + 2q_3 + 2q_{sc})$, we obtain the announced bound on B's advantage in solving the q-mBDHI problem. □

Lemma 12. *If a Type II adversary A_{II} has advantage ε against our CBSC scheme when asking at most q_{cu} queries to $O^{CreateUser}$,*

q_{sc} *queries to* $O^{Signcryption}$, q_{dsc} *queries to* $O^{Designcryption}$, *and* q_i *queries to random oracles* $H_1 \sim H_3$, *then there exists an algorithm B to solve the BDH problem with advantage*

$$\varepsilon' \geq \frac{\varepsilon}{q_{cu}(q_2 + 2q_3 + 2q_{sc})}\left(1 - q_{sc}\frac{q_2 + 2q_3 + 2q_{sc}}{2^k}\right)$$
$$\times \left(1 - \frac{q_{dsc}}{2^k}\right). \tag{5}$$

Proof. Assume that B is given a BDH instance (P, aP, bP, cP), where a, b, c are three random elements from Z_p^*. B interacts with A_{II} as follows.

In the setup phase, B randomly chooses $\alpha \in Z_p^*$, sets $Q = aP$, and computes $P_{\text{pub}} = \alpha P$ and $g = e(P, Q)$. Furthermore, it randomly chooses an index θ with $1 \leq \theta \leq q_{cu}$. Then, B starts *IND-CBSC-CCA2 Game-II* by supplying A_{II} with $msk = \alpha$ and $params = \{p, G, G_T, e, n, P, Q, P_{\text{pub}}, g, H_1, H_2, H_3\}$, where $H_1 \sim H_3$ are random oracles controlled by B. A_{II} can make queries on these random oracles at any time during the game.

Now, B starts to respond various queries as follows.

H_1 *Queries*. B maintains a list H_1List of tuples $\langle id_i, PK_i, h_{1,i} \rangle$. On input (id_i, PK_i), B does the following: if (id_i, PK_i) already appears on H_1List in a tuple $\langle id_i, PK_i, h_{1,i} \rangle$, then B returns $h_{1,i}$ to A_{II}; otherwise, it returns a random $h_{1,i} \in Z_p^*$ and inserts $\langle id_i, PK_i, h_{1,i} \rangle$ into H_1List.

H_2 *Queries*. B responds as in the proof of Lemma 11.

H_3 *Queries*. B responds as in the proof of Lemma 11.

$O^{CreateUser}$ *Queries*. B maintains a list KeyList of tuples $\langle id_i, PK_i, SK_i \rangle$. On input (id_i), B does the following: (1) if id_i already appears on KeyList in a tuple $\langle id_i, PK_i, SK_i \rangle$, B returns PK_i to A_{II}. (2) Else if $id_i = id_\theta$, B returns $PK_\theta = e(bP, Q) = e(bP, aP)$ to A_{II} and inserts $\langle id_\theta, PK_\theta, \perp \rangle$ into KeyList. Note that the private key for the identity id_θ is b which is unknown to B. (3) Else B randomly chooses $x_i \in Z_p^*$ as the private key SK_i for the identity id_i and computes the corresponding public key as $PK_i = g^{x_i}$. It then inserts $\langle id_i, PK_i, SK_i \rangle$ into KeyList and returns PK_i to A_{II}.

$O^{ExtractPrivateKey}$ *Queries*. On receiving such a query on id_i, B does the following: if $id_i = id_\theta$, then B aborts; otherwise, B searches id_i in KeyList to find the tuple $\langle id_i, PK_i, SK_i \rangle$ and returns SK_i to A_{II}.

$O^{Signcryption}$ *Queries*. On input (M, id_S, id_R), B does the following: if $id_S \neq id_\theta$, B can answer the query according to the specification of the Signcrypt algorithm since it knows the sender id_S's private key and certificate. Otherwise, B randomly chooses $r, h_3 \in Z_p^*$, $h_2 \in \{0,1\}^n$ and computes $U = r(H_1(id_\theta, PK_\theta)P + P_{\text{pub}}) - h_3(H_1(id_R, PK_R)bP + \alpha bP)$, $V = rCert_R$, $C = M \oplus h_2$, $R_1 = e(U, Cert_R)$, and $R_2 = e(U, Cert_R)^{SK_R}$, respectively. It is easy to verify that $e(H_1(id_\theta, PK_\theta)P + P_{\text{pub}}, V) \cdot (PK_\theta)^{-h_3} = e(U, Cert_R)$. It then inserts $\langle R_1, R_2, h_2 \rangle$ and $\langle M, U, R_1, R_2, id_\theta, PK_\theta, id_R, PK_R, h_3, C \rangle$

into H_2List and H_3List respectively, and returns the ciphertext $\sigma = (C, U, V)$ to A_{II}. Note that B fails if H_2List or H_3List is already defined in the corresponding value, but this only happens with probability smaller than $(q_2 + 2q_3 + 2q_{sc})/2^k$.

$O^{Designcryption}$ Queries. B responds as in the proof of Lemma 11.

In the challenge phase, A_{II} outputs $(M_0, M_1, id_S^*, id_R^*)$, on which it wants to be challenged. If $id_R^* \neq id_\theta$, then B aborts. Otherwise, B randomly chooses $C^* \in \{0,1\}^n$, $V^* \in G$, computes $U^* = (H_1(id_\theta, PK_\theta) + \alpha)cP$, and returns $\sigma^* = (C^*, U^*, V^*)$ to A_{II} as the challenge ciphertext. Observe that the decryption of C^* is $C^* \oplus H_2(e(U^*, Cert_\theta), e(U^*, Cert_\theta)^{SK_\theta})$.

In the guess phase, A_{II} outputs a bit, which is ignored by B. Note that A_{II} cannot recognize that σ^* is not a valid ciphertext unless it queries H_2 on $(e(U^*, Cert_\theta), e(U^*, Cert_\theta)^{SK_\theta})$ or H_3 on $(M_\beta, U^*, e(U^*, Cert_\theta), e(U^*, Cert_\theta)^{SK_\theta}, id_S^*, PK_S^*, id_\theta, PK_\theta)$, where $\beta \in \{0,1\}$. Standard arguments can show that a successful A_{II} is very likely to query H_2 on $(e(U^*, Cert_\theta), e(U^*, Cert_\theta)^{SK_\theta})$ or H_3 on $(M_\beta, U^*, e(U^*, Cert_\theta), e(U^*, Cert_\theta)^{SK_\theta}, id_S^*, PK_S^*, id_\theta, PK_\theta)$ if the simulation is indistinguishable from a real attack environment. To produce a result, B picks a random tuple $\langle R_1, R_2, h_2 \rangle$ or $\langle M, U, R_1, R_2, id_S, PK_S, id_R, PK_R, h_3, C \rangle$ from H_2List or H_3List. With probability $1/(q_2 + 2q_3 + 2q_{sc})$ (as H_2List, H_3List contain at most $q_2 + q_3 + q_{sc}$, $q_3 + q_{sc}$ tuples, resp.), the chosen tuple will contain the right element $R_2 = e(U^*, Cert_\theta)^{SK_\theta} = e(P,P)^{abc}$. B then returns R_2 as the solution to the given BDH problem.

We now derive B's advantage in solving the BDH problem. From the above construction, the simulation fails if any of the following events occurs: (1) E_1: in the challenge phase, B aborts because $id_R^* \neq id_\theta$; (2) E_2: A_{II} makes an $O^{ExtractPrivateKey}$ query on id_θ; (3) E_3: B aborts in answer A_{II}'s $O^{Signcryption}$ query because of a collision on H_2 or H_3; (4) E_4: B rejects a valid ciphertext at some point of the game.

We clearly have that $\Pr[\neg E_1] = 1/q_{cu}$ and $\neg E_1$ implies $\neg E_2$. We also already observed that $\Pr[E_3] \leq (q_2 + 2q_3 + 2q_{sc})/2^k$ and $\Pr[E_4] \leq q_{dsc}/2^k$. Thus, we have that

$$\Pr\left[\neg E_1 \wedge \neg E_2 \wedge \neg E_3 \wedge \neg E_4\right]$$
$$\geq \frac{1}{q_{cu}}\left(1 - q_{sc}\frac{q_2 + 2q_3 + 2q_{sc}}{2^k}\right)\left(1 - \frac{q_{dsc}}{2^k}\right). \quad (6)$$

Since B selects the correct tuple from H_2List or H_3List with probability $1/(q_2 + 2q_3 + 2q_{sc})$, we obtain the announced bound on B's advantage in solving the BDH problem. $\square$

Theorem 13. *The CBSC scheme above is EUF-CBSC-CMA secure under the hardness of the q-CAA and CDH problems in the random oracle model.*

This theorem can be proved by combining the following two lemmas.

Lemma 14. *If a Type I adversary A_I asks at most q_{cu} queries to $O^{CreateUser}$, q_{sc} queries to $O^{Signcryption}$, q_{dsc} queries to $O^{Designcryption}$, and q_i queries to random oracles $H_1 \sim H_3$ and produces a valid forgery with probability $\varepsilon \geq 10(q_{sc}+1)(q_{sc}+q_3)/2^k$, then there exists an algorithm B to solve the $(q_1 - 1)$-CAA problem with advantage $\varepsilon' \geq 1/(9q_1)$.*

Proof. Assume that B is given a q-CAA instance $(P, \alpha P, (\omega_1 + \alpha)^{-1}P, \ldots, (\omega_q + \alpha)^{-1}P, \omega_1, \ldots, \omega_q)$, where $q = q_1 - 1$. B interacts with A_I as follows.

In the setup phase, B randomly chooses $t \in Z_p^*$, sets $P_{pub} = \alpha P$, and computes $Q = tP$ and $g = e(P, Q)$. Furthermore, it randomly chooses a value $\omega^* \in Z_p^*$ such that $\omega^* \notin \{\omega_1, \ldots, \omega_q\}$ and an index $\theta \in [1, q_1]$. Then, B starts EUF-CBSC-CMA Game-I by supplying A_I with $params = \{p, G, G_T, e, n, P, Q, P_{pub}, g, H_1, H_2, H_3\}$, where $H_1 \sim H_3$ are random oracles controlled by B. Note that the corresponding master key is $msk = \alpha$ which is unknown to B.

In the query phase, B responds to various oracle queries as in the proof of Lemma 11.

Finally, in the forge phase A_I outputs a valid forgery $(\sigma^* = (C^*, U^*, V^*), id_S^*, id_R^*)$ with probability $\varepsilon \geq 10(q_{sc}+1)(q_{sc}+q_3)/2^k$ [29]. If $(id_S^*, PK_S^*) \neq (id_\theta, PK_\theta)$, B aborts. Otherwise, having the knowledge of SK_R^* and $Cert_R^*$, B runs the algorithm $Designcrypt(params, \sigma^*, id_S^*, PK_R^*, SK_R^*, Cert_R^*, id_\theta, PK_\theta)$ to obtain the message M^* and then simulates the random oracle H_3 on its own to obtain $h_3^* = H_3(M^*, U^*, e(U^*, Cert_R^*), e(U^*, Cert_R^*)^{SK_R^*}, id_\theta, PK_\theta, id_R^*, PK_R^*)$. Using the oracle replay technique [29], B replays A_I with the same random tape but with the different hash value $h_3^{*\prime}(\neq h_3^*)$ to generate one more valid ciphertext $\sigma^{*\prime} = (C^*, U^*, V^{*\prime})$ such that $V^{*\prime} \neq V^*$. Since $\sigma^* = (C^*, U^*, V^*)$ and $\sigma^{*\prime} = (C^*, U^*, V^{*\prime})$ are both valid ciphertexts for the same message M^* and the randomness r^*, we obtain the following relations:

$$V^* - V^{*\prime} = \left(h_3^* SK_\theta + r^*\right)Cert_\theta - \left(h_3^{*\prime} SK_\theta + r^*\right)Cert_\theta$$
$$= \left(h_3^* - h_3^{*\prime}\right)SK_\theta Cert_\theta. \quad (7)$$

Because $Cert_\theta = t(\omega^* + \alpha)^{-1}P$, B can compute $(\omega^* + \alpha)^{-1}P = [t(h_3^* - h_3^{*\prime})SK_\theta]^{-1}(V^* - V^{*\prime})$ as the solution to the given q-CAA problem.

We now derive B's advantage in solving the q-CAA problem. From the above construction, the simulation fails after A_I outputs a valid forgery if any of the following events occurs: (1) E_1: in the forge phase, B aborts because $(id_S^*, PK_S^*) \neq (id_\theta, PK_\theta)$; (2) E_2: B fails in using the oracle replay technique to generate one more valid ciphertext.

Clearly, $\Pr[\neg E_1] = 1/q_1$. Moreover, from the forking lemma [29], we know that $\Pr[\neg E_2] \geq 1/9$. Thus, we have that if A_I produces a forgery, then B will succeed in solving the q-CAA problem with probability $\varepsilon' = \Pr[\neg E_1 \wedge \neg E_2] \geq 1/(9q_1)$. $\square$

Lemma 15. *If a Type II adversary A_{II} asks at most q_{cu} queries to $O^{CreateUser}$, q_{sc} queries to $O^{Signcryption}$, q_{dsc} queries to $O^{Designcryption}$, and q_i queries to random oracles $H_1 \sim H_3$ and*

TABLE 2: Performance of the CBSC schemes.

Schemes	Signcryption cost	Designcryption cost	Ciphertext overhead
Ours	$2e + 3m + 3h$	$2p + 2e + 1m + 3h$	$2|G|$
[18]	$1p + 1e + 4m + 3h$	$3p + 1e + 1m + 3h$	$2|G|$
[20]	$1p + 5m + 4h$	$4p + 2m + 3h$	$3|G| + |id|$
[21]	$2p + 4e + 3m + 3h$	$3p + 4e + 3h$	$2|G| + 2|Z_p|$

produces a valid forgery with probability $\varepsilon \geq 10(q_{sc} + 1)(q_{sc} + q_3)/2^k$, *then there exists an algorithm B to solve the CDH problem with advantage* $\varepsilon' \geq 1/(9q_{cu})$.

Proof. Assume that B is given a random CDH instance (P, aP, bP) where a, b are two random elements from Z_p^*. B interacts with A_{II} as follows.

In the setup phase, B randomly chooses $\alpha \in Z_p^*$, sets $Q = aP$, and computes $P_{pub} = \alpha P$ and $g = e(P, Q)$. Furthermore, it randomly chooses an index θ with $1 \leq \theta \leq q_{cu}$. Then, B starts *EUF-CBSC-CMA Game-II* by supplying A_{II} with $msk = \alpha$ and $params = \{p, G, G_T, e, n, P, Q, P_{pub}, g, H_1, H_2, H_3\}$, where $H_1 \sim H_3$ are random oracles controlled by B.

In the query phase, B responds to various oracle queries as in the proof of Lemma 12.

Finally, in the forge phase A_I outputs a valid forgery $(\sigma^* = (C^*, U^*, V^*), id_S^*, id_R^*)$ with probability $\varepsilon \geq 10(q_{sc} + 1)(q_{sc} + q_3)/2^k$ [29]. If $id_S^* \neq id_\theta$, then B aborts. Otherwise, having the knowledge of SK_R^* and $Cert_R^*$, B runs the algorithm *Designcrypt*$(params, \sigma^*, id_R^*, PK_R^*, SK_R^*, Cert_R^*, id_\theta, PK_\theta)$ to obtain the message M^* and then simulates the random oracle H_3 on its own to obtain $h_3^* = H_3(M^*, U^*, e(U^*, Cert_R^*), e(U^*, Cert_R^*)^{SK_R^*}, id_\theta, PK_\theta, id_R^*, PK_R^*)$. Using the oracle replay technique [29], B replays A_{II} with the same random tape but with the different hash value $h_3^{*'} (\neq h_3^*)$ to generate one more valid ciphertext $\sigma^{*'} = (C^*, U^*, V^{*'})$ such that $V^{*'} \neq V^*$. Since $\sigma^* = (C^*, U^*, V^*)$ and $\sigma^{*'} = (C^*, U^*, V^{*'})$ are both valid ciphertexts for the same message M^* and randomness r^*, we obtain the following relations:

$$V^* - V^{*'} = \left(h_3^* SK_\theta + r^*\right) Cert_\theta - \left(h_3^{*'} SK_\theta + r^*\right) Cert_\theta$$
$$= \left(h_3^* - h_3^{*'}\right) SK_\theta (H_1(id_\theta, PK_\theta) + \alpha)^{-1} Q. \tag{8}$$

Then, we have the following relations:

$$e\left(H_1(id_\theta, PK_\theta)P + \alpha P, V^* - V^{*'}\right)$$
$$= e\left(P, \left(h_3^* - h_3^{*'}\right) SK_\theta Q\right). \tag{9}$$

Because $Q = aP$ and $SK_\theta = b$, B can compute $abP = SK_\theta Q = (h_3^* - h_3^{*'})^{-1}(H_1(id_\theta, PK_\theta) + \alpha)(V^* - V^{*'})$ as the solution to the given CDH problem.

We now derive B's advantage in solving the CDH problem. From the above construction, the simulation fails if any of the following events occurs: (1) E_1: in the forge phase, B aborts because $id_S^* \neq id_\theta$; (2) E_2: B fails in using the oracle replay technique to generate one more valid ciphertext. Clearly, $\Pr[\neg E_1] = 1/q_{cu}$. From the forking lemma [29], we

TABLE 3: Timings needed to perform atomic operations and representation of group elements in bits.

Curves	Relative timings (1 unit = 1 scalar multiplication in G)			Representation sizes (bits)	
	m	e	p	$|G|$	$|G_T|$
MNT/80	1	36	150	171	1026
SS/80	1	4	20	512	1024

know that $\Pr[\neg E_2] \geq 1/9$. Thus, we have that if A_{II} produces a valid forgery, then B will succeed in solving the CDH problem with probability $\varepsilon' = \Pr[\neg E_1 \wedge \neg E_2] \geq 1/(9q_{cu})$. □

5.3. Performance. To evaluate the performance of our new CBSC scheme, we compare our scheme with the previous CBSC schemes in terms of the computational cost and the communicational cost.

In the computational cost comparison, we consider four major operations: pairing, exponentiation in G_T, scalar multiplication in G, and hash. Among these operations, the pairing is considered as the heaviest time-consuming one in spite of the recent advances in the implementation technique. For simplicity, we denote these operations by p, e, m, and h, respectively. In the communicational cost comparison, ciphertext overhead represents the difference (in bits) between the ciphertext length and the message length, $|id|$ denotes the bit-length of user's identity, and $|G|$ and $|Z_p|$ denote the bit-length of an element in G and Z_p, respectively. Without considering precomputation, the performances of the compared CBSC schemes are listed in Table 2.

The efficiency of a pairing-based cryptosystem always depends on the chosen curve. Boyen [30] computes estimated relative timings for all atomic asymmetric operations (exponentiations and pairings) and representation sizes for group elements when instantiated in supersingular curves with 80-bit security (SS/80) and MNT curves with 80-bit security (MNT/80). In Table 3, we recall the data from [30].

To make a much clearer comparison, Table 4 gives the concrete values of the computational cost and the communicational cost for the compared CBE schemes according to the data in Table 3. As the hash operation is much more efficient than the multiplication in the group G, the costs of the hash operations are ignored.

In our proposed CBSC scheme, the *Signcrypt* algorithm does not require computing any time-consuming pairings. It only needs to compute two exponentiations in G_T, three scalar multiplications in G and three hashes in each signcryption operation. The *Designcrypt* algorithm needs to compute two pairings, two exponentiations in G_T, one

TABLE 4: Performance comparison of the CBSC schemes.

Schemes	Signcryption cost	Designcryption cost	Ciphertext overhead
MNT/80			
Ours	75	373	342
[18]	187	487	342
[20]	155	602	$513 + |id|$
[21]	447	594	2390
SS/80			
Ours	11	49	1024
[18]	28	65	1024
[20]	25	82	$1536 + |id|$
[21]	59	76	3072

scalar multiplication in G, and three hashes to designcrypt a ciphertext. From Tables 2 and 4, we can see that our scheme is more efficient than the previous CBSC scheme, especially in the computational efficiency. Actually, the computational performance of our scheme can be further optimized when $H_1(id_U, PK_U)P + P_{pub}$ can be precomputed. Such a precomputation enables us to additionally reduce one scalar multiplication computation in G and one hash computation in both the *Signcrypt* algorithm and the *Designcrypt* algorithm. In addition and most importantly, it is believed that our scheme is the first signcryption scheme in the certificate-based cryptographic setting that achieves security against both the public key replacement attacks and the insider attacks.

6. Conclusions

In this paper, we have introduced an improved security model of CBSC that captures both public key replacement attack and insider security. Our cryptanalysis has shown that Luo et al.'s CBSC scheme [20] is insecure in our security model. We have proposed a new CBSC scheme that resists both the key replacement attacks and the insider attacks. Compared with the previous CBSC schemes in the literature, the proposed scheme enjoys better performance, especially in the computation efficiency. However, a limitation of our schemes is that its security can only be achieved in the random oracle model [31]. Therefore, it would be interesting to construct a secure CBSC scheme without random oracles.

Conflict of Interests

The authors declare that there is no conflict of interests regarding the publication of this paper.

Acknowledgments

The authors would like to thank the anonymous referees for their helpful comments. This work is supported by the National Natural Science Foundation of China (Grant no. 61272542).

References

[1] A. Shamir, "Identity-based cryptosystems and signature schemes," in *Proceedings of the Advances in Cryptology (CRYPTO '84)*, pp. 47–53, 1984.

[2] S. S. Al-Riyami and K. G. Paterson, "Certificateless public key cryptography," in *Proceedings of the 9th International Conference on the Theory and Application of Cryptology and Information Security (ASIACRYPT '03)*, pp. 452–473, 2003.

[3] C. Gentry, "Certificate-based encryption and the certificate revocation problem," in *Proceedings of the International Conference on the Theory and Applications of Cryptographic Techniques (EUROCRYPT '03)*, pp. 272–293, 2003.

[4] S. S. Al-Riyami and K. G. Paterson, "CBE from CL-PKE: a generic construction and efficient schemes," in *Proceedings of the 8th International Workshop on Theory and Practice in Public Key Cryptography (PKC '05)*, pp. 398–415, January 2005.

[5] P. Morillo and C. Ràfols, "Certificate-based encryption without random oracles," Tech. Rep. 2006/12, Cryptology ePrint Archive, http://eprint.iacr.org/2006/012.pdf.

[6] D. Galindo, P. Morillo, and C. Ràfols, "Improved certificate-based encryption in the standard model," *Journal of Systems and Software*, vol. 81, no. 7, pp. 1218–1226, 2008.

[7] J. K. Liu and J. Zhou, "Efficient certificate-based encryption in the standard model," in *Proceedings of the 6th International Conference on Security and Cryptography for Networks*, pp. 144–155, 2008.

[8] Y. Lu, J. Li, and J. Xiao, "Constructing efficient certificate-based encryption with paring," *Journal of Computers*, vol. 4, no. 1, pp. 19–26, 2009.

[9] Z. Shao, "Enhanced certificate-based encryption from pairings," *Computers and Electrical Engineering*, vol. 37, no. 2, pp. 136–146, 2011.

[10] Y. Lu and J. Li, "Constructing certificate-based encryption secure against key replacement attacks," *ICIC Express Letters B: Applications*, vol. 3, no. 1, pp. 195–200, 2012.

[11] B. G. Kang, J. H. Park, and S. G. Hahn, "A certificate-based signature scheme," in *Proceedings of the Cryptographers' Track at the RSA Conference (CT-RSA '04)*, pp. 99–111, 2004.

[12] M. H. Au, J. K. Liu, W. Susilo, and T. H. Yuen, "Certificate based (linkable) ring signature," in *Proceedings of the 3rd Information Security Practice and Experience Conference*, pp. 79–92, 2007.

[13] J. Li, X. Huang, Y. Mu, W. Susilo, and Q. Wu, "Certificate-based signature: security model and efficient construction," in *Proceedings of the 4th European PKI Workshop*, pp. 110–125, 2007.

[14] J. K. Liu, J. Baek, W. Susilo, and J. Zhou, "Certificate based signature schemes without pairings or random oracles," in *Proceedings of the 11th International conference on Information Security*, pp. 285–297, 2008.

[15] W. Wu, Y. Mu, W. Susilo, and X. Huang, "Certificate-based signatures revisited," *Journal of Universal Computer Science*, vol. 15, no. 8, pp. 1659–1684, 2009.

[16] J. Li, X. Huang, Y. Mu, W. Susilo, and Q. Wu, "Constructions of certificate-based signature secure against key replacement attacks," *Journal of Computer Security*, vol. 18, no. 3, pp. 421–449, 2010.

[17] Y. Zheng, "Digital signcryption or how to achieve cost(signature & encryption) ≪ cost(signature) + cost(encryption)," in *Proceedings of the 17th Annual International Cryptology Conference (CRYPTO '97)*, pp. 165–179, 1997.

[18] F. Li, X. Xin, and Y. Hu, "Efficient certificate-based signcryption scheme from bilinear pairings," *International Journal of Computers and Applications*, vol. 30, no. 2, pp. 129–133, 2008.

[19] L. Chen and J. Malone-Lee, "Improved identity-based signcryption," in *Proceedings of the 8th International Workshop on Theory and Practice in Public Key Cryptography (PKC '05)*, pp. 362–379, January 2005.

[20] M. Luo, Y. Wen, and H. Zhao, "A certificate-based signcryption scheme," in *Proceedings of the International Conference on Computer Science and Information Technology (ICCSIT '08)*, pp. 17–23, September 2008.

[21] J. Li, X. Huang, M. Hong, and Y. Zhang, "Certificate-based signcryption with enhanced security features," *Computers and Mathematics with Applications*, 2012.

[22] J. H. Park and D. H. Lee, "On the security of status certificate-based encryption scheme," *IEICE Transactions on Fundamentals of Electronics, Communications and Computer Sciences*, vol. E90-A, no. 1, pp. 303–304, 2007.

[23] J. An, Y. Dodis, and T. Rabin, "On the security of joint signature and encryption," in *Proceedings of the International Conference on the Theory and Applications of Cryptographic Techniques (EUROCRYPT '02)*, pp. 83–107, 2002.

[24] J. Baek, R. Steinfeld, and Y. Zheng, "Formal proofs for the security of signcryption," *Journal of Cryptology*, vol. 20, no. 2, pp. 203–235, 2007.

[25] A. W. Dent, "Hybrid signcryption schemes with insider security," in *Proceedings of the 10th Australasian Conference on Information Security and Privacy (ACISP '05)*, pp. 253–266, July 2005.

[26] D. Boneh and M. Franklin, "Identity-based encryption from the weil pairing," *SIAM Journal on Computing*, vol. 32, no. 3, pp. 586–615, 2003.

[27] S. Mitsunari, R. Sakai, and M. Kasahara, "A new traitor tracing," *IEICE Transactions on Fundamentals of Electronics, Communications and Computer Sciences*, vol. E85-A, no. 2, pp. 481–484, 2002.

[28] S. D. Selvi, S. S. Vivek, D. Shukla, and P. R. Chandrasekaran, "Efficient and provably secure certificateless multi-receiver signcryption," in *Proceedings of the 2nd International Provable Security (ProvSec '08)*, pp. 52–67, 2008.

[29] D. Pointcheval and J. Stern, "Security arguments for digital signatures and blind signatures," *Journal of Cryptology*, vol. 13, no. 3, pp. 361–396, 2000.

[30] X. Boyen, "The BB_1 identity-based cryptosystem: a standard for encryption and key encapsulation," IEEE 1363.3, 2006, http://grouper.ieee.org/groups/1363/IBC/submissions/Boyen-bb1_ieee.pdf.

[31] M. Bellare and P. Rogaway, "Random oracles are practical: a paradigm for designing efficient protocols," in *Proceedings of the 1st ACM Conference on Computer and Communications Security*, pp. 62–73, November 1993.

16

Reputation-Based Secure Sensor Localization in Wireless Sensor Networks

Jingsha He,[1] **Jing Xu,**[2] **Xingye Zhu,**[1] **Yuqiang Zhang,**[2] **Ting Zhang,**[2] **and Wanqing Fu**[3]

[1] *School of Software Engineering, Beijing University of Technology, Beijing 100124, China*
[2] *College of Computer Science and Technology, Beijing University of Technology, Beijing 100124, China*
[3] *Information Center, SINOPEC Research Institute of Petroleum Processing, Beijing 100083, China*

Correspondence should be addressed to Jingsha He; jhe@bjut.edu.cn

Academic Editor: Yuxin Mao

Location information of sensor nodes in wireless sensor networks (WSNs) is very important, for it makes information that is collected and reported by the sensor nodes spatially meaningful for applications. Since most current sensor localization schemes rely on location information that is provided by beacon nodes for the regular sensor nodes to locate themselves, the accuracy of localization depends on the accuracy of location information from the beacon nodes. Therefore, the security and reliability of the beacon nodes become critical in the localization of regular sensor nodes. In this paper, we propose a reputation-based security scheme for sensor localization to improve the security and the accuracy of sensor localization in hostile or untrusted environments. In our proposed scheme, the reputation of each beacon node is evaluated based on a reputation evaluation model so that regular sensor nodes can get credible location information from highly reputable beacon nodes to accomplish localization. We also perform a set of simulation experiments to demonstrate the effectiveness of the proposed reputation-based security scheme. And our simulation results show that the proposed security scheme can enhance the security and, hence, improve the accuracy of sensor localization in hostile or untrusted environments.

1. Introduction

The technologies of wireless sensor networks (WSNs) are becoming popular along with the rapid advancement of wireless communication technology, more remarkable performance of integrated circuits as well as decrease in cost and increase in functionality of sensor nodes. Since WSNs are a kind of intelligence networks that are able to integrate data collection, fusion, and transmission, such networks have been widely used in fields such as military defense, industrial and agricultural control, urban management, environment monitoring, health care, emergency rescue, and disaster relief. In addition, sensor networks also have a broad prospect of applications in tracking logistics management and space exploration. Depending on different application scenarios in the above areas, researchers have put forward some new technology and strategy, such as sensor deployment methods suitable for underwater detection [1] and intelligent monitoring technologies in Smart Home scenarios [2]. In short,

the applications of WSNs are being developed to achieve ubiquity that can bring more convenience for human beings in many areas.

In most applications, sensor nodes are used to collect physical data, such as temperature, humidity, water level, pressure, and wind speed, that are sent along with the location information to the data center to ensure that the collected data have spatial meaning. Furthermore, the location information of sensor nodes can also serve as the basis for some network functions, such as network configuration and real-time statistics of network coverage. Therefore, in massively deployed WSNs, location information of sensor nodes is very important for enabling many applications, which makes sensor localization one of the basic services and a core technology for WSNs.

Since sensor localization in wireless sensor networks (WSNs) is a fundamental technical issue and is critical for monitoring applications and for most location-based routing protocols and services, research in sensor localization technology has generated a wide spread interest and various issues on different aspects have been studied, which include efficiency [3], accuracy [4], and security [5], among many hot issues in sensor localization.

Current algorithms for sensor localization fall into two categories: range-free algorithms [6] and range-based algorithms [7]. In a range-free algorithm, such as Centroid [8] or CTDV-Hop [9], a node estimates its location using information of connectivity between different nodes. In a range-based algorithm, a sensor node estimates its own location based on information about distances or angles between sensor nodes and through using techniques such as time of arrival (TOA) [10], time difference of arrival (TDOA) [11], received signal strength indicator (RSSI) [12], and angle of arrival (AOA) [13] as well as methods such as trilateration, triangulation, or maximum likelihood estimation [14]. Among the many different sensor localization algorithms, RSSI-based positioning technology is perhaps the most popular due to its low cost and easy implementation. On the other hand, sensor localization results can be greatly affected by malicious nodes in hostile or untrusted environments. This is because sensor nodes can hardly perform accurate localization if they use location information that is provided by untrusted beacon nodes. Security in sensor localization has thus received a great deal of attention along with the development of sensor localization technologies for WSNs.

In the past few years, researchers have proposed several security strategies for sensor localization from different aspects. Some of the methods implement verification measures to reduce the impact of using unreliable or false location information [15] while some others apply a series of schemes in which temporal, spatial, and consistent properties are considered to deal with distance-consistent spoofing attacks [16]. However, in these schemes, sensor nodes are divided into just two types: secure and insecure sensor nodes through the mechanisms of comparing the nodes and their behavior against normal situations. However, such an approach cannot be very objective, which could cause many false positive and false negative results.

Meanwhile, some other researchers have proposed localization methods that are able to fight against attacks launched by compromised sensor nodes, a problem that is more difficult to deal with. Liu et al. proposed robust computing algorithms to improve the reliability of localization schemes [17]. Park and Shin proposed an attack-tolerant localization protocol that would perform adaptive management of a profile for normal localization behavior [18]. However, the limitation of these schemes is that they did not consider the security of sensor localization when sensor nodes are joining and leaving the network along with the passage of time. In addition, they did not pay enough attention to secure sensor localization in dynamic wireless networks.

As an effective means of ensuring security, the notion of reputation has been introduced and some reputation-based schemes have since been proposed for sensor localization. Srinivasan et al. proposed a distributed reputation-based beacon trust system [19] and Xu et al. proposed a reputation-based revising scheme for sensor localization which would incur high computation cost [20]. However, complicated reputation evaluation in the above schemes for sensor localization makes it necessary to further improve the efficiency of evaluation for beacon nodes. Any sensor localization method that can achieve good performance should ensure the reliability of location information before such information can be actually used for sensor localization.

In real applications, there may be other types of sensor localization methods to fit different application scenarios. Therefore, specific localization methods in real applications need to be continuously developed and improved based on orientation methods in order to adapt basic sensor localization schemes to the many different network scenarios. Consequently, in order to develop effective sensor localization methods, we should analyze and understand the main characteristics of specific networks and develop proper performance metrics that can be used to measure the performance of sensor localization schemes. In addition, we should also consider limitations of wireless sensor networks such as constrained energy supply in the sensor nodes as well as the complexity of network environments in the development of effective sensor localization methods.

In this paper, we propose a novel reputation-based secure sensor localization scheme to improve the accuracy of sensor localization for WSNs in hostile environments. In the proposed reputation model, the reputation of each beacon node is evaluated by each other to ensure that sensor nodes will get credible location information to perform sensor localization. The proposed scheme can therefore effectively reduce the impact of malicious beacon nodes on the localization of regular sensor nodes by relying on the security mechanism of beacon node evaluation. Our simulation results show that the proposed reputation-based secure sensor localization scheme can improve the accuracy of sensor localization in hostile or untrusted environments. In addition, the proposed secure sensor localization scheme possesses the desirable characteristics of expandability and flexibility since it can be used in both static and dynamic networks.

The remainder of this paper is structured as follows. In Section 2, we present a reputation model in which we first describe the network model and then propose a reputation evaluation model. In Section 3, we present our sensor localization scheme which is based on the evaluation of the reputation of beacon nodes. In Section 4, we describe the simulation that we have performed and present the simulation results. Finally, in Section 5, we conclude this paper in which we also discuss some future work.

2. The Reputation Model

In hostile network environments, which most current WSN deployments would assume, regular sensor nodes need to be confronted with security threats during the process of sensor localization. If a sensor node can identify the security and credibility of location information that it receives and subsequently use the information appropriately, the accuracy of sensor localization can be greatly improved or ensured in such environments. Therefore, in order to develop effective sensor localization schemes, we should understand the main characteristics of the specific networks as well as the performance goals of the localization schemes. To achieve the above objective, we need to consider such characteristics as resource constraints in the sensor nodes and the complexity of the environment where the sensor nodes are deployed. Any sensor localization scheme must be effectively working in a specific WSN after the above-mentioned factors are considered in the design.

To achieve the above goal, we first propose a reputation scheme to be used in the sensor localization scheme we will propose later in this paper to deal with a hostile deployment environment in which malicious nodes can be dropped into the network at will and regular sensor nodes can also be easily compromised to make them behave in a malicious manner. We call the scheme that we propose the reputation-based localization scheme (RBL). The main characteristics of the reputation model and the RBL are as follows.

(1) The proposed secure sensor localization scheme is developed based on a reputation model and on the evaluation of all the beacon nodes for deriving a reputation value for each and every beacon node.

(2) In the reputation model, the reputation of each beacon node is evaluated and consequently used by regular sensor nodes to determine the credibility of the location information provided by the beacon node.

(3) In the reputation model, the reputation of each beacon node is updated continuously with the passage of time if sensor localization needs to be carried out from time to time.

In the following sections, we will first describe the network model followed by the threat model and the reputation model.

2.1. The Network Model. The WSN under consideration is composed of beacon nodes and regular sensor nodes. Beacon nodes are capable of positioning themselves (e.g., by determining their positions through GPS) while the regular sensor nodes need to locate their own positions based on position information from other nodes, especially from the beacon nodes.

Our sensor localization method in this paper requires that a regular sensor node first estimates its relative position to some of the beacon nodes through the means of receiving signals from creditable beacon nodes and by computing the distances between them using a signal attenuation formula. Then, the sensor node estimates its position using the maximum likelihood estimation method [21] after it has collected enough position information.

2.2. The Threat Model. An analysis of the network model described above indicates that position information received from beacon nodes and the estimation of relative positions between a regular sensor node and the referenced beacon nodes can determine the accuracy of sensor localization. There are, however, two primary types of security threats for the network model as described below.

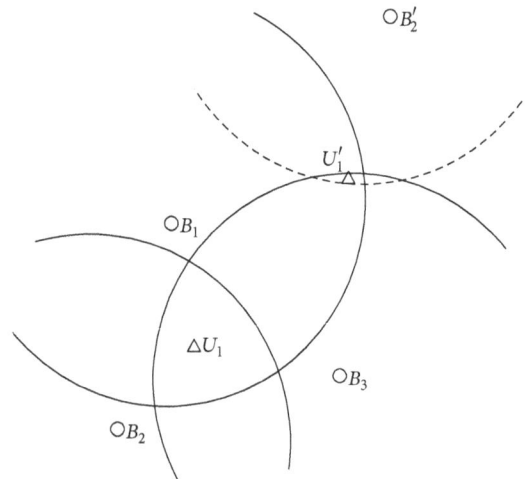

FIGURE 1: An example of sensor localization in hostile environments.

(1) Sending false beacon information: if malicious beacon nodes send false position information, such information received by regular sensor nodes may not be accurate. Then, the estimated position of a regular sensor node cannot be guaranteed to be accurate and will lose its credibility because it is calculated based on received false information from beacon nodes. The impact to sensor localization from this type of attacks is shown in Figure 1. We can see from the figure that B_2' is the false position of beacon node B_2, which makes sensor node U_1 receive a false localization result U_1'.

(2) Obstructing physical property: if malicious nodes interfere with normal signals from beacon nodes, no regular sensor node would be able to estimate its relative position to the beacon nodes accurately by the means of signal attenuation, leading to reduced accuracy for sensor localization.

Consequently, the scheme that we propose in this paper needs to deal with the potential threats that result from the above two types of attacks in order to improve the accuracy of sensor localization in WSNs.

2.3. The Proposed Reputation Model. To deal with the above security threats, we propose a novel reputation model for sensor localization in WSNs. In the reputation model, beacon nodes evaluate each other using information such as the characteristics about the perception of positions and provide the evaluation results to the regular sensor nodes. The regular sensor nodes use the evaluation results provided by the beacon nodes to rank the beacon nodes and base the credibility of the location information provided by beacon nodes on such ranking.

First, let us make the following assumptions in our reputation model.

(1) The reputation value for each and every beacon node is a number between 0 and 1, indicating values from the lowest to the highest reputations.

(2) The reputation value for each and every beacon node is initialized to be 0.5, a medium value to start with.

(3) In the reputation model, each beacon node performs evaluation only on its neighboring beacon nodes, that is, the beacon nodes that are one hop away from it.

Pseudocode 1 contains the pseudocode for our proposed reputation model for beacon node B_j and sensor node U_m.

The details of the evaluation procedure in the proposed reputation model are as follows.

(1) Beacon node B_i sends its coordinate (x_i, y_i) to its neighboring beacon nodes.

(2) Each neighboring beacon node to B_i will calculate its distance to B_i using the received coordinate information and the signal strength information independently. Let $l_{B_{ji}}$ denote the distance between B_j and B_i based on the coordinate information and let $d_{B_{ji}}$ denote the distance based on the signal strength information. B_j can then calculate $l_{B_{ji}}$ using the coordinate information from B_i and calculate $d_{B_{ji}}$ through a signal strength ranging algorithm based on the strength of the signals received from B_i.

(3) All the neighboring beacon nodes evaluate the reputation of B_i. The value of reputation evaluation is determined using (1) in which $R^t_{B_{ji}}$ and $R^{t+\Delta t}_{B_{ji}}$ denote the reputation values on B_i by B_j at times t and $t + \Delta t$, respectively, and Δt denotes the time interval of two reputation values. Let Δd be the threshold for the distance, that is, the error that can be tolerated for the distance, and let α be the weight of the evaluation value which is determined using (2)

$$R^{t+\Delta t}_{B_{ji}} = \alpha \times R^t_{B_{ji}} + (1 - \alpha), \quad \left| l_{B_{ji}} - d_{B_{ji}} \right| \leq \Delta d$$
$$R^{t+\Delta t}_{B_{ji}} = (1 - \alpha) \times R^t_{B_{ji}}, \quad \left| l_{B_{ji}} - d_{B_{ji}} \right| > \Delta d, \tag{1}$$

$$\alpha = \frac{\left| l_{B_{ji}} - d_{B_{ji}} \right|}{l_{B_{ji}} + d_{B_{ji}}}. \tag{2}$$

(4) The neighboring sensor nodes get the reputation values for B_i from the beacon nodes. Each regular sensor node collects the evaluation values from all the neighboring beacon nodes and computes the average reputation value using (3) in which $R^{t+\Delta t}_{U_m, B_i}$ and $R^{t+\Delta t}_{B_{ki}}$ denote the reputation value on beacon node B_i from a sensor node U_m and that on B_i evaluated by B_k at time $t + \Delta t$, where B_k is a neighboring beacon node to B_i and n is the number of such neighboring nodes. Consider

$$R^{t+\Delta t}_{U_m, B_i} = \frac{\sum_{k=1}^{n} R^{t+\Delta t}_{B_{ki}}}{n}. \tag{3}$$

(5) Every regular sensor node ranks the neighboring beacon nodes from high to low based on the received reputation values.

3. The Sensor Localization Scheme

The sensor localization scheme in this paper uses the proposed reputation evaluation scheme described above in which the reputation model is relied upon by the beacon nodes to evaluate each other. In the illustration below, we use the RSSI ranging technology for sensor localization although the same reputation scheme can be applied equally to TOA, TDOA, and AOA ranging methods in practical applications.

After receiving the evaluation results, a regular sensor node will select credible beacon nodes based on the reputation values. Afterwards, the sensor node will measure the distance to the credible beacon nodes using the RSSI ranging technology and estimate its location through maximum likelihood estimation. The main steps in our localization scheme are described as follows.

(i) Every beacon node provides its location information to all the neighbor nodes. As shown in Figure 2, beacon node B_1 sends its position coordinate (x_1, y_1) to all the neighboring nodes.

(ii) Beacon nodes in the network will evaluate each other using the proposed reputation model and each will send its evaluation results to all the neighboring nodes. As shown in Figure 2, beacon nodes B_2, B_3, B_4, and B_5 evaluate the reputation of B_1 after receiving the location information from B_1 using the proposed reputation model and each will send the evaluation result to their neighboring sensor nodes including node U_1.

(iii) Each regular sensor node will select credible beacon nodes based on the results from the reputation evaluation. Sensor node U_1 computes the reputation value for B_1 and collects the reputation values from neighboring beacon nodes using (3). Then, U_1 ranks the neighboring beacon nodes according to the reputation values in the order of high to low, based on which it selects the credible beacon nodes accordingly.

(iv) Regular sensor nodes estimate their relative positions to the credible beacon nodes using the signal attenuation formula in RRSI [12].

(v) Regular sensor nodes calculate their coordinates using maximum likelihood estimation. Suppose that the number of credible neighboring beacon nodes around U_1 is p with coordinates (x_1, y_1), $(x_2, y_2), \ldots, (x_p, y_p)$, respectively, and the distances between $U_1(x_{U_1}, y_{U_1})$ and the beacon nodes are $d_1, d_2, \ldots, d_p$, respectively; then the position of U_1 can be calculated using the following:

$$\left(x_{U_1} - x_i \right)^2 + \left(y_{U_1} - y_i \right)^2 = d_i^2, \quad i = 1, 2, \ldots, p. \tag{4}$$

In addition, p distance equations about U_1 and the p beacon nodes are listed as in (5) that result from subtracting

```
// Beacon nodes evaluate each other between neighbor beacon nodes
R_{B_{ji}}^t = 0.5
while(true)
        α = |l_{B_{ji}} − d_{B_{ji}}| / (l_{B_{ji}} + d_{B_{ji}})
        if |l_{B_{ji}} − d_{B_{ji}}| ≤ Δd
                R_{B_{ji}}^{t+Δt} = α × R_{B_{ji}}^t + (1 − α)
        else
                R_{B_{ji}}^{t+Δt} = (1 − α) × R_{B_{ji}}^t
        sleep(Δt)
                // Beacon nodes send the reputation value to their neighbor sensor nodes
                Send (node B_j, node U_m, reputation value)
                // Sensor nodes compute the reputation value of their neighbor beacon nodes
        R_{U_m,B_i}^{t+Δt} = (Σ_{k=1}^n R_{B_{ki}}^{t+Δt}) / n
```

PSEUDOCODE 1: Pseudocode for the reputation model.

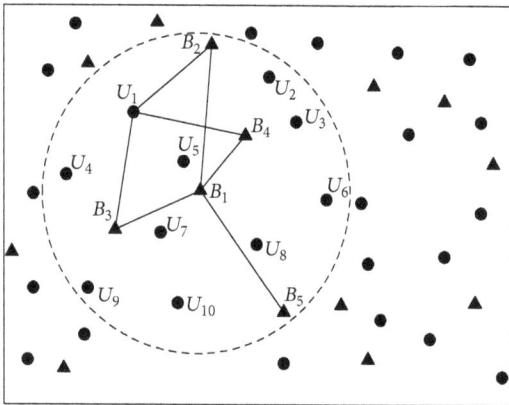

FIGURE 2: The network topology.

the last equation from each of the first $p − 1$ equations. Consider

$$x_1^2 - x_p^2 - 2\left(x_1 - x_p\right)x_{U_1} + y_1^2 - y_p^2$$
$$- 2\left(y_1 - y_p\right)y_{U_1} = d_1^2 - d_p^2$$
$$\cdots \tag{5}$$
$$x_{p-1}^2 - x_p^2 - 2\left(x_{p-1} - x_p\right)x_{U_1} + y_{p-1}^2 - y_p^2$$
$$- 2\left(y_{p-1} - y_p\right)y_{U_1} = d_{p-1}^2 - d_p^2.$$

$U_1(x_{U_1}, y_{U_1})$ can then be calculated using the following:

$$U_1 = A^{-1}b. \tag{6}$$

The matrices in (6) can then be expressed as the following expressions:

$$A = 2\begin{bmatrix} x_1 - x_p & y_1 - y_p \\ \cdots & \\ x_{p-1} - x_p & y_{p-1} - y_p \end{bmatrix},$$

$$b = \begin{bmatrix} x_1^2 - x_p^2 + y_1^2 - y_p^2 - d_1^2 + d_p^2 \\ \cdots \\ x_{p-1}^2 - x_p^2 + y_{p-1}^2 - y_p^2 - d_{p-1}^2 + d_p^2 \end{bmatrix}, \tag{7}$$

$$U_1 = \begin{bmatrix} x_{U_1} \\ y_{U_1} \end{bmatrix}.$$

The final solution to (6) can be obtained using the following:

$$U = \left(A^T A\right)^{-1} A^T b. \tag{8}$$

From the above steps, we can see that a regular sensor node would treat the location information from neighboring beacon nodes differently according to the result of reputation evaluation. There is no need to determine the position relationship between regular sensor nodes and beacon nodes that have low reputation values, which are required in the signal attenuation formula, resulting in reducing a certain amount of computational overhead.

4. Simulation and Analysis

We have performed some simulation on wireless sensors localization with the proposed RBL to evaluate the performance of the scheme.

The network configuration for the simulation is set up as follows. The regular sensor nodes and beacon nodes are deployed randomly in an area of 650 m × 600 m. The transmission radius of each beacon and sensor node is set

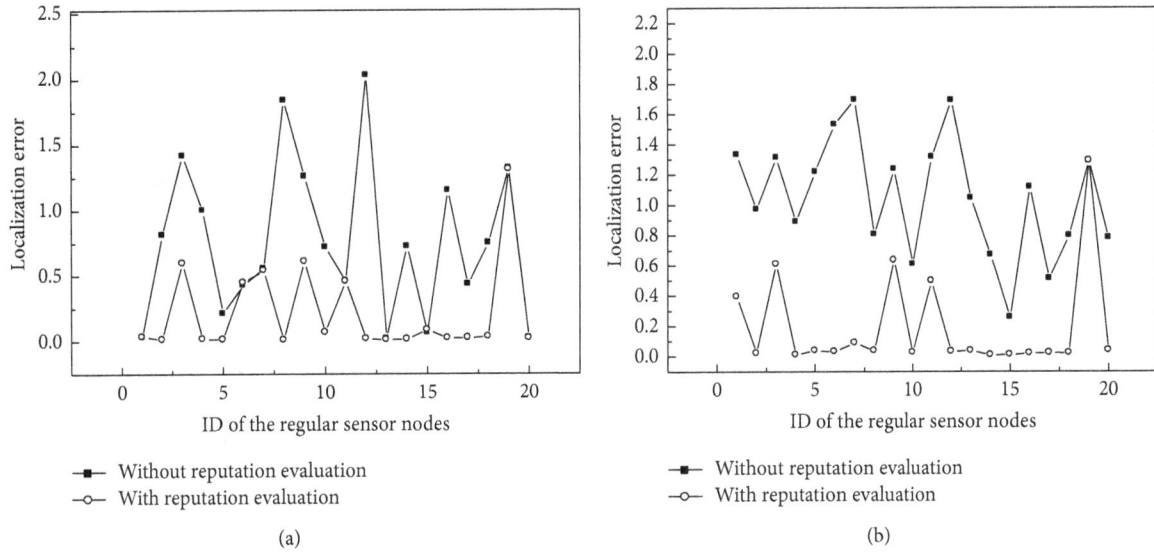

FIGURE 3: Sensor localization error with a different number of malicious beacon nodes: (a) 10 normal and 5 malicious beacon nodes; (b) 10 normal and 10 malicious beacon nodes.

at 200 m. There exist some malicious beacon nodes that randomly send out false location information.

Localization error is one important indicator of the performance in sensor localization for WSNs, which is calculated using (9). In the formula, (x_{U_m}, y_{U_m}) and (x'_{U_m}, y'_{U_m}) denote the measured coordinates and the actual coordinates for node U_m, respectively, R denotes the transmission radius of the nodes, and e_m is the localization error. Consider

$$e_m = \frac{\sqrt{\left(x_{U_m} - x'_{U_m}\right)^2 + \left(y_{U_m} - y'_{U_m}\right)^2}}{R}.$$ (9)

The localization error from the simulation for 20 sensor nodes is shown in Figure 3. We can see from the figure that reputation evaluation is effective for reducing localization error in hostile environments and the improvement is more significant as the number of malicious beacon nodes increases.

There are two types of threats in sensor localization: attacks targeted at the nodes and attacks targeted at the location information. RBL evaluates the credibility of beacon nodes by evaluating the location information that beacon nodes provide in order to reduce the influence of compromised beacon nodes on localization results and to resist the threat of location information tampering by the malicious beacon nodes. To measure the capability of RBL on countering the above security threats, in our evaluation, we deploy 40 regular sensor nodes to expand the scale of our experiment in which we measure the average localization error using (10) where N denotes the number of regular sensor nodes in the network. The average localization error from our simulation on a network in which there exist one or more compromised beacon nodes is shown in Figure 4. We can see from the figure that although the average localization error fluctuates with the number and the locations of the regular sensor nodes,

the result of RBL is much better than that of the primary localization scheme (PLS) using RRSI in which no evaluation of beacon nodes is performed. Consider

$$\bar{e} = \frac{\sum_{i=1}^{N} e_i}{N}.$$ (10)

Since WSNs possess the characteristics of dynamic network topology, an advanced secure sensor localization scheme should not only be able to ensure the security of sensor localization in a static network, but also be able to handle the cases of nodes joining, leaving, and removing from the network. We have performed some simulations on sensor localization for the above scenarios.

In order to expand the coverage of beacon nodes in a network so as to make more regular sensor nodes the neighbors of the beacon nodes in the network, consequently improving the utilization of beacon information, we can increase the signal transmission power of the beacon nodes to effectively expand the signal transmission radius of the beacon nodes.

In the simulations, we first deploy 20 regular nodes and 4 normal beacon nodes in the area. Then, we add more beacon nodes into the network at the rate of one node per minute starting at the moment of 1.5 min with normal and malicious beacon nodes being added alternately. Malicious beacon nodes that are added into the network would send out false position information randomly while normal beacon nodes always send out their real position information. Figure 5 shows the average localization error for regular sensor nodes during the first seven minutes from which we can see that the average localization error for regular sensor nodes fluctuates noticeably in the primary localization scheme but exhibits a good performance in our proposed RBL.

We have also performed some simulations to evaluate the impact of nodes leaving the network on sensor localization.

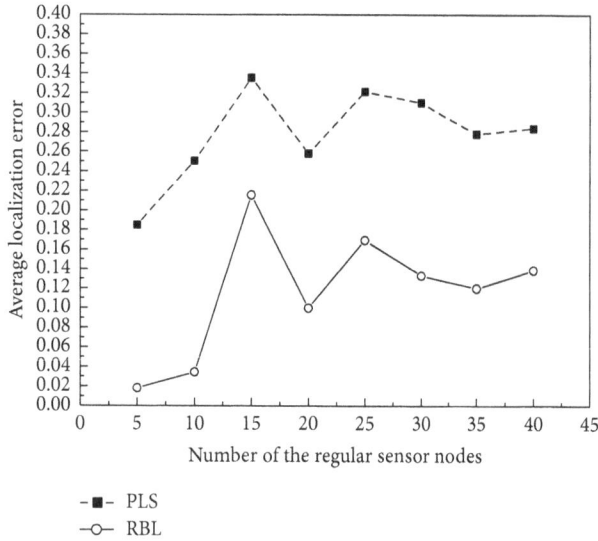

FIGURE 4: Average sensor localization error with a varying number of regular sensor nodes.

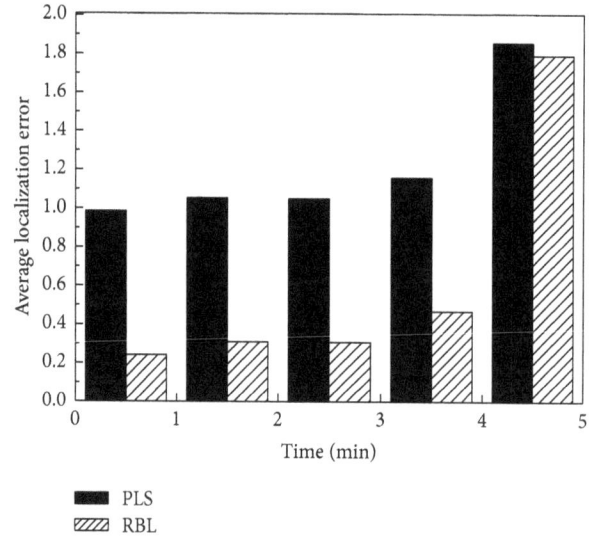

FIGURE 6: Average sensor localization error as beacon nodes are removed from the network.

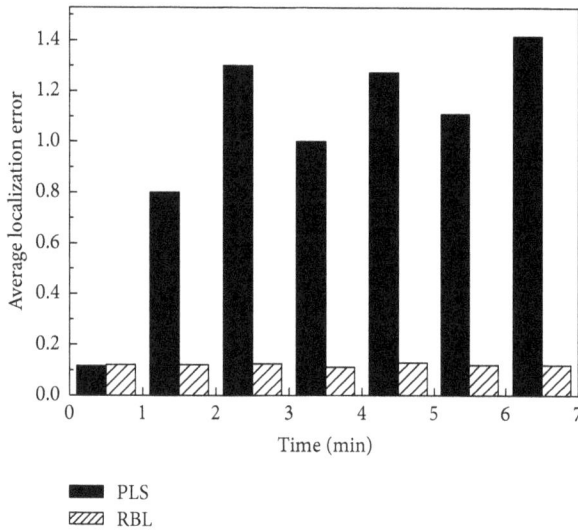

FIGURE 5: Average sensor localization error as beacon nodes are added into the network.

TABLE 1: State situations for the beacon nodes.

Categories	1	2	3	4
Situations	$RP_1 \cap CP_2$	$RP_2 \cap CP_1$	$RP_2 \cap CP_2$	$RP_2 \cap CP_3$

In the simulation, we first deploy 20 regular nodes, 4 normal beacon nodes, and 4 malicious beacon nodes in the area. Then, we remove the beacon nodes from the network at the rate of one node per 1 minute starting at the moment of 1.5 min with normal and malicious beacon nodes being removed alternately. Again, normal beacon nodes always claim their real positions while malicious beacon nodes would send out false position information randomly. Figure 6 shows the average sensor localization error for regular sensor nodes during the process from which we can see that the average localization error of RBL is noticeably lower than that of PLS in most cases. However, when the number of normal beacon nodes falls below three in the whole network,

the advantage would disappear, which seems to be a limitation of the current RBL.

Lastly, we evaluate the impact of status change among the existing beacon nodes on regular sensor nodes under the assumption that the total number of beacon nodes remains the same. Four possibilities exist for such status change as illustrated in Table 1 in which RP_1 and RP_2 represent the cases in which a beacon node does not change its real position and changes its real position, respectively, and CP_1, CP_2, and CP_3 represent the cases in which a beacon node does not change its claimed position information, changes its claimed position information randomly, and changes its claimed position information consistently, respectively.

We perform evaluations for four scenarios. In the first evaluation, we deploy 20 regular sensor nodes and 8 normal beacon nodes in the area and then change the status of 4 beacon nodes to the state that corresponds to situation 1 in Table 1 gradually during a 4-minute time period starting at the moment of 1.5 min. In the second evaluation, we deploy 20 regular sensor nodes and 8 normal beacon nodes in the area and then change the status of 4 beacon nodes to the state that corresponds to situation 2 in Table 1 gradually during a 4-minute time period starting at the moment of 1.5 min. In the third evaluation, we deploy 20 regular sensor nodes and 8 normal beacon nodes in the area and then change the status of 4 beacon nodes to the state that corresponds to situation 3 in Table 1 gradually during a 4-minute time period starting at the moment of 1.5 min. In the last evaluation, we deploy 20 regular sensor nodes and 8 normal beacon nodes in

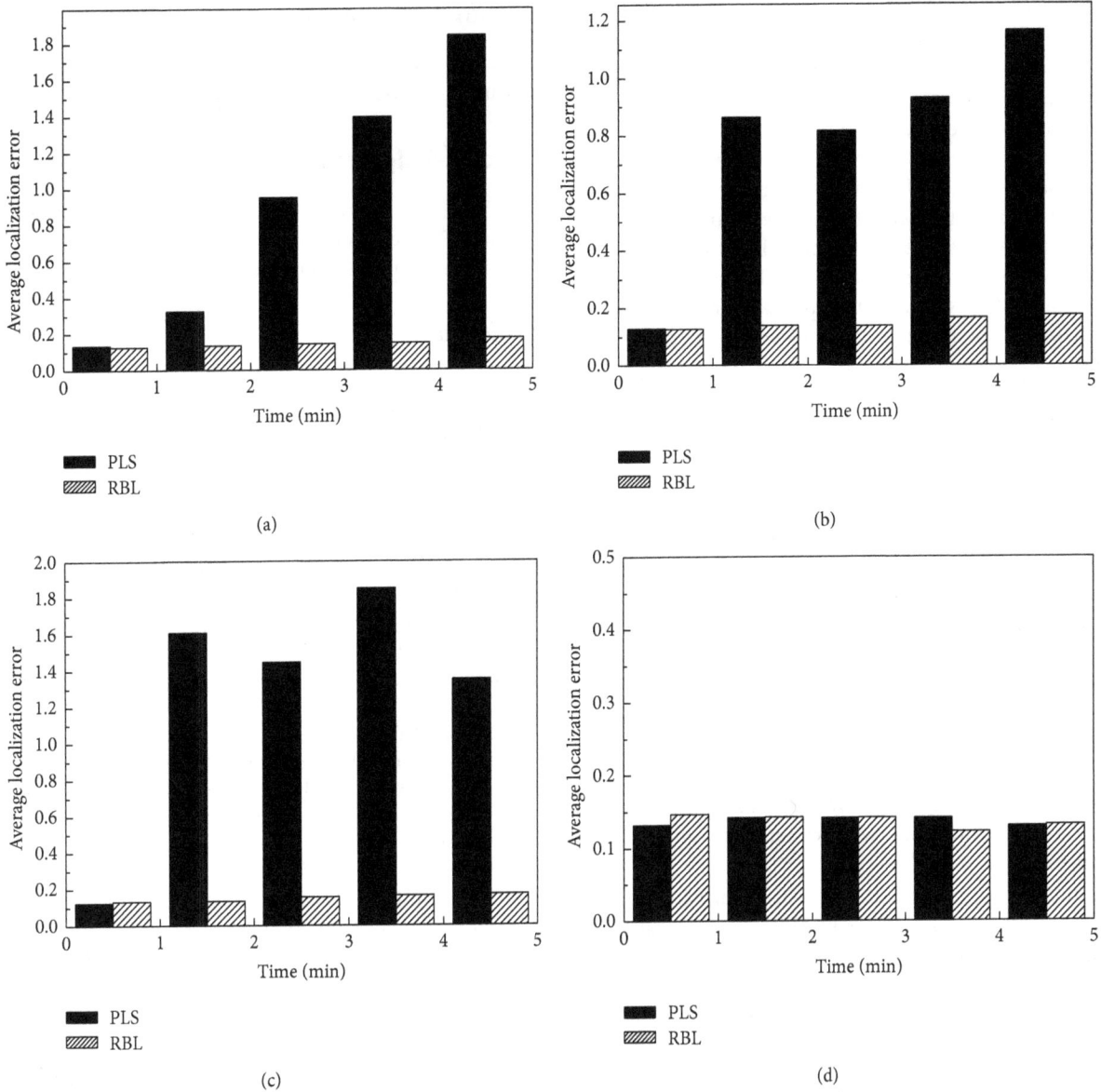

FIGURE 7: Average sensor localization error as beacon nodes change their status in various scenarios.

the area and then change the status of 4 beacon nodes to the state that corresponds to situation 4 in Table 1 gradually during a 4-minute time period starting at the moment of 1.5 min.

Figures 7(a), 7(b), 7(c), and 7(d) show the results of the evaluations that correspond to the above four evaluation scenarios. We can see from the figure that RBL can effectively filter out abnormal (or malicious) beacon nodes when some of the beacon nodes change their status in an unpredictable manner, which demonstrates that RBL is an effective scheme for secure sensor localization for WSNs, which clearly shows that RBL can improve the accuracy of sensor localization in hostile or untrusted environments.

In summary, so far, we have performed three sets of simulation experiments to verify the performance and

the effectiveness of the proposed security scheme for sensor localization in hostile or untrusted environments. In the first one, we evaluated the performance of localizing regular sensor nodes in the presence of a varying number of malicious beacon nodes. In the second one, we evaluated the average sensor localization error for different numbers of regular sensor nodes in hostile or untrusted environment. In the third one, we evaluated sensor localization results; when new beacon nodes join the network, existing beacon nodes leave the network and existing beacon nodes change their status that determines how they would make claims on their positions. It is clear that the purpose of the last experiment is to evaluate the influence on the localization of regular sensor nodes due to changes on the credibility of the beacon nodes. That is, the first two experiments are mainly aimed at showing the performance of RBL on secure sensor localization in static WSNs while the third one is aimed at verifying the effectiveness of RBL on secure sensor localization in dynamic WSNs. All the simulation results that we have obtained clearly show that RBL can reduce the effect of malicious beacon nodes on the localization of regular sensor node, thus allowing us to conclude that RBL can effectively improve the security and the accuracy of sensor localization in WSNs. The experiments also indicate that RBL can scale well with the size of the network and can be applied in dynamic WSNs, especially when new sensor nodes can join and existing sensor nodes can leave networks with the passage of time.

5. Conclusion

In this paper, we proposed a novel reputation model for regular sensor nodes to evaluate the credibility of beacon nodes in sensor localization. In the model, beacon nodes first evaluate each other and then provide the evaluation results to regular sensor nodes for them to determine the credibility of beacon nodes to ensure that they will receive and use credible position information from the beacon nodes in locating their own positions. The proposed security scheme can improve the accuracy of sensor localization in hostile or untrusted environments. The scheme can help to ensure the reliability of received location information under the scenario of signal attenuation by minimizing the effects of false location information as well as interfering signals caused by malicious beacon nodes.

In the future, we will extend our security scheme to counter other types of malicious attacks in sensor localization without incurring too much additional computational cost and communication overhead and to apply our reputation-based sensor localization scheme to different network environments to further verify and improve the scheme. We will also study the impact on evaluation due to other factors of sensor nodes to further improve the performance and usability of our secure sensor localization scheme in WSNs.

Notations

B_i: Beacon node i
U_m: Sensor node m
Δt: The time interval of the two reputation values
$l_{B_{ji}}$: The distance between beacon node j and i based on the coordinate information
$d_{B_{ji}}$: The distance between beacon node j and i based on the ranging techniques (such as received signal strength indicator)
$R^t_{B_{ki}}$: The reputation value on beacon node i from a beacon node k
$R^t_{U_m,B_i}$: The reputation value on beacon node i from a sensor node m at time t
(x_p, y_p): The coordinate of beacon node p
(x_{U_m}, y_{U_m}): The coordinate of sensor node m.

Conflict of Interests

The authors declare that there is no conflict of interests regarding the publication of this paper.

Acknowledgments

The work in this paper has been supported by National Natural Science Foundation of China (Grant no. 61272500) and Beijing Natural Science Foundation (Grant no. 4142008).

References

[1] E. F. Golen, B. Yuan, and N. Shenoy, "An evolutionary approach to underwater sensor deployment," *International Journal of Computational Intelligence Systems*, vol. 2, no. 10, pp. 184–201, 2009.

[2] J. C. Augusto, J. Liu, P. McCullagh, H. Wang, and J.-B. Yang, "Management of uncertainty and spatio-temporal aspects for monitoring and diagnosis in a smart home," *International Journal of Computational Intelligence Systems*, vol. 1, no. 4, pp. 361–378, 2008.

[3] S.-K. Yang and K.-F. Ssu, "An energy efficient protocol for target localization in wireless sensor networks," *World Academy of Science, Engineering and Technology*, vol. 56, no. 8, pp. 398–407, 2009.

[4] M. Boushaba, A. Hafid, and A. Benslimane, "High accuracy localization method using AoA in sensor networks," *Computer Networks*, vol. 53, no. 18, pp. 3076–3088, 2009.

[5] R. Sugihara and R. K. Gupta, "Sensor localization with deterministic accuracy guarantee," in *Proceedings of the IEEE INFOCOM*, pp. 1772–1780, April 2011.

[6] J. Park, Y. Lim, K. Lee, and Y.-H. Choi, "A polygonal method for ranging-based localization in an indoor wireless sensor network," *Wireless Personal Communications*, vol. 60, no. 3, pp. 521–532, 2011.

[7] Y. W. E. Chan and B. H. Soong, "A new lower bound on range-free localization algorithms in wireless sensor networks," *IEEE Communications Letters*, vol. 15, no. 1, pp. 16–18, 2011.

[8] N. Bulusu, J. Heidemann, and D. Estrin, "GPS-less low-cost outdoor localization for very small devices," *IEEE Personal Communications*, vol. 7, no. 5, pp. 28–34, 2000.

[9] H. Wu, M. Deng, L. Xiao, W. Wei, and A. Gao, "Cosine theorem-based DV-hop localization algorithm in wireless sensor networks," *Information Technology Journal*, vol. 10, no. 2, pp. 239–245, 2011.

[10] I. Güvenç and C.-C. Chong, "A survey on TOA based wireless localization and NLOS mitigation techniques," *IEEE Communications Surveys and Tutorials*, vol. 11, no. 3, pp. 107–124, 2009.

[11] A. Savvides, C.-C. Han, and M. B. Strivastava, "Dynamic fine-grained localization in ad-hoc networks of sensors," in *Proceedings of the 7th Annual International Conference on Mobile Computing and Networking*, pp. 166–179, July 2001.

[12] P. Bahl and V. N. Padmanabhan, "RADAR: an in-building RF-based user location and tracking system," in *Proceedings of the 19th Annual Joint Conference of the IEEE Computer and Communications Societies*, pp. 775–784, March 2000.

[13] D. Niculescu and B. Nath, "Ad hoc positioning system (APS) using AOA," in *Proceedings of the 22nd Annual Joint Conference on the IEEE Computer and Communications Societies*, pp. 1734–1743, April 2003.

[14] Y. Zhang, L. Bao, S.-H. Yang, M. Welling, and D. Wu, "Localization algorithms for wireless sensor retrieval," *Computer Journal*, vol. 53, no. 10, pp. 1594–1605, 2010.

[15] S. Čapkun, K. B. Rasmussen, M. Čagalj, and M. Srivastava, "Secure location verification with hidden and mobile base stations," *IEEE Transactions on Mobile Computing*, vol. 7, no. 4, pp. 470–483, 2008.

[16] H. Chen, W. Lou, and Z. Wang, "A novel secure localization approach in wireless sensor networks," *EURASIP Journal on Wireless Communications and Networking*, vol. 2010, pp. 1–12, 2010.

[17] D. Liu, P. Ning, A. Liu, C. Wang, and W. K. Du, "Attack-resistant location estimation in wireless sensor networks," *ACM Transactions on Information and System Security*, vol. 11, no. 7, pp. 22–39, 2008.

[18] T. Park and K. G. Shin, "Attack-tolerant localization via iterative verification of locations in sensor networks," *Transactions on Embedded Computing Systems*, vol. 8, no. 12, pp. 1–24, 2008.

[19] A. Srinivasan, J. Teitelbaum, and W. Jie, "DRBTS: distributed reputation-based beacon trust system," in *Proceedings of the 2nd IEEE International Symposium on Dependable, Autonomic and Secure Computing (DASC '06)*, pp. 277–283, October 2006.

[20] X. Xu, H. Jiang, L. Huang, H. Xu, and M. Xiao, "A reputation-based revising scheme for localization in wireless sensor networks," in *Proceedings of the IEEE Wireless Communications and Networking Conference (WCNC '10)*, pp. 1–6, April 2010.

[21] R.-I. Rusnac and A. Ş. Gontean, "Maximum Likelihood Estimation Algorithm evaluation for wireless sensor networks," in *Proceedings of the 12th International Symposium on Symbolic and Numeric Algorithms for Scientific Computing (SYNASC '10)*, pp. 95–98, September 2010.

17

On the Security of a Novel Probabilistic Signature Based on Bilinear Square Diffie-Hellman Problem and Its Extension

Zhenguo Zhao[1] and Wenbo Shi[2]

[1] School of Water Conservancy, North China University of Water Resources and Electric Power, Zhengzhou 450045, China
[2] Department of Electronic Engineering, Northeastern University at Qinhuangdao, Qinhuangdao 066004, China

Correspondence should be addressed to Zhenguo Zhao; zhenguozhao2013@163.com

Academic Editor: Junghyun Nam

Probabilistic signature scheme has been widely used in modern electronic commerce since it could provide integrity, authenticity, and nonrepudiation. Recently, Wu and Lin proposed a novel probabilistic signature (PS) scheme using the bilinear square Diffie-Hellman (BSDH) problem. They also extended it to a universal designated verifier signature (UDVS) scheme. In this paper, we analyze the security of Wu et al.'s PS scheme and UDVS scheme. Through concrete attacks, we demonstrate both of their schemes are not unforgeable. The security analysis shows that their schemes are not suitable for practical applications.

1. Introduction

Signature scheme is an important modern cryptographic mechanism of the public key cryptosystem. In the signature scheme, the signer uses his private key to sign a message and generate a signature, which could be verified by other users using the signer's public key. The signature could provide integrity, authenticity, and nonrepudiation; then it could be used in modern electronic commerce [1–5].

The undeniable signature (US) scheme is a variation of the signature scheme, which was first introduced by Chaum and van Antwerpen [6]. In the US scheme, the verifier should get the signer's cooperation to finish the verification. In order to remove the complicated cooperation between the signer and the verifier, Jakobsson et al. [7] introduced the concept of the designated verifier signature (DVS) scheme and proposed a concrete DVS scheme. However, Wang [8] found that there is serious security vulnerability in Jakobsson et al.'s scheme. Later, Steinfeld et al. [9, 10] introduced the concept of the universal designated verifier signature (UDVS) scheme to generate the concept of the DVS scheme. In the UDVS scheme, the signer could generate a signature and only the designated verifier could verify the signature using his private key.

Later, Zhang et al. [11] used Diffie-Hellman problem to construct a UDVS scheme and demonstrated that their scheme is provably secure in the standard model. Unfortunately, Cheon [12] found that Zhang et al.'s scheme had a security flaw. To enhance security, Huang et al. [13] presented a new UDVS scheme using the gap bilinear Diffie-Hellman problem. In order to satisfy applications in identity-based systems, Chen et al. [14] proposed the first identity-based UDVS scheme. In order to improve efficiency, Wu and Lin [15] proposed a probabilistic signature (PS) scheme using the bilinear square Diffie-Hellman (BSDH) problem. Then, they extended this PS scheme to a UDVS scheme. They also demonstrated that both of their schemes are provably secure in the random oracle. In this paper, we analyze the security of both Wu and Lin's PS scheme and UDVS scheme. Through concrete attacks, we show that neither of their schemes is unforgeable. We will also propose efficient countermeasures to withstand those attacks.

The organization of the paper is sketched as follows. Section 2 gives a brief review of Wu et al.'s PS scheme and UDVS scheme. Section 3 presents our attacks against Wu et al.'s PS scheme and UDVS scheme. Section 4 presents our countermeasures to withstand the proposed attacks. At last, Section 5 presents some conclusion of the paper.

2. Review of Wu and Lin's Schemes

In this section, we will give the details of Wu et al.'s PS scheme and UDVS scheme.

2.1. Review of Wu and Lin's PS Scheme. There are two participants in Wu and Lin's PS scheme, that is, a signer and a verifier, where the signer generates a publicly verifiable signature (PV-signature) using his private key and the verifier could verify the validity of the PV-signature using the signer's public key. There are three algorithms in Wu and Lin's PS scheme, that is, *Setup, PV-Signature-Generation*, and *PV-Signature-Verification*.

Setup. Taking a security parameter k as input, the system authority (SA) runs the following steps to generate system parameters. Besides, the user U_i registers his public key.

(1) SA chooses a random number q and selects two multiplicative groups $(G_1, \times)$ and $(G_2, \times)$ with the same order q, where the bit length of q is k.

(2) SA chooses a generator P of the group $(G_1, \times)$ and a bilinear pairing $e : G_1 \times G_1 \to G_2$.

(3) SA chooses two secure hash functions h_1 and h_2, where $h_1 : G_1 \to G_1$ and $h_2 : \{0, 1\}^* \times G_1 \to Z_q$.

(4) SA publishes the system parameters *params* $= \{G_1, G_2, q, P, e, h_1, h_2\}$.

(5) U_i chooses a random number $x_i \in Z_q$ as his private key and registers his public key $Y_i = P^{x_i}$.

PV-Signature-Generation. Upon receiving the message m, the signer U_s runs the following steps to generate a PV-signature Ω.

(1) U_s chooses a random number $r \in Z_q$ and computes $R = P^r$, $T = h_1(R)^{x_s}$, and $\rho = h_2(m, R)x_s^2 - r \bmod q$.

(2) U_s outputs $\Omega = (R, T, \rho)$ as the PV-signature of the message m.

PV-Signature-Verification. Upon receiving the message m, the PV-signature Ω, and the signer's public key Y_s, the verifier U_v runs the following steps to verify the validity of the PV-signature.

(1) U_v checks whether the equation $e(P^\rho R, h_1(R)) = e(Y_s, T)^{h_2(m,R)}$ holds.

(2) If the equation holds, U_v confirms the PV-signature is valid; otherwise, U_v confirms that the PV-signature is not valid.

2.2. Review of Wu and Lin's UDVS Scheme. There are two participants in Wu and Lin's UDVS scheme, that is, a signer and a verifier, where the signer generates a designated verifiable signature (DV-signature) using his private key and only the designated verifier could verify the validity of the DV-signature using the signer's public key. There are five

algorithms in Wu and Lin's UDVS scheme, that is, *Setup, PV-Signature-Generation, PV-Signature-Verification, DV-Signature-Generation*, and *DV-Signature-Verification*. Because the first three algorithms are the same as those in PS scheme, only the last two algorithms will be described in detail.

DV-Signature-Generation. Upon receiving a message m and the designated verifier U_v's public key Y_v, the signer U_s runs the following steps to generate a DV-signature Ω.

(1) U_s chooses a random number $r \in Z_q$ and computes $R = P^r$, $T = h_1(R)^{x_s}$, $\rho = h_2(m, R)x_s^2 - r \bmod q$, and $W = Y_v^\rho$.

(2) U_s outputs $\Omega = (R, T, W)$ as the DV-signature of the message m.

DV-Signature-Verification. Upon receiving a message m, the DV-signature Ω, and the signer's public key Y_s, the designated verifier U_v runs the following steps to verify the validity of the DV-signature.

(1) U_v checks whether the equation $e(W^{x_v^{-1}}R, h_1(R)) = e(Y_s, T)^{h_2(m,R)}$ holds.

(2) If the equation holds, U_v confirms the DV-signature is valid; otherwise, U_v confirms that the PV-signature is not valid.

3. Security Analysis of Wu and Lin's Schemes

In this section, we will give the security analysis of Wu et al.'s PS scheme and UDVS scheme.

3.1. Security Analysis of Wu and Lin's PS Scheme. Wu and Lin claimed that their PS scheme was unforgeable against various attacks. Through concrete attack, we will show that an adversary without the signer U_s's private key could forge a legal PV-signature of any message. Given a message m, the adversary A could forge a legal PV-signature through the following steps.

(1) A generates a random number $\rho \in Z_q$ and computes $R = Y_s P^{-\rho}$ and $T = h_1(R)^{h_2(m,R)^{-1} \bmod q}$.

(2) A outputs $\Omega = (R, T, \rho)$ as the PV-signature of the message m.

Since $R = Y_s P^{-\rho}$ and $T = h_1(R)^{h_2(m,R)^{-1} \bmod q}$, we could get

$$e\left(P^\rho R, h_1(R)\right)$$
$$= e\left(P^\rho Y_s P^{-\rho}, h_1(R)\right)$$
$$= e\left(Y_s, h_1(R)\right),$$
$$e(Y_s, T)^{h_2(m,R)}$$
$$= e\left(Y_s, h_1(R)^{h_2(m,R)^{-1} \bmod q}\right)^{h_2(m,R)}$$

$$= e(Y_s, h_1(R))^{h_2(m,R) \cdot h_2(m,R)^{-1} \bmod q}$$

$$= e(Y_s, h_1(R)).$$

$$(1)$$

From (1), we know that the equation $e(P^\rho R, h_1(R)) = e(Y_s, T)^{h_2(m,R)}$ holds. Then, the PV-signature generated by the adversary could pass the verifier's check. Therefore, the adversary could forge a legal PV-signature.

3.2. Security Analysis of Wu and Lin's UDVS Scheme. Wu and Lin claimed that their UDVS scheme was unforgeable against various attacks. Through concrete attack, we will show that an adversary without the signer U_s's private key could forge a legal DV-signature of any message. Given a message m and the designated verifier U_v's public key Y_v, the adversary A could forge a legal DV-signature through the following steps.

(1) A generates a random number $\rho \in Z_q$ and computes $R = Y_s P^{-\rho}$, $T = h_1(R)^{h_2(m,R)^{-1} \bmod q}$ and $W = Y_v^\rho$.

(2) A outputs $\Omega = (R, T, \rho)$ as the DV-signature of the message m.

Since $R = Y_s P^{-\rho}$, $T = h_1(R)^{h_2(m,R)^{-1} \bmod q}$, $W = Y_v^\rho$, and $Y_v = P^{x_v}$, we could get

$$e\left(W^{x_v^{-1}} R, h_1(R)\right)$$

$$= e\left((Y_v^\rho)^{x_v^{-1}} Y_s P^{-\rho}, h_1(R)\right)$$

$$= e\left(((P^{x_v})^\rho)^{x_v^{-1}} Y_s P^{-\rho}, h_1(R)\right)$$

$$= e\left(P^{x_v \rho x_v^{-1}} Y_s P^{-\rho}, h_1(R)\right)$$

$$= e\left(P^\rho Y_s P^{-\rho}, h_1(R)\right)$$

$$= e\left(Y_s, h_1(R)\right),$$

$$e(Y_s, T)^{h_2(m,R)}$$

$$= e\left(Y_s, h_1(R)^{h_2(m,R)^{-1} \bmod q}\right)^{h_2(m,R)}$$

$$= e(Y_s, h_1(R))^{h_2(m,R) \cdot h_2(m,R)^{-1} \bmod q}$$

$$= e(Y_s, h_1(R)).$$

From (2), we know that the equation $e(W^{x_v^{-1}} R, h_1(R)) = e(Y_s, T)^{h_2(m,R)}$ holds. Then, the DV-signature generated by the adversary could pass the verifier's verification. Therefore, the adversary could forge a legal DV-signature.

4. Countermeasures

4.1. Countermeasure for Wu and Lin's PS Scheme. From the details of Wu and Lin's PS scheme, we know that the value T has no relation with the value of ρ. Then the adversary could

choose the value T freely to remove the relation between R and ρ. To withstand the attack described in Section 3.1, we just need to modify Wu and Lin's PS scheme slightly.

DV-Signature-Generation. Upon receiving a message m, the signer U_s runs the following steps to generate a PV-signature Ω.

(1) U_s chooses a random number $r \in Z_q$ and computes $R = P^r$, $T = h_1(R)^{x_s}$, and $\rho = h_2(m, R, T)x_s^2 - r \bmod q$.

(2) U_s outputs $\Omega = (R, T, \rho)$ as the PV-signature of the message m.

DV-Signature-Verification. Upon receiving a message m, the PV-signature Ω, and the signer's public key Y_s, the verifier U_v runs the following steps to verify the validity of the PV-signature.

(1) U_v checks whether the equation $e(P^\rho R, h_1(R)) = e(Y_s, T)^{h_2(m,R,T)}$ holds.

(2) If the equation holds, U_v confirms the PV-signature is valid; otherwise, U_v confirms that the PV-signature is not valid.

After the modification, the adversary A could generate a random number $\rho \in Z_q$ and compute $R = Y_s P^{-\rho}$, $T = h_1(R)^{h_2(m,R)^{-1} \bmod q}$. However, the equation $e(P^\rho R, h_1(R)) = e(Y_s, T)^{h_2(m,R,T)}$ never holds since the adversary cannot use $h_2(m, R)^{-1} \bmod q$ to remove the function $h_2(m, R, T)^{-1} \bmod q$. Then, the modified scheme is secure against the attack described in Section 3.1.

4.2. Countermeasure for Wu and Lin's UDVS Scheme. From the details of Wu and Lin's UDVS scheme, we know that the value T has no relation to the value of W. Then the adversary could choose the value T freely to remove the relation between R and W. To withstand the attack described in Section 3.2, we just need to modify Wu and Lin's UDVS scheme slightly.

DV-Signature-Generation. Upon receiving a message m and the designated verifier U_v's public key Y_v, the signer U_s runs the following steps to generate a DV-signature Ω.

(1) U_s chooses a random number $r \in Z_q$ and computes $R = P^r$, $T = h_1(R)^{x_s}$, $\rho = h_2(m, R, T)x_s^2 - r \bmod q$, and $W = Y_v^\rho$.

(2) U_s outputs $\Omega = (R, T, W)$ as the DV-signature of the message m.

DV-Signature-Verification. Upon receiving a message m, the DV-signature Ω, and the signer's public key Y_s, the designated verifier U_v runs the following steps to verify the validity of the DV-signature.

(1) U_v checks whether the equation $e(W^{x_v^{-1}} R, h_1(R)) = e(Y_s, T)^{h_2(m,R,T)}$ holds.

(2) If the equation holds, U_v confirms the DV-signature is valid; otherwise, U_v confirms that the PV-signature is not valid.

After the modification, the adversary A could generate a random number $\rho \in Z_q$ and compute $R = Y_s P^{-\rho}$, $T = h_1(R)^{h_2(m,R)^{-1} \bmod q}$ and $W = Y_v^{\rho}$. However, the equation $e(W^{x_v^{-1}} R, h_1(R)) = e(Y_s, T)^{h_2(m,R,T)}$ never holds since the adversary cannot use $h_2(m,R)^{-1} \bmod q$ to remove the function $h_2(m,R,T)^{-1} \bmod q$. Then, the modified scheme is secure against the attack described in Section 3.2.

5. Conclusion

Recently, Wu and Lin proposed a PS scheme using the bilinear square Diffie-Hellman problem and extended it to a UDVS scheme. They also demonstrated that their scheme is provably secure in the random oracle. Through concrete attacks, we demonstrate that neither of their schemes is unforgeable against common adversary. To improve security, we also propose efficient countermeasures to withstand the proposed attacks.

Conflict of Interests

The authors declare that they have no conflict of interests.

Acknowledgments

The authors thank Professor Junghyun Nam and anonymous reviewers for their valuable comments. This study was supported by the International S&T Cooperation Program from the Ministry of Science and Technology of China (no. 2012DFA91530), the "Twelfth five-year-plan" Support Plan Projects (no. 2011BAD25B01), the introduction of high-level Talents Foundation of North China University of Water Resources and Electric Power (no. NCWU201248), the Key Technique Program of the Education Department of Henan Province (13A570704), and the National Science foundation of China (no. 61202447).

References

[1] N. Tiwari and S. Padhye, "Analysis on the generalization of proxy signature," *Security and Communication Networks*, vol. 6, no. 5, pp. 549–566, 2013.

[2] L. Yi, G. Bai, and G. Xiao, "Proxy multi-signature scheme: a new type of proxy signature scheme," *Electronics Letters*, vol. 36, no. 6, pp. 527–528, 2000.

[3] C. Hsu, T. Wu, and T. Wu, "New nonrepudiable threshold proxy signature scheme with known signers," *Journal of Systems and Software*, vol. 58, no. 2, pp. 119–124, 2001.

[4] Z. Shao, "Proxy signature schemes based on factoring," *Information Processing Letters*, vol. 85, no. 3, pp. 137–143, 2003.

[5] D. He, B. Huang, and J. Chen, "New certificateless short signature scheme," *IET Information Security*, vol. 7, no. 2, pp. 113–117, 2013.

[6] D. Chaum and H. van Antwerpen, "Undeniable signature," in *Proceedings of the 10th Annual International Cryptology Conference (CRYPTO '90)*, Advances in Cryptology, pp. 212–216, Springer, Berlin, Germany, 1990.

[7] M. Jakobsson, K. Sako, and R. Impagliazzo, "Designated verifier proofs and their applications," in *Proceedings of the International Conference on the Theory and Application of Cryptographic Techniques (EUROCRYPT '96)*, Advances in Cryptology, pp. 143–154, Springer, Berlin, Germany, 1996.

[8] G. Wang, "An attack on not-interactive designated verifier proofs for undeniable signatures," Cryptology ePrint archive, 2003, http://eprint.iacr.org/2003/243.

[9] R. Steinfeld, L. Bull, H. Wang, and J. Pieprzyk, "Universal designated-verifier signatures," in *Proceedings of the 9th International Conference on the Theory and Application of Cryptology and Information Security (ASIACRYPT '03)*, Advances in Cryptology, pp. 523–542, Springer, Berlin, Germany, 2003.

[10] R. Steinfeld, H. Wang, and J. Pieprzyk, "Efficient extension of standard Schnorr/RSA signatures into universal designated-verifier signatures," in *Proceedings of the Public Key Cryptography (PKC '04)*, pp. 86–100, Springer, Berlin, Germany, 2004.

[11] R. Zhang, J. Furukawa, and H. Imai, "Short signature and universal designated verifier signature without random oracles," in *Proceedings of the 3rd International Conference on Applied Cryptography and Network Security (ACNS '05)*, vol. 3531, pp. 483–498, Springer, Berlin, Germany, June 2005.

[12] J. H. Cheon, "Security analysis of the strong Diffie- Hellman problem," in *Proceedings of the International Conference on the Theory and Application of Cryptographic Techniques (EUROCRYPT '96)*, Advances in Cryptology, pp. 1–11, Springer, Berlin, Germany, 2006.

[13] X. Huang, W. Susilo, Y. Mu, and W. Wu, "Secure universal designated verifier signature without random oracles," *International Journal of Information Security*, vol. 7, no. 3, pp. 171–183, 2008.

[14] X. Chen, G. Chen, F. Zhang, B. Wei, and Y. Mu, "Identity-based universal designated verifier signature proof system," *International Journal of Network Security*, vol. 8, no. 1, pp. 52–58, 2009.

[15] T. Wu and H. Lin, "A novel probabilistic signature based on bilinear square Diffie-Hellman problem and its extension," *Security and Communication Networks*, vol. 6, no. 6, pp. 757–764, 2013.

18

A Provably Secure Revocable ID-Based Authenticated Group Key Exchange Protocol with Identifying Malicious Participants

Tsu-Yang Wu,[1,2] Tung-Tso Tsai,[3] and Yuh-Min Tseng[3]

[1] *Shenzhen Graduate School, Harbin Institute of Technology, Shenzhen 518055, China*
[2] *Shenzhen Key Laboratory of Internet Information Collaboration, Shenzhen 518055, China*
[3] *Department of Mathematics, National Changhua University of Education, Jin-De Campus, Changhua City 500, Taiwan*

Correspondence should be addressed to Tsu-Yang Wu; wutsuyang@gmail.com

Academic Editor: Fei Yu

The existence of malicious participants is a major threat for authenticated group key exchange (AGKE) protocols. Typically, there are two detecting ways (passive and active) to resist malicious participants in AGKE protocols. In 2012, the revocable identity- (ID-) based public key system (R-IDPKS) was proposed to solve the revocation problem in the ID-based public key system (IDPKS). Afterwards, based on the R-IDPKS, Wu et al. proposed a revocable ID-based AGKE (RID-AGKE) protocol, which adopted a passive detecting way to resist malicious participants. However, it needs three rounds and cannot identify malicious participants. In this paper, we fuse a noninteractive confirmed computation technique to propose the first two-round RID-AGKE protocol with identifying malicious participants, which is an active detecting way. We demonstrate that our protocol is a provably secure AGKE protocol with forward secrecy and can identify malicious participants. When compared with the recently proposed ID/RID-AGKE protocols, our protocol possesses better performance and more robust security properties.

1. Introduction

In the past, group-oriented applications, such as collaboration works and teleconference, were popularly and widely used in the Internet. Authenticated group key exchange (AGKE) protocol [1] is a cryptographic primitive which provides secure group communications for users in cooperative and distributed applications. During executing the protocol, group participants not only cooperatively generate a common key which is used to encrypt the transmitted messages but also authenticate the participants' identities.

The existence of malicious participants is a major threat for AGKE protocols. The goal of malicious participants is to disturb the establishing of common keys. Hence, how to resist malicious participants in AGKE protocols becomes a critical research. Typically, there are two detecting ways to resist malicious participants. (I) Passive detection [2–4]: it involves an explicit key confirmation approach in AGKE protocols. The resulted protocols only detect the existence of malicious participants and an additional round is required.

(II) Active detection [5, 6]: it adopts a noninteractive confirmed computation technique into AGKE protocols. The resulted protocols can identify the identities of malicious participants without additional round. However, the computational cost of active detection is time-consuming than the one of passive detection.

Quite recently, the revocable identity- (ID-) based public key system (R-IDPKS) was proposed to solve the revocation problem of users in the ID-based public key system (IDPKS). The concept of IDPKS was introduced by Shamir [7] in 1984 and was practiced by Bonch and Franklin [8] in 2001. Indeed, they [8] had suggested a solution that the private key generator (PKG) renews these nonrevoked users' private keys periodically to answer the revocation problem in the IDPKS. The approach can be used to revoke the compromised or misbehaving users. Nevertheless, the heavy workload arose from the PKG for renewing users' private keys periodically.

In 2008, Boldyreva et al. [9] proposed a revocable ID-based encryption (RIBE) scheme by using binary tree. Their scheme can reduce the PKG's workload mentioned in the

Boneh-Franklin solution [8]. However, this scheme is based on a weak security model called the relaxed selective-ID model [10]. In 2009, Libert and Vergnaud [11] relied on Boldyreva et al.'s RIBE to present a secure RIBE scheme under an adaptive-ID model. Recently, Seo and Emura [12] demonstrated Boldyreva et al.'s scheme [9] is vulnerable to decryption key exposure and then proposed a provably secure tree based RIBE scheme. Subsequently, Seo and Emura [13] presented a hierarchical RIBE scheme to solve the open problem mentioned in [11].

In 2011, Tseng and Tsai [14] proposed a practical RIBE scheme over a public channel. The key construction of the Tseng-Tsai scheme is different from the previous schemes [9, 11–13]. In [14], each user's private key consists of a fixed initial private key and an update key, where the update key is renewed along with the current period. For an honest (nonrevoked) user, the PKG periodically issues new update key and sends it to the user via a public channel. Upon receiving the new updating key, the user can renew her/his private key by herself/himself. To revoke a malicious user, the PKG only stops issuing the new update key in current period. Thus, the user cannot compute the newest private key. In other words, she/he cannot execute any cryptographic behaviors in later periods. Later on, several revocable ID-based cryptographic schemes based on the Tseng-Tsai R-IDPKS [14] were presented such as encryption [15], signature [16, 17], authenticated group key exchange (AGKE) [4], and signcryption [18].

In 2012, Wu et al. [4] proposed the first provably secure revocable ID-based AGKE (RID-AGKE) protocol. Their protocol adopted a passive detecting way to resist malicious participants. However, it requires three rounds and cannot identify the identities of malicious participants. In this paper, we fuse the key construction of Tseng-Tsai R-IDPKS [14] and a noninteractive confirmed computation technique [6] to present a two-round RID-AGKE protocol with identifying malicious participants. In our protocol, each group participant can confirm whether the broadcast values are correctly computed by other participants. Based on the detecting approach, our protocol can easily identify the participants who maliciously broadcast the incorrect values to disturb the common key establishing. The framework and security notions for RID-AGKE protocols are defined to formalize possible threats and attacks. We demonstrate the security of our protocol in the random oracle model [19] and under two mathematical assumptions (the computational Diffie-Hellman and the decisional bilinear Diffie-Hellman). Finally, we make the comparisons between our protocol and the recently proposed ID/RID-AGKE protocols to show the advantages of the proposed protocol.

The rest of this paper is organized as follows. We briefly review the concepts of bilinear pairings and related mathematical problems in Section 2. The security model and notions of RID-AGKE are presented in Section 3. We propose a concrete RID-AGKE protocol in Section 4. Security analysis of the proposed RID-AGKE protocol is demonstrated in Section 5. We make the performance analysis and comparisons in Section 6. Conclusions are drawn in Section 7.

2. Preliminaries

In this section, we briefly review the properties of bilinear pairings and related mathematical problems. For the details, a reader can refer to [8, 20, 21] for full descriptions.

2.1. Bilinear Pairings. Let G_1 and G_2 be two groups of a large prime order q, where G_1 is an additive cyclic group and G_2 is a multiplicative cyclic group. A bilinear pairing e is a map defined by $e : G_1 \times G_1 \rightarrow G_2$ and satisfies the following three conditions.

(1) *Bilinearity*: for all $P, Q \in G_1$ and $a, b \in Z_q$, $e(aP, bQ)$ $= e(P, Q)^{ab}$.

(2) *Nondegeneracy*: there exist $P, Q \in G_1$ such that $e(P, Q)$ $\neq 1$.

(3) *Computability*: for all $P, Q \in G_1$, there exists an algorithm to compute $e(P, Q)$.

2.2. Mathematical Hard Problems and Assumptions. Here, we present two mathematical hard problems and define the corresponding assumptions as follows.

(1) *Computational Diffie-Hellman (CDH) problem*: given $P, aP, bP \in G_1$ for some $a, b \in Z_q^*$, the CDH problem is to compute $abP \in G_1$.

(2) *Decisional bilinear Diffie-Hellman (DBDH) problem*: given $P, aP, bP, cP, dP \in G_1$ for some $a, b, c, d \in Z_q^*$, the DBDH problem is to distinguish $(P, aP, bP, cP, dP, e(P, P)^{abc})$ from $(P, aP, bP, cP, dP, e(P, P)^d)$.

Definition 1 (CDH assumption). Given $P, aP, bP \in G_1$ for some $a, b \in_R Z_q^*$, there does not exist a probabilistic polynomial-time algorithm A with a nonnegligible probability to compute $abP \in G_1$. The advantage of A within running time t is defined as $Adv_{CDH}(t) = \Pr[A(P, aP, bP) = abP \mid P, aP, bP \in G_1]$.

Definition 2 (DBDH assumption). Given $P, aP, bP, cP, dP \in G_1$ for some $a, b, c, d \in_R Z_q^*$, there does not exist a probabilistic polynomial-time algorithm A with nonnegligible probability to distinguish $(P, aP, bP, cP, dP, e(P, P)^{abc})$ from $(P, aP, bP, cP, dP, e(P, P)^d)$. The advantage of A within running time t is defined as $Adv_{DBDH}(t) = \Pr[A(e(P, P)^{abc}, e(P, P)^d) = 1 \mid P, aP, bP, cP, dP \in G_1]$.

3. Model and Notions

In this section, we define the model and notions for RID-AGKE protocol. Note that some of the following definitions and notations are referred to in [4, 6, 22–24].

Initialization. The initialization of RID-AGKE protocol has three algorithms.

(1) *Setup Algorithm.* This algorithm is a probabilistic algorithm which takes as input a security parameter k and a total

number z of periods. It returns a system private key s and public parameters *param*. Note that the whole life time of the system is divide into distinct periods $1, 2, \ldots, z$. Here, *param* is made public.

(2) *Initial Key Extract Algorithm.* This algorithm is a deterministic algorithm which takes as input the system private key s and a participant's identity ID. It returns the participant's initial private key DID.

(3) *Key Update Algorithm.* This algorithm is a deterministic algorithm which takes as input the system private key s, a participant's identity ID, and a period index j, where $1 \leq j \leq z$. It returns the participant's update key TID_j.

Here, note that the participant's private key for period j is defined by $DID_j = DID + TID_j$.

Related Notions. For simplicity, there is a fixed set $G = \{U_1, U_2, \ldots, U_n\}$ with polynomial size of potential participants. Assume that each participant U_i has a unique identity $ID_i \in \{0, 1\}^*$. Any subset of G may run a RID-AGKE protocol many times (possibly concurrently) in some period index j to establish a group session key, where $1 \leq j \leq z$ and z is a total number of periods. Note that the set of participants' identities, $\mathbf{ID} = \{ID_1, ID_2, \ldots, ID_n\}$ is known by all participants (including adversary).

An instance t of participant U in period j is denoted by $\Pi_U^{j(t)}$, where t is a positive integer. Each instance $\Pi_U^{j(t)}$ has associated with seven variables as follows.

(i) $state_U^{j(t)}$: it presents the current state of instance $\Pi_U^{j(t)}$.

(ii) $acc_U^{j(t)}$ and $term_U^{j(t)}$: they take Boolean values to demonstrate whether $\Pi_U^{j(t)}$ *has accepted* or terminated. Informally, we say that an instance *has accepted* meaning that it does not detect any incorrect behavior. An instance is called terminating if it has sent and received messages. Note that a terminated instance may also possibly accept.

(iii) $used_U^{j(t)}$: it indicates whether $\Pi_U^{j(t)}$ is used in a RID-AGKE protocol.

(iv) $pid_U^{j(t)}$: the partner ID of instance $\Pi_U^{j(t)}$ is a set which contains the identities of participants in the group with whom $\Pi_U^{j(t)}$ wants to establish a group session key (including U itself).

(v) $sid_U^{j(t)}$: the session ID of instance $\Pi_U^{j(t)}$ is a concatenation of all messages sent and received by the instance in a given execution of RID-AGKE protocol.

(vi) $sk_U^{j(t)}$: a group session key which is accepted by instance $\Pi_U^{j(t)}$.

In the following definitions, we will only focus on the three variables $pid_U^{j(t)}$, $sid_U^{j(t)}$, and $sk_U^{j(t)}$. The remaining variables will be left implicit. We say that two instances $\Pi_U^{j(t)}$ and $\Pi_{U'}^{j(v)}$ are *partnered* if (1) they have accepted the same group session key; (2) $sid_U^{j(t)} = sid_{U'}^{j(v)}$; and (3) $pid_U^{j(t)} = pid_{U'}^{j(v)}$.

Adversarial Model. An adversary A can be viewed as a probabilistic polynomial-time algorithm. Here, we assume that A can potentially control all communications in a RID-AGKE protocol. The interaction between A and instances of participants in the protocol is modeled by the following oracles.

(i) *Execute* (V, j): when A makes *Execute query* on (V, j), it executes the RID-AGKE protocol between the unused instances of participants in V for period index j and then returns a transcript of the execution, where V is a subset of G. Here, *Execute query* is used to model passive attacks.

(ii) *Inextract* (ID_U): when A makes *Initial key extract query* on identity ID_U, it generates an initial private key DID_U corresponding to ID_U and returns it to A, where $ID_U \notin \mathbf{ID}$.

(iii) *Kupdate* (ID_U, j): when A makes *Key update query* on (ID_U, j), it generates an update key $TID_{U,j}$ corresponding to (ID_U, j) and returns it to A, where $ID_U \notin \mathbf{ID}$ and j is a period index.

(iv) *Send* (U, j, t, M): when A makes *Send query* on (U, j, t, M), it sends message M to instance $\Pi_U^{j(t)}$ and then returns the reply generated by this instance according to procedures of RID-AGKE protocol.

(v) *Reveal* (U, j, t): when A makes *Reveal query* on (U, j, t), it returns a group session key $sk_U^{j(t)}$ for a terminated instance $\Pi_U^{j(t)}$. Here, *Reveal query* is used to model known session key attacks.

(vi) *Corrupt* (ID_U, j): when A makes *Corrupt query* on (ID_U, j), it returns a private key $DID_{U,j}$ of ID_U in period j. Note that *Corrupt query* models the corruption of this participant at a time in which it is not currently executing the protocol. We say that a participant U is honest if and only if no *Corrupt query* has been made by A.

(vii) *Test* $(U, j, t,)$: at any time, the adversary A makes *Test query* only once to this oracle during A's execution. In this moment, a random coin $b \in \{0, 1\}$ is selected. If $b = 1$, a group session key $sk_U^{j(t)}$ is retuned. Otherwise, a random value is retuned. Here, *Test query* is used to model the semantic security of group session key.

According to the above adversarial model, we define two types of adversaries. A *passive adversary* is allowed to make *Execute*, *Reveal*, *Corrupt*, and *Test* queries. An *active adversary* is allowed to make the above all queries. In order to get more precise analysis, we still use *Execute query* though it can be substituted by making *Send query* repeatedly.

Remark 3. According to the adversarial model above, the adversary A can compute the participant U's private key $DID_{U,j} = DID_U + TID_{U,j}$ for period index j while A makes both *Initial key extract query* on ID_U and *Key update query* on (ID_U, j) simultaneously. Hence, we disallow A to make both queries in the same time.

Correctness. A RID-AGKE protocol is called *correct* if the following three conditions hold.

(1) All participants are honest and all messages are delivered honestly.

(2) $acc_U^{j(t)} = acc_{U'}^{j(v)} =$ "True" and $sk_U^{j(t)} = sk_{U'}^{j(v)}$.

(3) $sid_U^{j(t)} = sid_{U'}^{j(v)}$ and $pid_U^{j(t)} = pid_{U'}^{j(v)}$ for all participants $U, U' \in V \subseteq G$ with instances $\Pi_U^{j(t)}$ and $\Pi_{U'}^{j(v)}$.

Freshness. We say that an instance $\Pi_U^{j(t)}$ is called *fresh* (or called holding a *fresh* group session key $sk_U^{j(t)}$) if the following three conditions hold.

(1) $\Pi_U^{j(t)}$ has accepted a group session key $sk_U^{j(t)}$.

(2) Neither $\Pi_U^{j(t)}$ nor its partners have been made *Reveal query*.

(3) No *Corrupt query* has been made on $ID_V \in pid_U^{j(t)}$ before *Send query* to $\Pi_U^{j(t)}$ or *Send query* to $\Pi_{U'}^{j(v)}$, where $ID_{U'} \in pid_U^{j(t)}$.

Here, we assume all instances are *fresh*. Note that the notion of *freshness* is defined appropriately for the purpose of forward secrecy.

Secure RID-AGKE. A secure RID-AGKE protocol contains the following four parts.

(1) *Freshness.*

(2) *Security of RID-AGKE Protocol.* The security of RID-AGKE protocol is defined in the following game played between an active adversary A and a set of instances:

(a) *initialization:* the system private key, public parameters, and participants' private keys are generated in this phase;

(b) *query:* A may make different types of queries to oracles and gets back the answers corresponding to the RID-AGKE protocol;

(c) *guess:* finally, the adversary A outputs its guess for the coin b in *Test query* and terminates.

In this game, the goal of A is to distinguish a group session key from a random value. Let *Succ* be the event that A correctly guesses the coin b in *Test query*. The advantage of A in attacking a RID-AGKE protocol Ψ is defined by $Adv_{A,\Psi}(k) = |2 \cdot \Pr[Succ] - 1|$. We say that the protocol Ψ is secure, if the advantage $Adv_{A,\Psi}(k)$ is negligible.

(3) *Forward Secrecy.* We say that a RID-AGKE protocol Ψ provides *forward secrecy*. It means that though an adversary A obtains participants' private keys in Ψ, the previous establishing group session keys is preserved. The advantage of A in attacking the protocol Ψ within running time t is defined by $Adv_\Psi^{RIDAGKE\text{-}fs}(t, q_{ex}, q_s)$, where q_{ex} and q_s are the maximum numbers of making *Execute* and *Send* queries, respectively.

(4) *Authentication.* We say that a RID-AGKE protocol Ψ provides *implicit key authentication* if all participants in Ψ are guaranteed that nobody other than their partners can learn the session key. In other words, any adversary should not learn the key. Note that this security property does not guarantee that the partners have computed the key.

Malicious Participant. A participant U_m is called *malicious* in a RID-AGKE protocol Ψ if he is a legal participant but is fully controlled by adversary. The goal of malicious participant is to disturb the group key establishing in Ψ.

4. Concrete Protocol

In this section, we propose a concrete RID-AGKE protocol with identifying malicious participants. Our protocol fuses the Tseng-Tsai R-IDPKS [14] and a noninteractive confirmed computation technique [6]. In the initialization phase, given a security parameter k and a total number z of periods, a private key generator (PKG) executes *Setup algorithm* to generate the system private key s and the public parameters $param = \{G_1, G_2, e, q, P, P_{pub}, H_1, H_2, H_3, H_4\}$ defined in Notations section at the end of the paper.

When a participant U with identity $ID_U \in \{0, 1\}^*$ wants to obtain her/his initial private key DID_U, the PKG runs *Initial key extract algorithm* to compute $DID_U = s \cdot H_1(ID_U) = s \cdot QID_U$ and returns it to U via a secure channel. For a nonrevoked participant U with identity ID_U in time period j, the PKG runs *Key update algorithm* to compute her/his update key $TID_{U,j} = s \cdot H_2(ID_U, j) = s \cdot RID_{U,j}$ and returns it to U via a public channel, where $1 \leq j \leq z$. Hence, any nonrevoked participant U can update her/his private key $DID_{U,j} = DID_U + TID_{U,j}$ by itself in period j.

Let $G = \{U_1, U_2, \ldots, U_n\}$ be a set of participants who want to establish a group session key SK_j in period j. We assume that each U_i has a unique identity $ID_i \in \{0, 1\}^*$ as public key and U_i's private key is $DID_{i,j} = DID_i + TID_{i,j}$ for period j. Note that the indices are subject to modulo n; that is, U_{n+1} and U_0 denote U_1 and U_n, respectively. Finally, $m \in \{0, 1\}^*$ is a preknown common message by all participants. The details of proposed RID-AGKE protocol are described as follows.

Round 1. Each participant U_i randomly selects a secret value $a_i \in Z_q^*$ and computes $P_i = a_i \cdot P$, $h_i = H_3(ID_i, PID_j, j, m, P_i)$, and $V_i = DID_{i,j} + a_i \cdot h_i \cdot P_{pub}$, where PID_j denotes the concatenation of all participants' identities in period j; that is, $PID_j = ID_1 \| ID_2 \| \cdots \| ID_n$. Finally, each U_i broadcasts (ID_i, j, P_i, V_i) to other participants.

Round 2. Upon receiving $(ID_{i-1}, j, P_{i-1}, V_{i-1})$ and $(ID_{i+1}, j, P_{i+1}, V_{i+1})$, each U_i first verifies them by checking

$$e\left(P, \sum_{k \in \{-1,1\}} V_{i+k} \right)$$
$$\stackrel{?}{=} e\left(P_{pub}, \sum_{k \in \{-1,1\}} H_1(ID_{i+k}) + H_2(ID_{i+k}, j) \right. \tag{1}$$
$$\left. + h_{i+k} \cdot P_{i+k} \right),$$

where $h_{i+k} = H_3(ID_{i+k}, PID_j, j, m, P_{i+k})$. If the verification is true, each U_i uses her/his secret value a_i to compute $D_i = e(a_i \cdot (P_{i+1} - P_{i-1}), P_{pub})$. Then, U_i randomly selects a value $r_i \in Z_q^*$ and computes a tuple $(ID_i, j, D_i, \alpha_i, \beta_i, \gamma_i)$, where $\alpha_i = r_i \cdot P$, $\beta_i = r_i \cdot (P_{i+1} - P_{i-1})$, $\gamma_i = r_i \cdot P_i + w_i \cdot a_i \cdot P_{pub}$, $w_i = H_4(ID_i \parallel PID_j \parallel j \parallel D_i \parallel S_j, P_{i+1} - P_{i-1}, \alpha_i, \beta_i)$, and $S_j = P_1 \parallel P_2 \parallel \cdots \parallel P_n$. Finally, U_i sends this tuple to all other participants.

Group Key Computation. Upon receiving all $(ID_k, j, D_k, \alpha_k, \beta_k, \gamma_k)$ for $k = 1, 2, \ldots, n$ except i, each U_i verifies them by checking

$$e\left(P, \sum_{k=1, k \neq i}^{n} \gamma_k\right) = \prod_{k=1, k \neq i}^{n} e\left(P_k, \alpha_k + w_k \cdot P_{pub}\right),$$

$$e\left(P_{k+1} - P_{k-1}, \gamma_k\right) \overset{?}{=} e(\beta_k, P_k) \cdot D_k^{w_k}. \tag{2}$$

If the two verifications hold, U_i can confirm that each D_k is computed by U_k using her/his secret a_k honestly for $k = 1, 2, \ldots, n$ except i. Finally, in period j, each participant U_i can compute the group session key $SK_j = e(a_i \cdot P_{i-1}, P_{pub})^n \cdot D_i^{n-1} \cdot D_{i+1}^{n-2} \cdots D_{i-2}$.

Identifying Malicious Participant. When a malicious participant U_m tries to send a wrong tuple $(ID_m, D_m, j, \alpha_m, \beta_m, \gamma_m)$ to disrupt the establishment of group session key, he will be identified as a malicious participant by using the following two verifying equations: $e(P, \gamma_k) \overset{?}{=} e(P_k, \alpha_k + w_k \cdot P_{pub})$ and $e(P_{k+1} - P_{k-1}, \gamma_k) \overset{?}{=} e(\beta_k, P_k) \cdot D_k^{w_k}$. Later on, U_m will be deleted from the participant set G and other honest participants may rerun the protocol.

5. Security Analysis

In this section, we prove the security of the proposed RID-AGKE protocol in the random oracle model [19] and under the CDH and DBDH assumptions.

ID and Forgery Attacks

Theorem 4. *The proposed RID-AGKE protocol is secure against ID and forgery attacks.*

Proof. Note that we adopt a revocable ID-based signature (RIDS) scheme [16] in Round 1 and a pairing-based signature scheme [6] in Round 2, respectively. The two signature schemes had been proven secure against ID and forgery attacks for single signature and multiple signatures with batch verification. Therefore, the proposed RID-AGKE protocol Ψ is secure against ID and forgery attacks.

Secure RID-AGKE Providing Forward Secrecy. Now, we demonstrate that the proposed RID-AGKE protocol Ψ is a secure RID-AGKE providing forward secrecy. Note that we use a similar technique in [3, 4, 6] to prove Theorem 5. □

Theorem 5. *Assume that four hash functions H_1, H_2, H_3, and H_4 are random oracles. Then, the proposed RID-AGKE protocol Ψ is a secure RID-AGKE providing forward secrecy under the decisional bilinear Diffie-Hellman (DBDH) and the computational Diffie-Hellman (CDH) assumptions. Concretely,*

$$Adv_\Psi^{RIDAGKE\text{-}fs}(t, q_{ex}, q_s)$$
$$\leq 2nq_{ex} \cdot Adv_{DBDH}(t) + Adv_\Psi^{forge}(t), \tag{3}$$

where q_{ex} and q_s are total numbers of making Execute and Send queries, respectively. Note that $Adv_\Psi^{forge}(t)$ denotes the advantage of any forgers successfully attacking the protocol Ψ.

Proof. Assume that A is an active adversary in attacking the proposed RID-AGKE protocol Ψ with a nonnegligible advantage. Now, we consider the two possible cases. The first case is that A with the advantage can impersonate a participant (i.e., forging authentication transcripts). Another case is that A with the advantage can break the protocol Ψ without modifying any transcripts.

Case 1. We assume that the adversary A with an adaptive impersonation ability can break the RID-AGKE protocol Ψ. Using A, we would like to construct a forger F which can return valid signature tuples (ID, j, aP, V) and $(ID, j, D, \alpha, \beta, \gamma)$ with respect to the proposed protocol Ψ as follows. The forger F first generates all needed system parameters and keys. Then, F simulates the oracle queries made by A. This simulation is called perfect indistinguishable from A's oracle queries except that A makes Corrupt query on (ID, j), where j is a period index. If it occurs, F fails and stops. Otherwise, when A generates two signature tuples (ID, j, aP, V) and $(ID, j, D, \alpha, \beta, \gamma)$, F returns the tuples (ID, j, aP, V) and $(ID, j, D, \alpha, \beta, \gamma)$. Let *Forge* be the event that the adversary A successfully generates two valid signature tuples. Then, the probability that F successfully returns two valid signature tuples is bounded by $Pr_A[Forge] \leq Adv_{F,\Psi}^{forge}(t) \leq Adv_\Psi^{forge}(t)$.

Case 2. We assume that the adversary A can break the proposed RID-AGKE protocol without modifying any transcripts. We first focus on the case that A makes Execute query once on $(ID_1, ID_2, \ldots, ID_n, j)$ and then extends this to the case that A makes multiple Execute queries, where the number of participants n and period j are selected by A. The real execution of Ψ is given by

$$param = \left\{ \begin{array}{c} (G_1, G_2, e) \leftarrow \text{PKG}; P \leftarrow G_1; s \leftarrow Z_q^*; P_{pub} = s \cdot P; \\ QID_1, \ldots, QID_n, RID_{1,j}, \ldots, RID_{n,j} \leftarrow G_1; \\ DID_{1,j} = (QID_1 + RID_{1,j}) \cdot s, \ldots, DID_{n,j} = (QID_n + RID_{n,j}) \cdot s : \\ (G_1, G_2, e, P, P_{pub}, PID) \end{array} \right\},$$

$$Real = \left\{ \begin{array}{c} a_1,\ldots,a_n,h_1,\ldots,h_n,r_1,\ldots,r_n,w_1,\ldots,w_n \longleftarrow Z_q^*; \\ P_1 = a_1 P,\ldots,P_n = a_n P; V_1 = DID_{1,j} + a_1 h_1 P_{pub},\ldots,V_n = DID_{n,j} + a_n h_n P_{pub}; \\ D_1 = e(P_2 - P_n, P_{pub})^{a_1},\ldots,D_n = e(P_1 - P_{n-1}, P_{pub})^{a_n}; \\ \alpha_1 = r_1 P,\ldots,\alpha_n = r_n P; \beta_1 = r_1(P_2 - P_n),\ldots,\beta_n = r_n(P_1 - P_{n-1}); \\ \gamma_1 = r_1 P_1 + w_1 a_1 P_{pub},\ldots,\gamma_n = r_n P_n + w_n a_n P_{pub}; \\ T = (P_1,\ldots,P_n,V_1,\ldots,V_n,D_1,\ldots,D_n,\alpha_1,\ldots,\alpha_n,\beta_1,\ldots,\beta_n,\gamma_1,\ldots,\gamma_n); \\ SK_j = e(a_1 P_n, P_{pub})^n \cdot D_1^{n-1} \cdot D_2^{n-2} \cdots D_{n-1} : (j, T, SK_j) \end{array} \right\}, \tag{4}$$

where T denotes the transcript and SK_j is the group session key for period j.

In *Real*, each $D_i = e(P_{i+1} - P_{i-1}, P_{pub})^{a_i} = e(a_i P_{i+1}, P_{pub})/e(a_i P_{i-1}, P_{pub}) = e(a_i a_{i+1} P, P_{pub})/e(a_{i-1} a_i P, P_{pub})$. by

the bilinear pairing operations. We can use a random value $d_{1,2} \in Z_q^*$ to substitute $a_1 \cdot a_2$. Thus, a new distribution *Fake₁* is obtained as follows:

$$Fake_1 = \left\{ \begin{array}{c} d_{1,2},a_1,\ldots,a_n,h_1,\ldots,h_n,r_1,\ldots,r_n,w_1,\ldots,w_n \longleftarrow Z_q^*; \\ P_1 = a_1 P,\ldots,P_n = a_n P; \\ V_1 = DID_{1,j} + a_1 h_1 P_{pub},\ldots,V_n = DID_{n,j} + a_n h_n P_{pub}; \\ D_1 = \dfrac{e(d_{1,2} P, P_{pub})}{e(a_n a_1 P, P_{pub})}, D_2 = \dfrac{e(a_2 a_3 P, P_{pub})}{e(d_{1,2} P, P_{pub})},\ldots,D_n = \dfrac{e(a_n a_1 P, P_{pub})}{e(a_{n-1} a_n P, P_{pub})}; \\ \alpha_1 = r_1 P,\ldots,\alpha_n = r_n P; \beta_1 = r_1(P_2 - P_n),\ldots,\beta_n = r_n(P_1 - P_{n-1}); \\ \gamma_1 = r_1 P_1 + w_1 a_1 P_{pub},\ldots,\gamma_n = r_n P_n + w_n a_n P_{pub}; \\ T = (P_1,\ldots,P_n,V_1,\ldots,V_n,D_1,\ldots,D_n,\alpha_1,\ldots,\alpha_n,\beta_1,\ldots,\beta_n,\gamma_1,\ldots,\gamma_n); \\ SK_j = e(a_1 P_n, P_{pub})^n \cdot D_1^{n-1} \cdot D_2^{n-2} \cdots D_{n-1} : (j, T, SK_j) \end{array} \right\}. \tag{5}$$

Note that A can obtain all private keys $DID_{i,j}$ and hash values h_i by making *Corrupt* and *Hash* queries. It means that A can compute all $a_i \cdot P_{pub} = h_i^{-1} \cdot (V_i - DID_{i,j})$ for $i = 1, 2,\ldots,n$. Since the discrete logarithm assumption in G_1 is intractable, A cannot obtain some information about a_i from $a_i \cdot P_{pub}$ for $i = 1, 2,\ldots,n$.

In the following claim, we want to show that to distinguish two distributions *Real* from *Fake₁* can be reduced to solve the decisional bilinear Diffie-Hellman (DBDH) problem. Let $\varepsilon(t) = Adv_{DBDH}(t)$.

Claim. For any algorithm A with running time t, we have

$$\left| \Pr\left[(j, T, SK_j) \longleftarrow Real \mid A(j, T, SK_j) = 1 \right] \right. $$
$$\left. - \Pr\left[(j, T, SK_j) \longleftarrow Fake_1 \mid A(j, T, SK_j) = 1 \right] \right|$$
$$\leq \varepsilon(t). \tag{6}$$

Proof. As mentioned above, each $D_i = e(a_i a_{i+1} P, P_{pub})/e(a_{i-1} a_i P, P_{pub}) = e(P, P_{pub})^{a_i a_{i+1}} e(P, P_{pub})^{a_{i-1} a_i}$. Here, we use $\Gamma_{i,i+1}$ to substitute $e(P, P_{pub})^{a_i a_{i+1}}$ and then each D_i can be written into $\Gamma_{i,i+1}/\Gamma_{i-1,i}$ for $i = 1, 2,\ldots,n$. Hence, the group session key SK_j also can be written into $(\Gamma_{n,1})^n \cdot D_1^{n-1} \cdot D_2^{n-2} \cdots D_{n-1}$, where $(\Gamma_{n,1})^n = e(P, P_{pub})^{n a_n a_1} = e(a_1 P_n, P_{pub})^n$.

To solve the DBDH problem, we use a technique to dispose the related parameter. Considering the following algorithm D which inputs $P_a = aP, P_b = bP$, and $P_c = cP \in G_1$ for some $a, b, c \in_R Z_q^*$. D first generates (j, T, SK_j) according to the distribution *Dist¹*. Then, D runs $A(j, T, SK_j)$ and outputs whatever A outputs. The distribution *Dist¹* is defined as follows:

$$Dist^1 = \left\{ \begin{array}{c} a_1,\ldots,a_n,h_1,\ldots,h_n,u_1,\ldots,u_{n-2},r_1,\ldots,r_n,w_1,\ldots,w_n \longleftarrow Z_q^*; \\ P_1 = a_1 P,\ldots,P_n = a_n P; V_1 = DID_{1,j} + a_1 h_1 P_{pub},\ldots,V_n = DID_{n,j} + a_n h_n P_{pub}; \\ \Gamma_{1,2} = g_{sab} \in G_2, \Gamma_{2,3} = e(P_b, P_{pub})^{u_1}, \Gamma_{i,i+1} = e(P, P_{pub})^{u_{i-2} u_{i-1}} \text{ for } i = 3 \text{ to } n-1 \\ \Gamma_{n,1} = e(P_a, P_{pub})^{u_{n-2}}; D_1 = \dfrac{\Gamma_{1,2}}{\Gamma_{n,1}},\ldots,D_n = \dfrac{\Gamma_{n,1}}{\Gamma_{n-1,n}}; \\ \alpha_1 = r_1 P,\ldots,\alpha_n = r_n P; \beta_1 = r_1(P_2 - P_n),\ldots,\beta_n = r_n(P_1 - P_{n-1}); \\ \gamma_1 = r_1 P_1 + w_1 a_1 P_{pub},\ldots,\gamma_n = r_n P_n + w_n a_n P_{pub}; \\ T = (P_1,\ldots,P_n,V_1,\ldots,V_n,D_1,\ldots,D_n,\alpha_1,\ldots,\alpha_n,\beta_1,\ldots,\beta_n,\gamma_1,\ldots,\gamma_n); \\ SK_j = (\Gamma_{n,1})^n \cdot D_1^{n-1} \cdot D_2^{n-2} \cdots D_{n-1} : (j, T, SK_j) \end{array} \right\}. \tag{7}$$

Note that this distribution depends on P_a, P_b, and P_c.

By the above distribution $Dist^1$, let $\Gamma_{1,2} = e(P, P_{pub})^{ab} = e(P, P)^{sab}$. Then, we can obtain another distribution called $Dist^1_{DBDH}$. Obviously, $Dist^1_{DBDH}$ is identical to $Real$ because

$$
\begin{aligned}
SK_j &= (\Gamma_{n,1})(\Gamma_{1,2})(\Gamma_{2,3})\cdots(\Gamma_{n-2,n-1})(\Gamma_{n-1,n}) \\
&= e(P_a, P_{pub})^{u_{n-2}} \cdot e(P, P)^{sab} \\
&\quad \cdot e(P_b, P_{pub})^{u_1} \cdots e(P, P_{pub})^{u_{n-4}u_{n-3}} \\
&\quad \cdot e(P, P_{pub})^{u_{n-3}u_{n-2}} \\
&= e(P, P)^{su_{n-2}a + sab + sbu_1 + \cdots + su_{n-4}u_{n-3} + su_{n-3}u_{n-2}}.
\end{aligned}
\tag{8}
$$

Similarly, let $\Gamma_{1,2} = e(P_c, P_{pub}) = e(P, P)^{sc}$ for some $c \neq ab \in Z_q^*$. Then, we can obtain another distribution called $Dist^1_{Random}$. Obviously, $Dist^1_{Random}$ is identical to $Fake_1$ because

$$
\begin{aligned}
SK_j &= (\Gamma_{n,1})(\Gamma_{1,2})(\Gamma_{2,3})\cdots(\Gamma_{n-2,n-1})(\Gamma_{n-1,n}) \\
&= e(P_a, P_{pub})^{u_{n-2}} \cdot e(P, P)^{sc} \\
&\quad \cdot e(P_b, P_{pub})^{u_1} \cdots e(P, P_{pub})^{u_{n-4}u_{n-3}} \\
&\quad \cdot e(P, P_{pub})^{u_{n-3}u_{n-2}} \\
&= e(P, P)^{su_{n-2}a + sc + sbu_1 + \cdots + su_{n-4}u_{n-3} + su_{n-3}u_{n-2}}.
\end{aligned}
\tag{9}
$$

Therefore, we have

$$
\begin{aligned}
&\Big| \Pr\Big[(j, T, SK_j) \longleftarrow Real \,\Big|\, A(j, T, SK_j) = 1 \Big] \\
&\quad - \Pr\Big[(j, T, SK_j) \longleftarrow Fake_1 \,\Big|\, A(j, T, SK_j) = 1 \Big] \Big| \\
&\leq \varepsilon(t).
\end{aligned}
\tag{10}
$$

This completes the proof of claim.

Using the same process in $Fake_1$, we can define other distributions $Fake_i$ for $i = 2, 3, \ldots, n$. By a similar approach in claim, we can obtain the following n-1 equations in (11) for any adversary A with running time t

$$
\begin{aligned}
&\Big| \Pr\Big[(j, T, SK_j) \longleftarrow Fake_1 \,\Big|\, A(j, T, SK_j) = 1 \Big] \\
&\quad - \Pr\Big[(j, T, SK_j) \longleftarrow Fake_2 \,\Big|\, A(j, T, SK_j) = 1 \Big] \Big| \\
&\leq \varepsilon(t), \\
&\qquad\qquad\qquad \vdots \\
&\Big| \Pr\Big[(j, T, SK_j) \longleftarrow Fake_{n-1} \,\Big|\, A(j, T, SK_j) = 1 \Big] \\
&\quad - \Pr\Big[(j, T, SK_j) \longleftarrow Fake_n \,\Big|\, A(j, T, SK_j) = 1 \Big] \Big| \\
&\leq \varepsilon(t).
\end{aligned}
\tag{11}
$$

This implies

$$
\begin{aligned}
&\Big| \Pr\Big[(j, T, SK_j) \longleftarrow Real \,\Big|\, A(j, T, SK_j) = 1 \Big] \\
&\quad - \Pr\Big[(j, T, SK_j) \longleftarrow Fake_n \,\Big|\, A(j, T, SK_j) = 1 \Big] \Big| \\
&\leq n \cdot \varepsilon(t).
\end{aligned}
\tag{12}
$$

In $Fake_n$, the values $d_{1,2}, d_{2,3}, \ldots, d_{n-1,n}, d_{n,1}$ are constrained by T according to the following n equations:

$$
\begin{aligned}
\log_g D_1 &= s \cdot (d_{1,2} - d_{n,1}), \\
\log_g D_2 &= s \cdot (d_{2,3} - d_{1,2}), \ldots, \\
\log_g D_n &= s \cdot (d_{n,1} - d_{n-1,n}),
\end{aligned}
\tag{13}
$$

where $g = e(P, P)$. Since SK_j can be expressed as $e(P, P)^{sd_{1,2} + sd_{2,3} + \cdots + sd_{n,1}}$, we can obtain $\log_g SK_j = sd_{1,2} + sd_{2,3} + \cdots + sd_{n,1}$. Because $sd_{1,2} + sd_{2,3} + \cdots + sd_{n,1}$ is linear and independent from the set $\{\log_g D_i = s \cdot (d_{i,i+1} - d_{i-1,i}) \mid i = 1, 2, \ldots, n\}$, it implies that SK_j is independent for the transcript T. In other words, for any adversary A

$$
\begin{aligned}
&\Pr\Big[(j, T, SK_{j,0}) \longleftarrow Fake_n, SK_{j,1} \longleftarrow G_2 \mid A(j, T, SK_{j,b}) \\
&= 1, b \longleftarrow \{0, 1\} \Big] = \frac{1}{2}.
\end{aligned}
\tag{14}
$$

Therefore, the advantage of A on the event $\neg Forge$ is bounded by $2n \cdot Adv_{DBDH}(t)$. Combining the two cases, the advantage of A is bounded by

$$
Adv_\Psi^{RIDAGKE\text{-}fs}(t, 1, q_s) \leq 2n \cdot Adv_{DBDH}(t) + Adv_\Psi^{forge}(t).
\tag{15}
$$

Finally, a standard hybrid argument immediately demonstrates that

$$
\begin{aligned}
Adv_\Psi^{RIDAGKE\text{-}fs}(t, q_{ex}, q_s) &\leq 2n q_{ex} \cdot Adv_{DBDH}(t) \\
&\quad + Adv_\Psi^{forge}(t) \text{ for } q_{ex} > 1.
\end{aligned}
\tag{16}
$$

$\square$

Under the decisional bilinear Diffie-Hellman (DBDH) assumption, the advantage $Adv_{DBDH}(t)$ is negligible. By Theorem 4, the advantage $Adv_\Psi^{forge}(t)$ is also negligible. Hence, we can obtain that the advantage $Adv_\Psi^{RIDAGKE\text{-}fs}(t, q_{ex}, q_s)$ is negligible according to the result in Theorem 5. It implies that the proposed RID-AGKE protocol Ψ is a secure RID-AGKE providing forward secrecy.

Identifying Malicious Participant

Theorem 6. *The proposed RID-AGKE protocol can identify malicious participants.*

TABLE 1: Comparisons between our protocol and the previously proposed AGKE protocols.

	Tseng's AGKE [25]	Choi et al.'s ID-AGKE [26]	Wu et al.'s ID-AGKE [6]	Wu et al.'s RID-AGKE [4]	Our protocol
Public key setting	ElGmal	IDPKS	IDPKS	R-IDPKS	R-IDPKS
Certificate management	Required	Not required	Not required	Not required	Not required
Rounds	2	2	2	3	2
Computational cost for each participant	$(8n-2)T_{exp} + (n+1)T_{inv}$	$6TG_e + (n+11)TG_{mul} + (n+3)TG_H$	$(3n+3)TG_e + (n+10)TG_{mul} + 3TG_H + (n-1)T_{exp}$	$8TG_e + (2n+8)TG_{mul} + 4nTG_H$	$(3n+2)TG_e + (n+9)TG_{mul} + 4TG_H + (n-1)T_{exp}$
Security	Provably secure	Existing attack [27]	Provably secure	Provably secure	Provably secure
Revocation functionality	Using CRL [28]	No	No	Yes	Yes
Resistant to malicious participants	Yes (confirmed computation)	No	Yes (confirmed computation)	Yes (explicit key confirmation)	Yes (confirmed computation)
Identifying malicious participants	Yes	No	Yes	No	Yes

Proof. Note that in Round 2 a noninteractive confirmed computation technique is involved in adopted pairing-based signature scheme. The security of confirmed computation had been proven in [6]. Concretely, each participant U_i can confirm the broadcasted value D_k is computed by U_k using her/his secret a_k after passing two verifying equations for $k = 1, 2, \ldots, n$ except i. Hence, if there is a participant U_m who broadcasts a wrong D_m to disturb the group session key establishing, he will be identified as a malicious participant. In other words, the proposed RID-AGKE protocol can identify malicious participants by using the confirmed computation technique. □

6. Performance Analysis and Comparisons

For convenience to evaluate the computational cost, we focus on the time-consuming pairing-based operations as follows:

(i) TG_e: the time of executing a bilinear map operation $e : G_1 \times G_1 \rightarrow G_2$;

(ii) TG_{mul}: the time of executing a point scalar multiplication operation in G_1;

(iii) TG_H: the time of executing a map-to-point hash function $H_1, H_2 : \{0, 1\}^* \rightarrow G_1$;

(iv) T_{exp}: the time of executing a modular exponentiation operation over a finite field F_p, where p is a large prime number;

(v) T_{inv}: the time of executing a modular multiplicative inverse operation over a finite field F_p, where p is a large prime number.

Here, we first analyze the computational cost of our protocol. In Round 1, $2TG_{mul}$ is required to compute (P_i, V_i). In Round 2, each participant requires $3TG_e + 7TG_{mul} + 4TG_H$ to verify $(ID_{i+k}, j, P_{i+k}, V_{i+k})$ for $k \in \{-1, 1\}$ and to generate $(D_i, \alpha_i, \beta_i, \gamma_i)$. In the group key computation phase, $(3n-1)TG_e + nTG_{mul} + (n-1)T_{exp}$ is required to verify all $(ID_i, j, D_i, \alpha_i, \beta_i, \gamma_i)$ and to compute a group key SK_j.

Note that to evaluate $SK_j = e(a_i \cdot P_{i-1}, P_{pub})^n \cdot D_i^{n-1} \cdot D_{i+1}^{n-2} \cdots D_{i-2}$ is required $TG_e + TG_{mul}$ since $SK_j = A_{i-1} \cdot A_i \cdot A_{i+1} \cdots A_{i-2}$, where $A_{i-1} = e(a_i \cdot P_{i-1}, P_{pub})$, $A_i = A_{i-1} \cdot D_i$, $A_{i+1} = A_i \cdot D_{i+1}, \ldots$, and $A_{i-2} = A_{i-3} \cdot D_{i-2}$. As a result, each participant requires $(3n+2)TG_e + (n+9)TG_{mul} + 4TG_H + (n-1)T_{exp}$ in our protocol.

In Table 1, we compare our RID-AGKE protocol with four previously proposed AGKE protocols which include Tseng's AGKE protocol [25], Choi et al.'s ID-AGKE protocol [26], Wu et al.'s ID-AGKE protocol [6], and Wu et al.'s RID-AGKE protocol [4] in terms of the public key setting, number of rounds, computational cost, and security properties. One recent non-ID-based and non-RID-based AGKE protocol with identifying malicious participants was proposed by Tseng [25]. Since Tseng's protocol is based on the ElGmal system [29], each participant must verify the other participants' certificates for participant authentication. It will increase the required computational costs for verifying certificates, besides $(8n-2)T_{exp} + (n+1)T_{inv}$. On the contrary, Choi et al.'s ID-AGKE [26], Wu et al.'s ID-AGKE [6], Wu et al.'s RID-AGKE [4], and our protocol rely on the IDPKS system [8] or the R-IDPKS system [14]. Thus, they need not manage and verify the participants' certificates. However, Choi et al.'s ID-AGKE [26] suffered from an insider colluding attack demonstrated by Wu and Tseng [27].

For Wu et al.'s ID-AGKE [6], Wu et al.'s RID-AGKE [4], and our protocol, they are provably secure and are able to resist malicious participants. It is easy to see that our protocol is more efficient than Wu et al.'s ID-AGKE [6] even though both protocols can identify malicious participants via confirmed computation approach. More importantly, Wu et al.'s ID-AGKE protocol [6] does not provide a solution to revoke the compromised or misbehaving user in the group. It is very serious because these revoked participants should not be allowed to establish a common key with other legal (nonrevoked) participants. In another aspect, Wu et al.'s RID-AGKE [4] is a three-round protocol and adopts explicit key confirmation approach to resist malicious participants.

Though their protocol can detect the existence of malicious participants, it cannot still identify malicious participant. Our proposed RID-AGKE is a two-round protocol and provides an active detection mechanism to identify malicious participants. According to Table 1, the advantage of our protocol is demonstrated.

7. Conclusions

In this paper, we have fused the Tseng-Tsai R-IDPKS system and a noninteractive confirmed computation technique to propose the first RID-AGKE protocol with identifying malicious participants. The framework and security notions for RID-AGKE protocols have been defined to formalize the possible threats and attacks. When compared with the recently proposed ID/RID-AGKE protocols resistant to malicious participants, our protocol has better performance and provides an active detection way to identify malicious participants. In the random oracle model and under two mathematical assumptions (CDH and DBDH), we have proven that the proposed protocol is a secure RID-AGKE protocol with forward secrecy and identifying malicious participants.

Notations

e: A bilinear map, $e : G_1 \times G_1 \rightarrow G_2$, defined in Section 2.1

s: The system private key, $s \in Z_q^*$

P: A generator of group G_1

P_{pub}:   The system public key, $P_{pub} = s \cdot P$

ID_i: The identity of participant U_i

DID_i: The participant U_i's initial private key

$TID_{i,j}$:   The participant U_i's update key for period j

$DID_{i,j}$:   The participant U_i's private key for period j, $DID_{i,j} = DID_i + TID_{i,j}$

H_1:   A map-to-point hash function, $H_1 : \{0, 1\}^* \rightarrow G_1$

H_2:   A map-to-point hash function, $H_2 : \{0, 1\}^* \rightarrow G_1$

H_3:   A hash function, $H_3 : \{0, 1\}^* \times G_1 \rightarrow Z_q$

H_4:   A hash function, $H_4 : \{0, 1\}^* \times G_1^3 \rightarrow Z_q$.

Conflict of Interests

The authors declare that there is no conflict of interests regarding the publication of this paper.

Acknowledgments

The authors thank the referees for their valuable comments and constructive suggestions. This research was partially supported by Shenzhen Peacock Project of China (no. KQC201109020055A) and Shenzhen Strategic Emerging Industries Program of China (no. ZDSY20120613125016389).

References

[1] M. Burmester and Y. Desmedt, "A secure and efficient conference key distribution system," in *Advances in Cryptology—EUROCRYPT '94*, vol. 950 of *Lecture Notes in Computer Science*, pp. 275–286, 1995.

[2] J. Katz and J. S. Shin, "Modeling insider attacks on group key-exchange protocols," in *Proceedings of the12th ACM Conference on Computer and Communications Security (CCS '05)*, pp. 180–189, November 2005.

[3] T.-Y. Wu, Y.-M. Tseng, and C.-W. Yu, "A secure ID-Based authenticated group key exchange protocol resistant to insider attacks," *Journal of Information Science and Engineering*, vol. 27, no. 3, pp. 915–932, 2011.

[4] T.-Y. Wu, Y.-M. Tseng, and T.-T. Tsai, "A revocable ID-based authenticated group key exchange protocol with resistant to malicious participants," *Computer Networks*, vol. 56, no. 12, pp. 2994–3006, 2012.

[5] Y.-M. Tseng, "A robust multi-party key agreement protocol resistant to malicious participants," *The Computer Journal*, vol. 48, no. 4, pp. 480–487, 2005.

[6] T.-Y. Wu and Y.-M. Tseng, "Towards ID-based authenticated group key exchange protocol with identifying malicious participants," *Informatica*, vol. 23, no. 2, pp. 315–334, 2012.

[7] A. Shamir, "Identity-based cryptosystems and signature schemes," in *Identity-Based Cryptosystems and Signature Schemes*, vol. 196 of *Lecture Notes in Computer Science*, pp. 47–53, 1985.

[8] D. Boneh and M. Franklin, "Identity-based encryption from the weil pairing," *SIAM Journal on Computing*, vol. 32, no. 3, pp. 586–615, 2003, Preliminary version: in Advances in Cryptology—CRYPTO '01, vol. 2139 of Lecture Notes in Computer Science, pp. 213–229, 2001.

[9] A. Boldyreva, V. Goyal, and V. Kumart, "Identity-based encryption with efficient revocation," in *Proceedings of the 15th ACM conference on Computer and Communications Security (CCS '08)*, pp. 417–426, October 2008.

[10] R. Canetti, S. Halevi, and J. Katz, "A forward-secure public-key encryption scheme," *Journal of Cryptology*, vol. 20, no. 3, pp. 265–294, 2007, Preliminary version: in Advances in Cryptology—EUROCRYPT 2003, vol. 2656 of Lecture Notes in Computer Science , pp. 255–271, 2003.

[11] B. Libert and D. Vergnaud, "Adaptive-ID secure revocable identity-based encryption," in *Topics in Cryptology—CT-RSA 2009*, vol. 5473 of *Lecture Notes in Computer Science*, pp. 1–15, 2009.

[12] J. H. Seo and K. Emura, "Revocable identity-based encryption revisited: security model and construction," in *Public-Key Cryptography—PKC 2013*, vol. 7778 of *Lecture Notes in Computer Science*, pp. 216–234, 2013.

[13] J. H. Seo and K. Emura, "Efficient delegation of key generation and revocation functionalities in identity-based encryption," in *Topics in Cryptology—CT-RSA 2013*, vol. 7779 of *Lecture Notes in Computer Science*, pp. 343–358, 2013.

[14] Y.-M. Tseng and T.-T. Tsai, "Efficient revocable ID-based encryption with a public channel," *The Computer Journal*, vol. 55, no. 4, pp. 475–486, 2012.

[15] T.-T. Tsai, Y.-M. Tseng, and T.-Y. Wu, "A fully secure revocable ID-based encryption in the standard model," *Informatica*, vol. 23, no. 3, pp. 487–505, 2012.

[16] T.-Y. Wu, T.-T. Tsai, and Y.-M. Tseng, "Revocable ID-based signature scheme with batch verifications," in *Proceedings of the*

8th International Conference on Intelligent Information Hiding and Multimedia Signal Processing (IIH-MSP '12), pp. 49–54, July 2012.

[17] T. T. Tsai, Y. M. Tseng, and T. Y. Wu, "Provably secure revocable ID-based signature in the standard model," *Security and Communication Networks*, vol. 6, no. 10, pp. 1250–1260, 2013.

[18] T.-Y. Wu, T.-T. Tsai, and Y.-M. Tseng, "A revocable ID-based signcryption scheme," *Journal of Information Hiding and Multimedia Signal Processing*, vol. 3, no. 3, pp. 240–251, 2012.

[19] M. Bellare and P. Rogaway, "Random oracles are practical: a paradigm for designing efficient protocols," in *Proceedings of the 1st ACM Conference on Computer and Communications Security (CCS '93)*, pp. 62–73, November 1993.

[20] L. Chen, Z. Cheng, and N. P. Smart, "Identity-based key agreement protocols from pairings," *International Journal of Information Security*, vol. 6, no. 4, pp. 213–241, 2007.

[21] T.-Y. Wu and Y.-M. Tseng, "An ID-based mutual authentication and key exchange protocol for low-power mobile devices," *The Computer Journal*, vol. 53, no. 7, pp. 1062–1070, 2010.

[22] K. Y. Choi, J. Y. Hwang, and D. H. Lee, "Efficient ID-based group key agreement with bilinear maps," in *Public Key Cryptography—PKC 2004*, vol. 2947 of *Lecture Notes in Computer Science*, pp. 130–144, 2004.

[23] J. Katz and J. S. Shin, "Modeling insider attacks on group key-exchange protocols," in *Proceedings of the12th ACM Conference on Computer and Communications Security (CCS '05)*, pp. 180–189, November 2005.

[24] R. Steinwandt and A. S. Corona, "Attribute-based group key establishment," *Advances in Mathematics of Communications*, vol. 4, no. 3, pp. 381–398, 2010.

[25] Y.-M. Tseng, "A communication-efficient and fault-tolerant conference-key agreement protocol with forward secrecy," *Journal of Systems and Software*, vol. 80, no. 7, pp. 1091–1101, 2007.

[26] K. Y. Choi, J. Y. Hwang, and D. H. Lee, "ID-based authenticated group key agreement secure against insider attacks," *IEICE Transactions on Fundamentals of Electronics, Communications and Computer Sciences*, vol. E91-A, no. 7, pp. 1828–1830, 2008.

[27] T. Y. Wu and Y. M. Tseng, "Comments on an ID-based authenticated group key agreement protocol with withstanding insider attacks," *IEICE Transactions on Fundamentals*, vol. E92-A, no. 10, pp. 2638–2640, 2009.

[28] R. Housley, W. Polk, W. Ford, and D. Solo, "Internet X.509 public key infrastructure certificate and certificate revocation list (CRL) profile," RFC 3280, IETF, Anaheim, Calif, USA, 2002.

[29] T. ElGamal, "A public-key cryptosystem and a signature scheme based on discrete logarithms," *IEEE Transactions on Information Theory*, vol. 31, no. 4, pp. 469–472, 1985.

19

The Application of Baum-Welch Algorithm in Multistep Attack

Yanxue Zhang,[1] Dongmei Zhao,[2] and Jinxing Liu[3]

[1] *College of Mathematics and Information Science, Hebei Normal University, Shijiazhuang 050000, China*
[2] *College of Information Technology, Hebei Normal University, Shijiazhuang 050000, China*
[3] *The First Aeronautics College of PLAAF, Xinyang 464000, China*

Correspondence should be addressed to Dongmei Zhao; zhaodongmei666@126.com

Academic Editor: Yuxin Mao

The biggest difficulty of hidden Markov model applied to multistep attack is the determination of observations. Now the research of the determination of observations is still lacking, and it shows a certain degree of subjectivity. In this regard, we integrate the attack intentions and hidden Markov model (HMM) and support a method to forecasting multistep attack based on hidden Markov model. Firstly, we train the existing hidden Markov model(s) by the Baum-Welch algorithm of HMM. Then we recognize the alert belonging to attack scenarios with the Forward algorithm of HMM. Finally, we forecast the next possible attack sequence with the Viterbi algorithm of HMM. The results of simulation experiments show that the hidden Markov models which have been trained are better than the untrained in recognition and prediction.

1. Introduction

Currently, the network security situation is increasingly sophisticated and the multistep network attack has become the mainstream of network attack. 2012 Chinese Internet network security reports released by the National Computer Network Emergency Response Technical Team Coordination Center of China (CNCERT/CC) show that the two typical multistep attacks: warms and distributed denial of service (DDOS) [1] account for 60% of overall network attacks. Multistep attack [2] means that the attacks apply multiple attack steps to attack the security holes of the target itself and achieve the devastating blow to the target. There are three features of attack steps of multistep attack. (1) In the multistep attack, there is a casual relationship between multiple attack steps. (2) The attack steps of multistep attack have the property of time sequence [3]. (3) The attack steps of multistep attack have the characteristics of uncertainty [4].

Multistep attack is one of the main forms of network attack behaviors, recognizing and predicting multistep attack that laid the foundation of active defense, which is still one of the hot spots nowadays. Literature (application of hidden Markov models to detect multistep network attacks) proposed a method to recognize multistep attack based on hidden Markov model.

Markov model literature (improving the quality of alerts and predicting intruder's next goal with hidden colored Petri-net) introduced the concept of attack "observation," but both stayed in the specific attack behaviors, which have some limitations. Current research on the approaches to forecast multistep attack behaviors mainly includes four types: (1) the approach to forecasting multistep attack based on the antecedents and consequences of the attack [5]. It applies the precursor subsequent relationship of the event, to forecast the attacker wants to implement attacks in the near future. Because of the complexity and the diversity of the attack behaviors, this approach is difficult to achieve. (2) The approach to forecasting multistep attack based on hierarchical colored Petri-nets (HCPN) applies the raw alerts by Petri-nets and considers that the attack intention is inferred by raw alerts [4]. But this approach focuses on the intrusion detection of multistep attack behaviors. (3) The approach to forecasting multistep attack based on Bayes game theory could forecast the probability that the attackers choose to

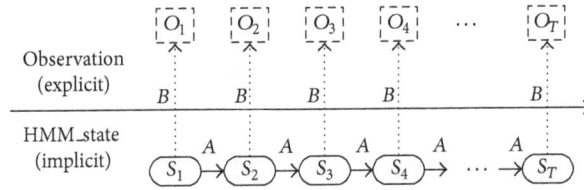

FIGURE 1: Model of recognizing and forecasting multistep attack based on hidden Markov model.

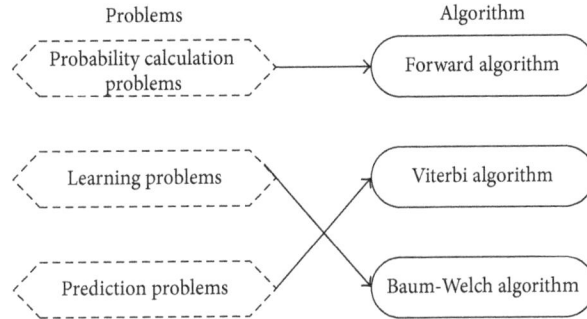

FIGURE 2: Correspondence between the problems and algorithms of hidden Markov model.

attack and the probability that the defenders choose to defend in the next stage rationally [6, 7]. However, in current study, only two-person game model is established, so this approach has some limitations. (4) The approach to forecasting multistep attack based on attack intention [3, 8] uses extended-directed graph to describe the logical relationship between attack behaviors and forecasts the next stage by logical relationship. The shortcoming of this approach is that it is difficult to determine the matching degree of the multistep attack. At the same time, there exists a certain degree of subjectivity in recognizing and forecasting multistep attack. In this regard, we integrate the attack intentions and hidden Markov model and propose a method to forecast multistep attack based on hidden Markov model. Firstly, we train the existing hidden Markov model(s) by the Baum-Welch algorithm of HMM. Then we recognize the alert belonging to attack scenarios with the Forward algorithm of HMM. Finally, we forecast the next possible attack sequence with the Viterbi algorithm of HMM. Simulation experiments results show that the hidden Markov models which have been trained are better than the untrained in recognition and prediction.

2. Hidden Markov Model

Hidden Markov model was first proposed by Baum and Petrie in 1966. It is a statistical model, which is used to describe a Markov process which contains a hidden parameter [9]. The research object of this model is a data sequence; each value of this data sequence is called an observation. Hidden Markov model assumes that there still exists another sequence which hides behind this data sequence; the other sequence consists of a series of states. Each observation occurs in a state, the state cannot be observed directly, and the features of the state can only be inferred from the observations.

A complete hidden Markov model (HMM) is usually represented by a triple $\lambda = (A, B, \pi)$, which includes the following five elements:

(1) a finite state, which is represented by the set S, where $S = \{s_1, s_2, \ldots, s_N\}$ and, at time t, the state is denoted by q_t;

(2) the set of observations, which is represented by the set O, where $O = \{o_1, o_2, \ldots, o_T\}$;

(3) the state transition matrix, which is represented by the matrix A, where $a_{ij} = p[q_{t+1} = s_j \mid q_t = s_j]$ and $1 \leq i, j \leq N$;

(4) the probability distribution of matrix A, which is represented by the matrix B, where $b_j(k) = p[o_k \mid q_t = s_j]$ and $1 \leq j \leq N, 1 \leq k \leq T$;

(5) the set of initial state probability distribution of HMM, which is represented by the set π, where $\pi_i = p[q_1 = s_i]$ and $1 \leq i \leq N$.

The model of recognizing and forecasting multistep attack based on hidden Markov model is shown in Figure 1.

There are three problems which can be solved by hidden Markov model well.

(1) *Probability Calculation Problems.* Calculate the probability $p(O \mid \lambda)$ under a given hidden Markov model $\lambda = (A, B, \pi)$ and the observation sequence $O = \{o_1, o_2, \ldots, o_T\}$.

(2) *Learning Problems.* Estimate the parameters of $\lambda = (A, B, \pi)$ when the observation sequence $O = \{o_1, o_2, \ldots, o_T\}$ is known, to maximize the probability $p(O \mid \lambda)$.

(3) *Prediction Problems.* Calculate the state sequence $I = \{i_1, i_2, \ldots, i_T\}$ under the maximum probability, when the hidden Markov model $\lambda = (A, B, \pi)$ and observation sequence $O = \{o_1, o_2, \ldots, o_T\}$ are given.

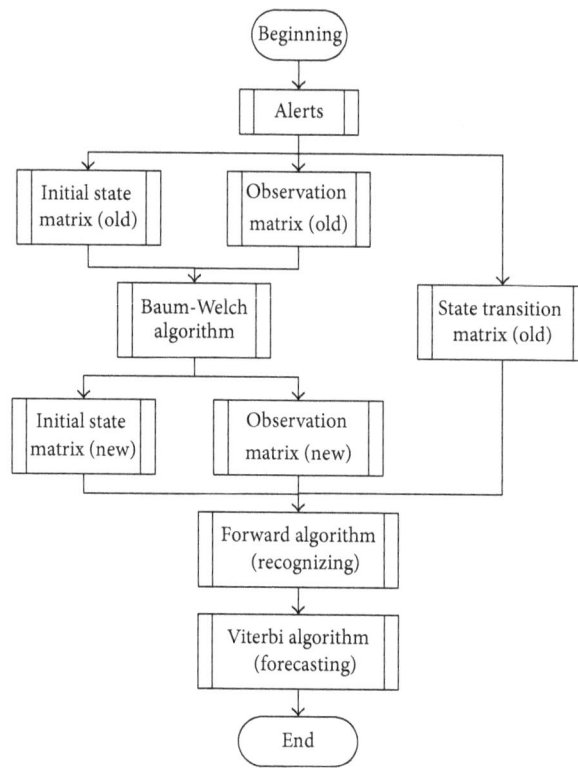

FIGURE 3: Flow chart of recognizing and forecasting multistep attack.

Input: alert sequence.
      $O = \{o_1, o_2, \ldots, o_T\}$;
Output: the parameters of hidden Markov model.
      $\lambda^{(n+1)} = (A^{(n+1)}, B^{(n+1)}, \pi^{(n+1)})$.
Step 1. Initialization.
      for $n = 0$, select $a_{ij}^{(0)}, b_j(k)^{(0)}, \pi_i^{(0)}$, we can obtain the initial model $\lambda^{(0)} = (A^{(0)}, B^{(0)}, \pi^{(0)})$.
Step 2. Iterative calculation.
for $n = 1, 2, \ldots$,
$$a_{ij}^{(n+1)} = \frac{\sum_{t=1}^{T-1} \xi_t(i, j)}{\sum_{t=1}^{T-1} \gamma_t(i)};$$
$$b_j(k)^{(n+1)} = \frac{\sum_{t=1, o_t=v_k}^{T} \gamma_t(j)}{\sum_{t=1}^{T} \gamma_t(j)};$$
$$\pi_i^{(n+1)} = \gamma_1(i).$$
where $\gamma_t(i) = \dfrac{\alpha_t(i)\beta_t(i)}{p(O \mid \lambda)} = \dfrac{\alpha_t(i)\beta_t(i)}{\sum_{j=1}^{N} \alpha_t(j)\beta_t(j)\sum}$;
$$\xi_t(i, j) = \frac{\alpha_t(i)a_{ij}b_j(o_{t+1})\beta_{t+1}(j)}{p(O \mid \lambda)} = \frac{\alpha_t(i)a_{ij}b_j(o_{t+1})\beta_{t+1}(j)}{\sum_{i=1}^{N}\sum_{j=1}^{N} \alpha_t(i)a_{ij}b_j(o_{t+1})\beta_{t+1}(j)}.$$
Step 3. Termination. We can obtain the parameters of hidden Markov model.
$\lambda^{(n+1)} = (A^{(n+1)}, B^{(n+1)}, \pi^{(n+1)})$.

ALGORITHM 1

Forward_Algorithm (λ, O):
Input: (1) alert sequence $O = \{alert_1, alert_2, \ldots, alert_T\}$;
      (2) hidden Markov model (HMM) λ.
Output: the probability $p(O \mid \lambda)$ generated by alert sequence $O = \{alert_1, alert_2, \ldots, alert_T\}$ of hidden Markov model.
Begin:
    (1) $\forall \text{int } ent_i \in \lambda, 1 \le i \le N$.
        // N is the number of attack intentions.
        calculate the probability of $alert_1$ generated by int ent_i: $\alpha_1(i) = \pi_i b_i(alert_1)$
    (2) calculate the probability of alert sequence $\{alert_1, alert_2, \ldots, alert_T\}$ and $q_{t+1} = \text{int } ent_j$.
        (a) at time t, calculate the probability of alert sequence $\{alert_1, alert_2, \ldots, alert_T\}$ and $q_t = \text{int } ent_j$: $\alpha_t(j)$.
        (b) at time $t+1$, calculate the probability of intent sequence $\{alert_1, alert_2, \ldots, alert_T\}$ generated by hidden Markov model (HMM): λ and

$$q_t = \text{int } ent_j: \alpha_{t+1}(j) = \left[\sum_{i=1}^{N} \alpha_t(i) a_{ij} \right] b_j(alert_{t+1}) \text{ where } 1 \le t \le T-1; 1 \le j \le N.$$

    (3) calculate the probability of the intent sequence $O = \{alert_1, alert_2, \ldots, alert_T\}$ generated by hidden Markov model (HMM): λ.

$$p(O \mid \lambda) = \sum_{i=1}^{N} \alpha_T(i).$$

    (4) Return $p(O \mid \lambda)$.
End;

ALGORITHM 2

Viterbi_Algorithm (λ, O):
Input: alert sequence $O = \{alert_1, alert_2, \ldots, alert_T\}$;
Output: (1) intent sequence: $Q = \{\text{int } ent_1, \text{int } ent_2, \ldots, \text{int } ent_T\}$.
        (2) the completed intent sequence and the next likely intent.
Begin:
    for $i = 1$ to HMM_m
    // HMM_m is the number of hidden Markov model(s)
    {
        Prob = Forward_Algorithm(hmm_i, O);
        // calculate the probability of alert sequence generated by each hidden Markov
        // model(s)
    }
    Most_likely_multi-step_attack_intention = maximum(Prob);
    Q = Viterbi_Algorithm(hmm_i', O);
    // Q is the completed intent sequence
    // hmm_i' is the maximum(Prob) of hmm_i
    $Q' = S - Q$ // the next likely intent
    // S is the intent sequence of hmm_i'
End;

ALGORITHM 3

Correspondence between the problems and algorithms of hidden Markov model are shown in Figure 2.

Hidden Markov model is usually used to deal with the problems related to the time sequence and it has been widely used in speech recognition, signal processing, bioinformation, and other fields. Based on the characteristics of the attack steps of hidden Markov model and the problems that hidden Markov model can be solved, we apply the hidden Markov model to the field of recognizing and forecasting multistep attack. Firstly, the improved Baum-Welch algorithm is used to train the hidden Markov model λ, and we get a new hidden Markov model λ'. Then we recognize the alert belonging to attack scenarios with the Forward algorithm of hidden Markov model. Finally, we forecast the next possible attack sequence with the Viterbi algorithm of hidden Markov model.

3. The Approach to Recognizing and Forecasting Multistep Attack

The steps of the approach to recognizing and forecasting multistep attack are as follows.

Step 1. Obtain the initial state matrix (old), state transition matrix (old), and observation matrix (old) of HMM (λ).

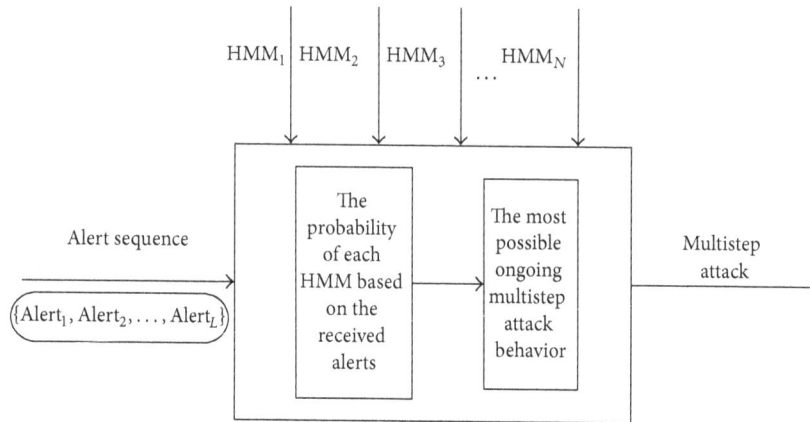

FIGURE 4: The structure of recognizing multistep attack with Forward algorithm.

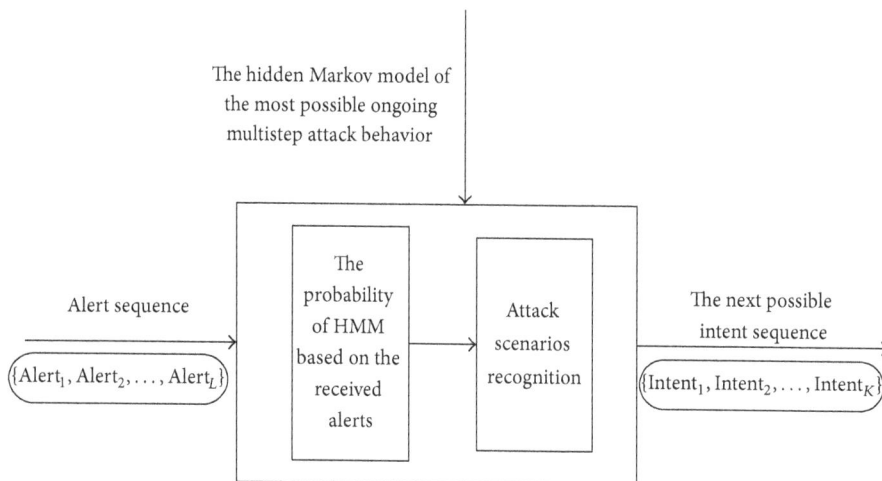

FIGURE 5: Forecasting multistep attack with Viterbi algorithm.

TABLE 1: The initial state matrix of DDoS_HMM.

State$_1$	State$_2$	State$_3$	State$_4$	State$_5$
0.250	0.750	0.000	0.000	0.000

TABLE 2: The state transition matrix of DDoS_HMM.

	State$_1$	State$_2$	State$_3$	State$_4$	State$_5$
State$_1$	0.000	1.000	0.000	0.000	0.000
State$_2$	0.000	0.177	0.823	0.000	0.000
State$_3$	0.000	0.228	0.688	0.028	0.056
State$_4$	0.000	0.000	0.000	0.750	0.250
State$_5$	0.000	0.000	0.000	0.000	0.000

Step 2. Use the improved Baum-Welch algorithm to train the initial state matrix (old) and observation matrix (old), and we get an initial state matrix (new), observation matrix (new), and a new HMM (λ').

Step 3. Recognize the alert belonging to attack scenarios with the Forward algorithm.

Step 4. Forecast the next possible attack sequence with the Viterbi algorithm.

The flow chart is shown in Figure 3.

3.1. The Introduction of Baum-Welch Algorithm. If we want to apply the hidden Markov model to the multistep attack, the biggest problem is to determine the observations of HMM. A better parameter can improve the efficiency of

calculation. Meanwhile, if the selection of observation is improper, this may result in a longer training time and even not complete the training. In this regard, we apply the Baum-Welch algorithm to train the given hidden Markov model. From the result of literature (accurate Baum-Welch algorithm free from overflow), we can learn that the most reliable algorithm to train the HMM is Baum-Welch algorithm. Baum-Welch algorithm can train the given hidden Markov model (λ) by an observation sequence and generate a new hidden Markov model (λ') for detection.

The steps of Baum-Welch algorithm are as in Algorithm 1.

TABLE 3: The observation matrix of DDoS_HMM.

	A1	A2	A3	A4	A5	A6	A7	A8	A9	A10	A11	A12	A13
S1	1.000	0.000	0.000	0.000	0.000	0.000	0.000	0.000	0.000	0.000	0.000	0.000	0.000
S2	0.000	0.490	0.490	0.020	0.000	0.000	0.000	0.000	0.000	0.000	0.000	0.000	0.000
S3	0.000	0.000	0.000	0.000	0.200	0.200	0.200	0.200	0.200	0.000	0.000	0.000	0.000
S4	0.000	0.000	0.000	0.000	0.000	0.000	0.000	0.000	0.000	1.000	0.000	0.000	0.000
S5	0.000	0.000	0.000	0.000	0.000	0.000	0.000	0.000	0.000	0.000	0.660	0.170	0.170

TABLE 4: The initial state matrix of DDoS_HMM$'$.

State$_1$	State$_2$	State$_3$	State$_4$	State$_5$
0.599	0.401	0.000	0.000	0.000

TABLE 5: The observation matrix of DDoS_HMM$'$.

	A1	A2	A3	A4	A5	A6	A7	A8	A9	A10	A11	A12	A13
S1	1.000	0.000	0.000	0.000	0.000	0.000	0.000	0.000	0.000	0.000	0.000	0.000	0.000
S2	0.000	0.499	0.499	0.002	0.000	0.000	0.000	0.000	0.000	0.000	0.000	0.000	0.000
S3	0.000	0.000	0.000	0.000	0.387	0.000	0.387	0.000	0.226	0.000	0.000	0.000	0.000
S4	0.000	0.000	0.000	0.000	0.000	0.000	0.000	0.000	0.000	1.000	0.000	0.000	0.000
S5	0.000	0.000	0.000	0.000	0.000	0.000	0.000	0.000	0.000	0.000	0.998	0.001	0.001

TABLE 6: DDoS_HMM.

STATE	ALERT
State$_1$	{Alert$_1$}
State$_2$	{Alert$_2$, Alert$_3$, Alert$_4$}
State$_3$	{Alert$_5$, Alert$_6$, Alert$_7$, Alert$_8$, Alert$_9$}
State$_4$	{Alert$_{10}$}
State$_5$	{Alert$_{11}$, Alert$_{12}$, Alert$_{13}$}

TABLE 7: FTP Bounce_HMM.

State	Alert
State$_1$	{Alert$_1'$, Alert$_2'$}
State$_2$	{Alert$_3'$, Alert$_4'$}
State$_3$	{Alert$_5'$, Alert$_6'$, Alert$_7'$}
State$_4$	{Alert$_8'$}
State$_5$	{Alert$_9'$, Alert$_{10}'$}

3.2. Forward Algorithm. The pseudocode of Forward algorithm is as in Algorithm 2.

Recognizing multistep attack is mainly based on the alert sequence. First, we calculate the probability of alert sequence generated by the given HMM(s). Then we decide that the attack which has the maximum is likely to be the ongoing attack. The structure of recognizing multistep attack with Forward algorithm is shown in Figure 4.

3.3. Viterbi Algorithm. The pseudocode of Viterbi algorithm is as in Algorithm 3.

Predicting the behavior of multistep attack is mainly to determine the intentions that the attackers have been completed and forecast the next possible attack intentions. The structure of forecasting multistep attack with Viterbi algorithm is shown in Figure 5.

4. The Simulation Experiment and Analysis

4.1. Baum-Welch Algorithm: Train the Given HMM(s). Based on the literature (approach to forecast multistep attack based on fuzzy hidden Markov model), we can obtain the initial state matrix, state transition matrix, and observation of DDoS_HMM, as is shown from Tables 1, 2, and 3.

The data set which is used in the simulation experiment is an attack scenario testing data set LLDOS1.0 (inside) provided by DARPA (Defense Advanced Research Projects Agency) in 2000. We extract two kinds of multistep attack from it; they are DDoS multistep attack and FTP Bounce multistep attack. While the calculation of the state transition matrix is completely the statistical calculations on data, we only train the initial state matrix and observation matrix of HMM. We can see that there are a large number of zeros in observation matrix clearly and the observation matrix is the sparse matrix. So we train the matrix(s) by block. We suppose that the number of observation sequences is S and the length of S is 32, where S multiplied by 32 equals the number of training data. And there is no corresponding sequence of state. In this regard, we can obtain the initial state matrix (new) and the observation matrix (new) of the DDoS_HMM$'$ (λ'), as is shown in Tables 4 and 5.

4.2. Forward Algorithm: Recognize the Alert Belonging to Attack Scenarios. The attack intentions and alerts of DDoS_HMM and FTP Bounce_HMM are shown in Tables 6 and 7, respectively.

When the alerts "Alert$_1$" and "Alert$_3$" were received, according to the Forward algorithm of hidden Markov model,

TABLE 8: The comparison of results.

	$p(\text{alerts} \mid \text{DDoS\_HMM})$	$p(\text{alerts} \mid \text{FTP Bounce\_HMM})$	$p(\text{alerts} \mid \text{DDoS\_HMM})$ $p(\text{alerts} \mid \text{FTP Bounce\_HMM})$
Before training	0.1225	0.0079	15.5
After training	0.2989	0.0036	83.0

we will obtain the probability based on DDoS_HMM$'$ and FTP Bounce_HMM$'$, respectively:

$$p(\text{alerts} \mid \text{DDoS\_HMM}) = 0.2989,$$

$$p(\text{alerts} \mid \text{FTP Bounce}) = 0.0036.$$

We can see from the above results, $p(\text{alerts} \mid \text{DDoS\_HMM}) > p(\text{alerts} \mid \text{FTP Bounce})$. That is to say, the ongoing multistep attack behavior is likely to be DDoS_HMM.

4.3. Viterbi Algorithm: Forecast the Next Possible Attack Sequence. When the alert sequence $\{\text{Alert}_1, \text{Alert}_3, \text{Alert}_7, \text{Alert}_8, \text{Alert}_{10}\}$ was received by the console, we can obtain the completed intent sequence $\{\text{State}_1, \text{State}_2, \text{State}_3, \text{State}_4\}$. That is to say, now completed intentions are the previous four attack intentions; the next intention will be state$_5$.

4.4. Comparison of Results. We compare the results between the untrained HMM(s) and the trained HMM(s) by Baum-Welch algorithm; the comparison of results are shown in Table 8.

5. Conclusion

The biggest difficulty of hidden Markov model applied in multistep attack is the determination of observations. Now the research of the determination of observations is still lacking, and it shows a certain degree of subjectivity. In this regard, we train the existing hidden Markov model(s) by the Baum-Welch algorithm of HMM based on several groups of observation sequence. And we can obtain a new hidden Markov model which is more objectively. Simulation experiments results show that the hidden Markov models which have been trained are better than the untrained in recognition and prediction.

Conflict of Interests

The authors declare that there is no conflict of interests regarding the publication of this paper.

Acknowledgments

The authors would like to thank the reviewers for their detailed reviews and constructive comments, which have helped in improving the quality of this paper. This work was supported by the National Natural Science Foundation of China no. 60573036, Hebei Science Fund under Grant no. F2013205193, and Hebei Science Supported Planning Projects no. 12213514D.

References

[1] B. L. Xie, S. Y. Jiang, and Q. S. Zhang, "Application-ialer DDoS attack detection based on request keywords," *Computer Science*, vol. 40, no. 7, pp. 121–125, 2013.

[2] C. Yuan, *Research on Multi-Step Attack Detection Method Based on GCT*, Jilin University, Jilin, China, 2010.

[3] C. Chen and B. Q. Yan, "Network attack forecast algorithm for multi-step attack," *Computer Engineering*, vol. 5, no. 37, pp. 172–174, 2011.

[4] G. Q. Zhai and S. Y. Zhou, "Construction and implementation of multistep attacks alert correlation model," *Journal of Computer Applications*, vol. 31, no. 5, pp. 1276–1279, 2011.

[5] Z. L. Wang and X. P. Cheng, "An Attack predictive algorithm based on the correlation of intrusion alerts in intrusion response," *Computer Science*, vol. 32, no. 4, pp. 144–146, 2005.

[6] H. Cao, Q. Q. Wang, Z. Y. Ma et al., "Attack Predition model based on dynamic bayesian games," *Computer Applications*, vol. 27, no. 6, pp. 1545–1547, 2007.

[7] H. Cao, Q. Q. Wang, Z. Y. Ma et al., "Attack predition model based on static Bayesian game," *Application Research of Computers*, vol. 24, no. 10, pp. 122–124, 2007.

[8] J.-W. Zhuge, X.-H. Han, Z.-Y. Ye, and W. Zou, "Network attack plan recognition algorithm based on the extended goal graph," *Chinese Journal of Computers*, vol. 29, no. 8, pp. 1356–1366, 2006.

[9] S. H. Zhang, *Research on Network Security Early Warning Technology Based on Hidden Markov Model*, PLA Information Engineering University, Henan, China, 2007.

20

A New Seamless Transfer Control Strategy of the Microgrid

Zhaoyun Zhang, Wei Chen, and Zhe Zhang

State Key Laboratory of Advanced Electromagnetic Engineering and Technology, Huazhong University of Science & Technology, Wuhan 430074, China

Correspondence should be addressed to Zhaoyun Zhang; zzy_zhaoyun@163.com

Academic Editors: Y. Mao, X. Meng, J. Zhou, and Z. Zhou

A microgrid may operate under two typical modes; the seamless transfer control of the microgrid is very important. The mode conversion controller is installed in microgrid and the control logic of master power is optimized for microgrid mode conversion. In the proposed scheme, master power is very important. The master-power is under the PQ control when microgrid is under grid-connected. And it is under V/F control when the microgrid is under islanding. The microgrid mode controller is used to solve the planned conversion. Three types of conversion are simulated in this paper. The simulation results show the correctness and validity of the mode control scheme. Finally, the implementation and application of the operation and control device are described.

1. Introduction

A microgrid is a low-voltage distribution grid comprising various controllable loads, storage devices, and distributed generators as a controlled entity that can either be isolated from or operate interconnectedly with the main grid. Distributed generation (DG) and the microgrid (MG) system have received increasing research attention [1–6]. At the same time, many demonstration projects of microgrids have been constructed in China, such as Dongfushan Island microgrid in Zhejiang and Zhangbei microgrid in Hebei. In general, demonstration projects involve the power such as photovoltaic, wind, and storage battery [7–9].

The microgrid may operate under two typical modes: it can connect with the main grid, known as grid-connected mode (GM), and it can operate without main grid, called islanding mode (IM). Mode conversion is one of the core issues of the microgrid control. The researches have focused on the grid-connected mode inverter control [10–12], but few research has been done on the mode conversion.

The microgrid control may be implemented under the master-slave control mode, droop control mode [13, 14], and so forth. However, many microgrids have been built or under construction adopting master-slave mode, mainly because this type of microgrids can keep the voltage and frequency of the microgrid near nominal point. On the other hand, the active power of the solar and wind energy usually is not controllable continuously, and PV systems and wind turbines often work at the maximum power point (MPP). So, control of solar and wind energy is in PQ control mode whether it is under islanding mode or grid-connected mode. Control logic is relatively simple. The reactive and active power of energy storage can be adjusted, and the energy storage system becomes the "master" power of microgrid when the microgrid is under islanding mode, and it is the frequency and voltage support of microgrid.

Frequency and voltage of the "master-slave" architecture microgrid can remain near the nominal point, control structures is clear, the control logic of the "slave" power is simple, and the "slave" power has the plug and play features. Because of these advantages, this microgrid architecture has been used in a wide range of applications. In papers [15–17], dual-mode inverter for this type of microgrid was researched, and the system is under PQ control when grid-connected mode is adopted; the system is under V/F control when islanding mode is adopted. The authors also proposed a microgrid mode conversion as a preliminary study but did not give more specific solutions.

The mode conversion of microgrid with "master-slave" architecture is discussed in this paper. The microgrid mode conversion includes the following four types:

(1) planned conversion from grid-connected mode to islanding mode,

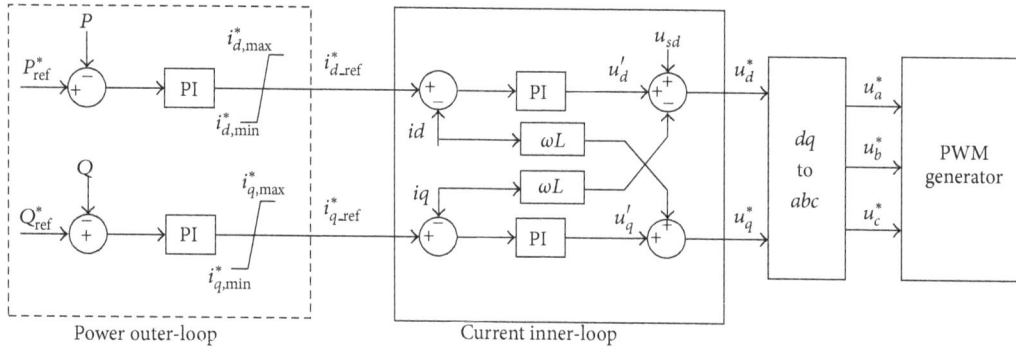

FIGURE 1: Schematic diagram of the PQ controller.

(2) unplanned conversion from grid-connected mode to islanding mode,

(3) planned conversion from islanding mode to grid-connected mode,

(4) unplanned conversion from islanding mode to grid-connected mode.

The fourth conversion can be avoided, but others cannot. In this study, for the microgrid mode conversion, microgrid sets a centralized mode controller and optimizes the master-power's control logic. The unplanned mode conversion is solved by the logic optimization of the master power, and the planned mode conversion is solved by the mode controller and the logic optimization of the master power.

A detailed program of the mode controller and an optimization scheme of the master power converter control system are presented. The energy storage system as an example of master power is described. The master power operates under PQ mode when microgrid works under grid-connected mode and master power operates under V/F mode when microgrid operates under islanding mode. The microgrid operating mode is detected through the microgrid information such as current, voltage, and digital input. The master power will change the operating mode, when the microgrid changes its operating mode. In the mode conversion process, a series of programs will be used to ensure microgrid stability.

2. Design of Control System

In a large number of the latest microgrid demonstration projects, a microgrid includes the photovoltaic power generation system, wind power systems, and energy storage system in which the "wind-solar-storage" mode is adopted. There are the photovoltaic power generation system, wind power systems, and energy storage system. Therefore, this study focuses on this type of microgrids. Without loss of generality, all of the PV systems are equivalent to one photovoltaic power generation system; all the wind systems in parallel are equivalent to one wind power system; all of the energy storage systems in parallel are equivalent to one storage system; load is distributed in microgrid.

In photovoltaic systems and direct drive wind power generation system of the microgrid, which are under PQ control mode, the maximum power tracking is always used and reactive power output is always 0.

Storage system is under the PQ control when microgrid is under grid-connected mode. The system is under V/F control, when the microgrid is under islanding.

To achieve the smooth transition between islanding mode and grid-connected mode, the control of the microgrid mode conversion includes two parts: one is the conversion control system between PQ control and V/F control for the storage system and the other is the mode controller for the mode conversion of microgrid.

2.1. Mode Conversion for the Storage System. The mode conversion between the PQ control and the V/F control for the storage system is a key component for the mode conversion of the microgrid.

2.1.1. PQ Control of the Storage System. When the storage system is under PQ control mode, the double loop control is used, which includes power outer-loop control and current inner-loop control. Figure 1 is the control diagram.

In power outer loop, the energy management system of the microgrid provides the active power reference value P_{ref}^* and reactive power reference value Q_{ref}^*, which depend on the state of the storage system and the load balance of microgrid.

The difference between the reference value and the actual value of active power is the input of the outer-loop PI regulator, for which the output is the reference values of the d-axis current i_d^* in the inner loop. Value of the q-axis current i_q^* is similar to i_d^*.

In current inner loop, the difference between reference value and actual value of the d-axis current is the input of the PI regulator, for which the output is the reference value of the d-axis voltage (u_d^*) of the inverter. The difference between reference value and actual value of the q-axis current is the input of the PI regulator, for which the output is the reference value of the d-axis voltage (u_q^*) of the inverter.

2.1.2. V/F Control of Storage System. When storage system is under V/F control mode, the voltage reference value is set. And frequency reference value is set, too. Voltage difference between reference value and the actual value is the input of PI regulator, for which the output is the reference values of the control voltage of the inverter. Figure 2 is the control diagram.

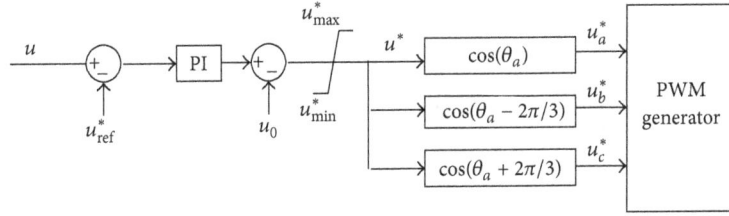

FIGURE 2: Schematic diagram of the V/F controller.

Where, u is the actual value of the voltage of the storage system output, u^* is the reference of microgrid voltage, f is the microgrid frequency, which can be set at 50 Hz (or 60 Hz), θ_0 is the initial phase angle, and u_0 is the initial voltage value.

To keep frequency and voltage of the microgrid stable, angle θ_0 and voltage u_0 will be adjusted.

2.1.3. Mode Conversion of the Storage System. Whether the system is under PQ control mode or V/F control mode, the last step is to get u_a^*, u_b^*, and u_c^* to PWM driver circuit and control IGBT turn-off and turn-on. In order to ensure a smooth transition, the reference voltages u_a^*, u_b^*, and u_c^* must be changed smoothly.

To ensure the continuity of the PWM reference voltage, it is necessary to ensure continuous amplitude and phase.

When the microgrid switches from grid-connected mode to islanding mode, storage system switch control includes the following.

(a) The amplitude u_a^* and phase angle θ_0 of PWM reference voltage are recorded when system is under grid-connected mode.

(b) When the microgrid switches from grid-connected mode to islanding mode, storage system is from the PQ control mode to the V/F control mode. The voltage reference value u_0 in V/F control mode is equal to the recording amplitude, when system is under PQ control mode, and the phase angle is

$$\theta_a = \theta_0 + 2\pi f \Delta t, \qquad (1)$$

where θ_0 is the recording phase angle when the storage system is under PQ mode, Δt is time after the storage system switches into the V/F control mode, and f is the voltage frequency value of the microgrid.

When microgrid operates continuously under V/F mode, the recursive algorithm is used in the phase angle calculation; the formula is

$$\theta_t = \theta_{t-\tau} + 2\pi f \tau, \qquad (2)$$

where θ_t is the phase angle of the current sampling point, $\theta_{t-\tau}$ is the phase angle of the latest sampling point, τ is the sampling interval, and f is frequency value of microgrid.

When microgrid switches from islanding mode to grid-connected mode, the synchronization function is fulfilled by the mode controller of microgrid. The storage system only

needs to change the response time of the PQ control mode (increasing K_i and reducing K_p) and the system can complete the smooth transition from the V/F control mode to the PQ control mode.

2.1.4. Inverter Parameters. The storage system inverter control includes the following parameters:

(a) the control mode: the PQ mode or V/F mode, which can through an external input, active detection, or the set of the inverter,

(b) the active power reference value: valid under PQ mode,

(c) the reactive power reference value: valid under the PQ mode,

(d) the voltage reference value: valid under V/F mode,

(e) the frequency reference value: valid under V/F mode,

(f) the phase angle reference values: valid under V/F mode; this generally is not for the external input, instead of the internal automatic continuous calculation.

2.2. The Mode Controller of Microgrid. With mode conversion of storage system, microgrid can smoothly switch between grid-connected mode and islanding mode. The microgrid mode controller is used to solve two problems and accomplish two goals as follows:

(a) planned conversion from grid-connected mode to islanding mode,

(b) planned conversion from islanding mode to grid-connected mode.

2.2.1. Planned Conversion from Grid-Connected Mode to Islanding Mode. Planned conversion from grid-connected mode to islanding mode does not trip the PCC's breaker immediately. A series of control methods are adopted via the mode controller of the microgrid, and the appropriate time for the trip breaker is selected, and then the system switches from the grid-connected mode to the islanding mode.

Control logic is as follows.

(a) After splitting command is produced, the storage system continues to work under PQ control mode. The mode controller begins to change the reference power of the storage

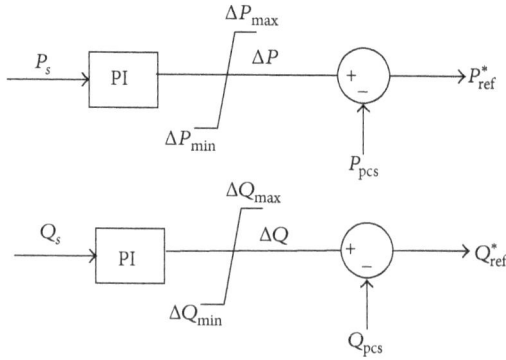

FIGURE 3: Schematic diagram of the power adjustment.

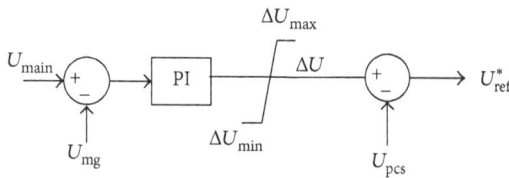

FIGURE 4: The voltage amplitude adjustment diagram.

system until the apparent power reaches 0. The control logic is as shown in Figure 3.

In Figure 3, P_s is the output active power from microgrid to main grid, Q_s is the output reactive power from microgrid to main grid, P_{pcs} is the actual output active power of PCS (storage system), Q_{pcs} is the actual output reactive power of PCS, P_{ref}^* is reference value of output active power of PCS, and Q_{ref}^* is reference value of output reactive power of PCS.

(b) If the active power and reactive power of microgrid are lower than the given values, the microgrid mode controller trips PCC breaker.

The microgrid mode controller monitors apparent power amplitude $S_s = P_s + jQ_s$ in real time. When the amplitude is continuously lower than the threshold value for a period of time (e.g., 0.1 s), the mode controller of the microgrid sends the command to trip the PCC breaker. The storage system continues to work under PQ control mode.

(c) After the trip command is received, PCC breaker turns off, and the microgrid works under the islanding mode. The storage system detects the islanding mode and starts to work under V/F control mode. Planned conversion is completed.

2.2.2. Planned Conversion from Islanding Mode to Grid-Connected Mode. With planned conversion from islanding mode to grid-connected mode, the synchronization problem between microgrid and main grid is resolved. Frequency and voltage adjustment is adjusted to allow microgrid to eventually meet the synchrony conditions.

The control logic is as follows:

(a) voltage adjustment: through adjusting the voltage reference value of storage system, the microgrid voltage becomes closer to main grid's voltage; the control logic is shown in Figure 4;

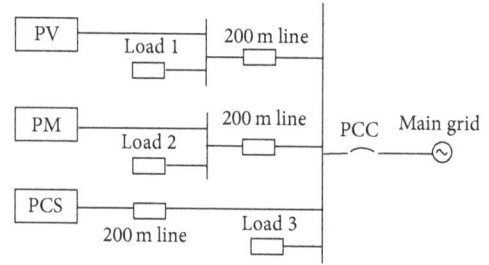

FIGURE 5: Schematic diagram of the microgrid.

(b) frequency adjustment: the mode controller detects the main grid's frequency and phase difference between the microgrid and main grid; then microgrid frequency is set as:

$$f_{ref} = f_{main} \pm \Delta f, \tag{3}$$

where f_{ref} is the storage system's reference frequency, f_{main} is the main grid's frequency, and Δf is the minor adjustment frequency, which can be set to 0.05 Hz;

the selection of addition or subtraction depends on the initial phase angle difference, which can be set as subtraction;

(c) synchronization check: mode controller makes checking synchronization feature to take effect; when the voltage difference and phase angle difference meet synchronization requests, the mode controller sends a close PCC breaker command;

(d) after the close command is received, the PCC break turns on, and microgrid operate under grid-connected mode; storage system detects this and works in PQ control mode; planned conversion is completed.

3. Simulation Analysis

3.1. Simulation Parameters. The simulation system is shown in Figure 5. Microgrid has three distributed sources, photovoltaic "PV," direct-drive wind power "PM," and battery energy storage system "PCS," and three groups of loads which are voltage-sensitive.

The photovoltaic generation "PV" works in PQ control mode, where the rated power factor is 1.0, the maximum active power track generation is adopted, and the rated active power is 150 kW. Direct-drive wind power "PM" works in PQ control, where the rated active power is 200 kW, the rated power factor is 1.0, and the maximum active power track is adopted.

In this example, all lines are 380 V line, $R = 0.642\ \Omega/\text{km}$, $X = 0.102\ \Omega/\text{km}$. Load is constant impedance load, expressed as $Z_{ld} = R_{ld} + jX_{ld}$. Load parameters are as follows:

$$Z_{ld1} = 0.922 + 0.218j\ \Omega,$$
$$Z_{ld2} = 1.296 + 0.466j\ \Omega, \tag{4}$$
$$Z_{ld3} = 1.294 + 0.466j\ \Omega.$$

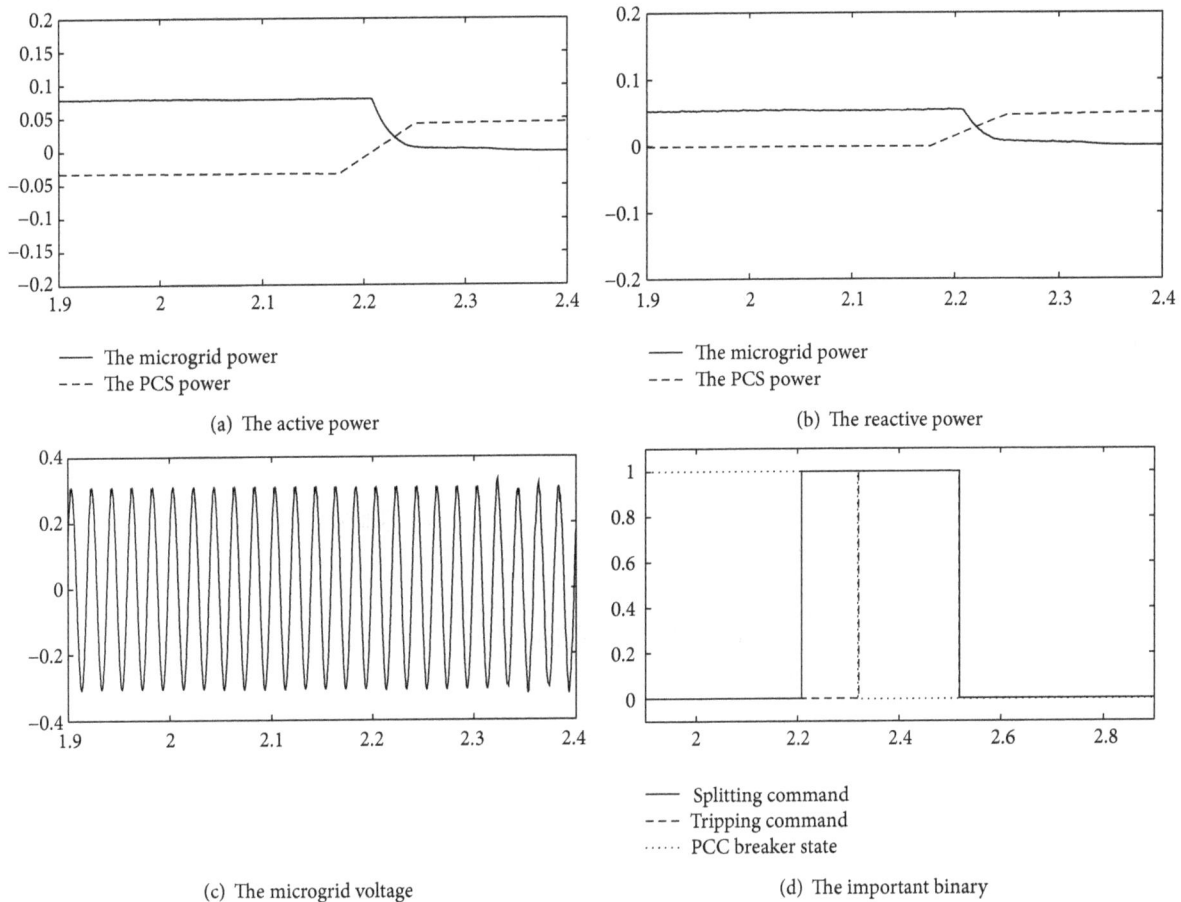

(a) The active power

— The microgrid power
--- The PCS power

(b) The reactive power

— The microgrid power
--- The PCS power

(c) The microgrid voltage

(d) The important binary

— Splitting command
--- Tripping command
······ PCC breaker state

FIGURE 6: Operation results (1).

3.2. Simulation Results

3.2.1. Planned Conversion from Grid-Connected Mode to Islanding. Figure 6 shows the operation results.

The microgrid works under the grid-connected mode before 2.2 seconds, and the switching reactive power is 50 kVA; the switching active power is 80 kW.

At 2.2 seconds, the mode controller received the splitting commands. The mode controller began to change the reference power of the storage system until the apparent power reached 0, as shown in Figures 6(a) and 6(b).

After adjustment for 150 milliseconds, the amplitude of apparent power is continuously lower than the threshold value for a period of time, and it meets the tripping conditions; the mode controller of the microgrid sends the command to trip the PCC breaker, as shown in Figure 6(d).

Before and after splitting, the voltage of the microgrid remained unchanged, as shown in Figure 6(c).

Through the simulation results, the mode controller of the microgrid is useful in this conversion and maintains stability of power system voltage before and after splitting.

3.2.2. Unplanned Conversion from Grid-Connected Mode to Islanding. Operation results are shown in Figure 7.

The microgrid works under grid-connected mode before 4.5 seconds, and the switching reactive power is 110 kVA; the switching active power is 170 kW.

At 4.5 seconds, the microgrid disconnected from the main grid.

After 200 milliseconds, storage system detected islanding mode, operated under V/F control mode, and adjusted the output power.

At moment of splitting, the voltage will have some changes. In this case, the voltage dropped a little. After the storage system works from PQ control mode to VF control model, the voltage of the microgrid gets gradually regeneration and ultimately achieves stable operation, as shown in Figure 7(c).

Through the simulation results, the storage system control logic is useful in this conversion and maintains stability of power system voltage before and after splitting.

3.2.3. Planned Conversion from Islanding Mode to Grid-Connected Mode. Operation results are shown in Figure 8.

The microgrid operates in islanding mode before 5.0 seconds.

The microgrid sends the reclosing command at 5.0 seconds. The voltage of microgrid and the main grid is shown

(a) The active power

(b) The reactive power

(c) The microgrid voltage

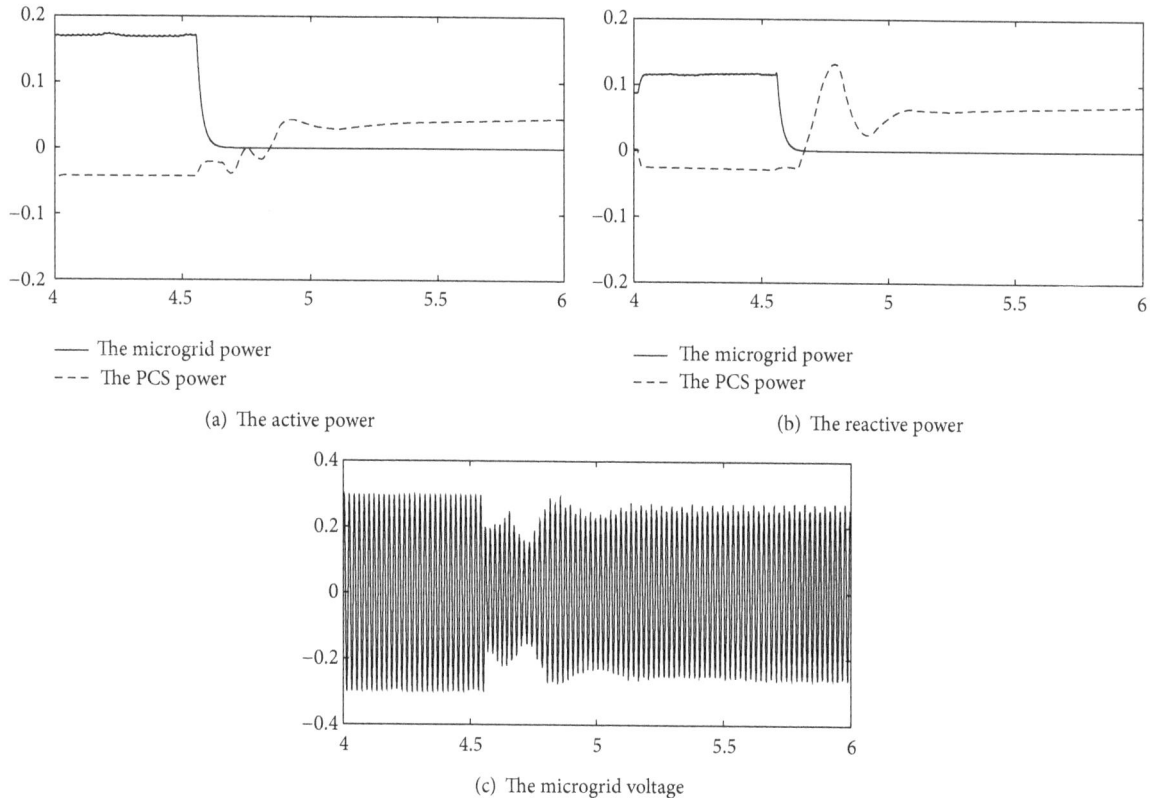

FIGURE 7: Operation results (2).

in Figure 8(a). In Figure 8(a), the phase difference between microgrid and main grid is about 40°.

After adjustment about 0.6 seconds, synchronization condition is met at 5.7 seconds, as shown in Figure 8(b). After the mode controller sends a close command, complete the conversion, as shown in Figure 8(e). Figures 8(c) and 8(d) are phase difference and amplitude difference in the process.

The simulation results show that the mode controller of microgrid is valid for synchronous reclosing.

Asynchronous reclosing is not detailed in this paper.

4. The Implementation of Device

The operation and control device is the important device of the microgrid, and the mode controller function is one of the important functions in the operation and control device. The characteristic of the operation and control device can influence the stability of the microgrid.

4.1. The Architecture of the Operation and Control Device. The architecture of the operation and control device is as shown in Figure 9.

The operation and control device included three layers as follows:

(1) hardware layer which includes CPU (Inter D525) and the necessary interface, such as RS-485, Ethernet, GPS, binary input, and binary output,

(2) operating system layer which is the operation and control device used LINUX operating system,

(3) software layer which includes operation and control function, mode controller function, power control function, energy management function, communication function, and so forth.

The core control strategies of the operation and control device including:

(1) microgrid black start,

(2) microgrid power control,

(3) microgrid optimization control,

(4) microgrid mode controller,

(5) microgrid energy management.

In order to adapt to different grid structures, different micropower types, multiterminals, and other different applications, the operation and control device needed be programmed by xml file, and then it is consistent with the energy storage device and the secondary control equipment to achieve and mode conversion smoothly.

4.2. The Communication of the Operation and Control Device. With the purpose of implementing control function, the device needs to communicate with the intelligent terminal in microgrid. The IEC61850 protocol, such as 9-2, GOOSE, and MMS, can be support by purposed device. The device

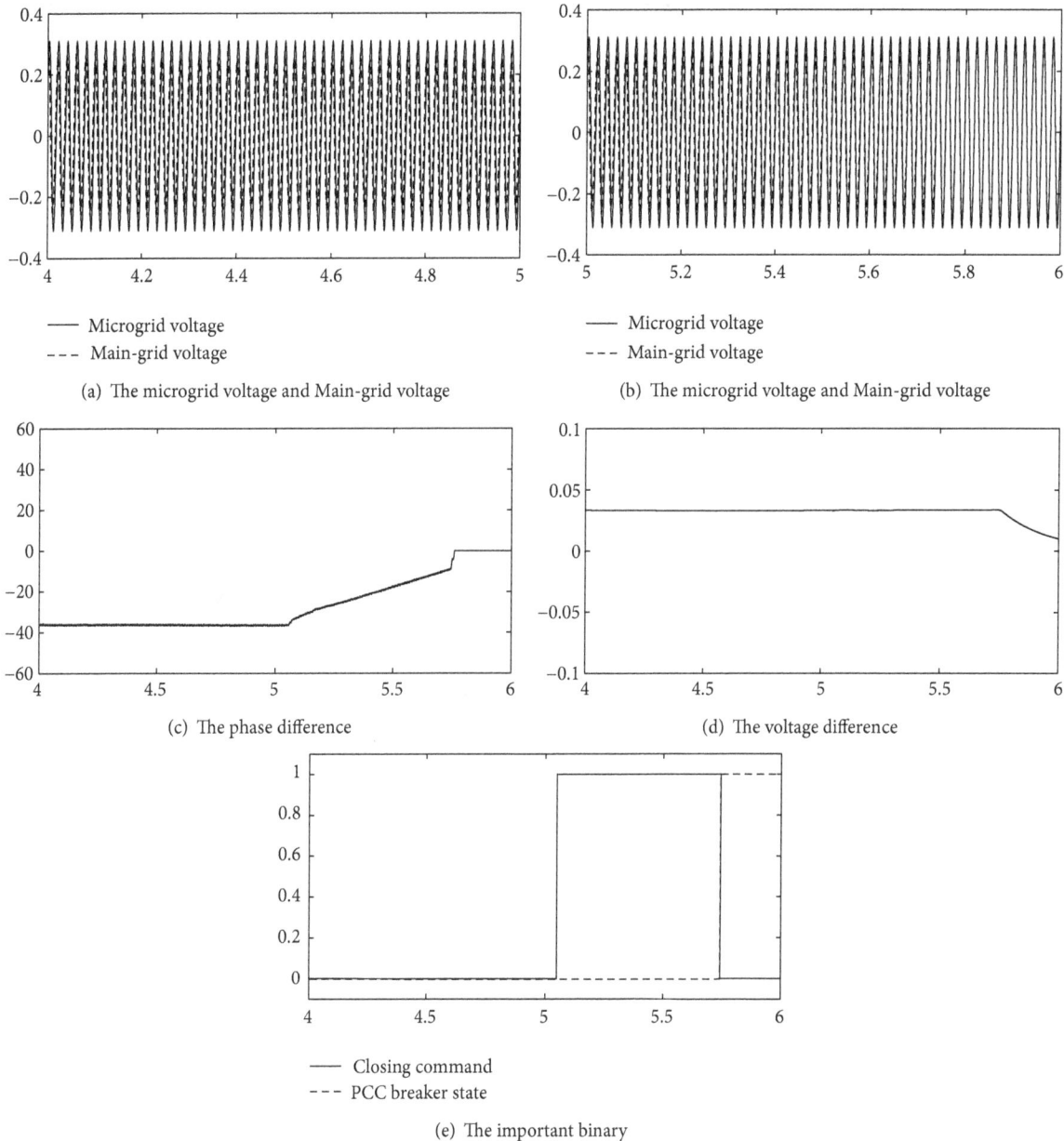

(a) The microgrid voltage and Main-grid voltage

(b) The microgrid voltage and Main-grid voltage

(c) The phase difference

(d) The voltage difference

(e) The important binary

FIGURE 8: Operation results (3).

adopts the 61850 communication architecture, and uses the SMV protocol and GOOSE protocol. The communication of microgrid can adopt the networking mode or the point-to-point mode between intelligent terminal and the operation and control device. The communication of the microgrid involves three kinds of network services, which are SMV, GOOSE, and time synchronization network, each of which can use a separate network, or share a common network. In practical applications, the IEEE 1588 Synchronous Ethernet technologies are adopted to build a shared network for SMV, GOOSE, and time synchronization, and optimized data flow distribution is achieved by using VLAN. A typical communication architecture of the microgrid based on IEC61850 is shown in Figure 10.

As shown in Figure 10, the operation and control system of microgrid mainly consists of the following:

(1) the operation and control device: each microgrid sets up one operation and control device,

(2) intelligent terminal: the intelligent terminal is mainly responsible for collecting AC quantities and acquiring information from local circuit breakers and executing the control commands issued from the central unit,

(3) synchronization clock source: the synchronization clock source mainly provides clock synchronization of the intelligent terminal.

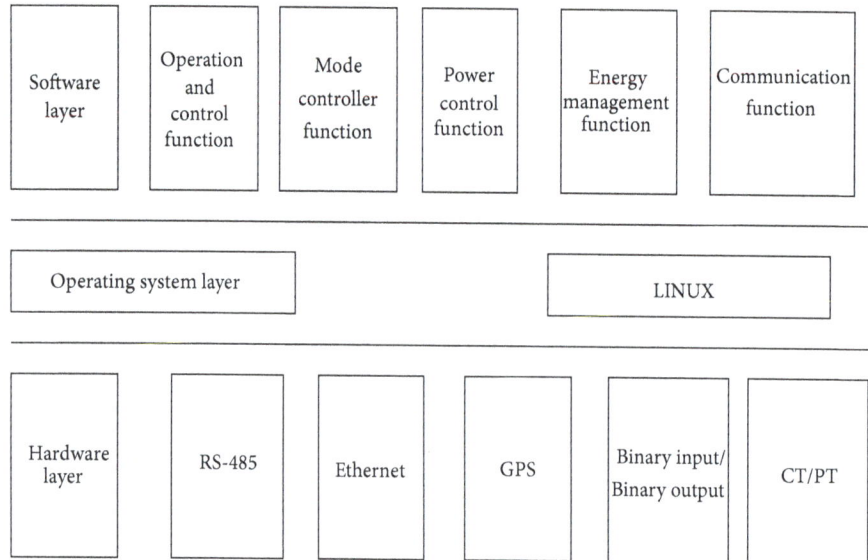

FIGURE 9: The operation and control device of microgrid.

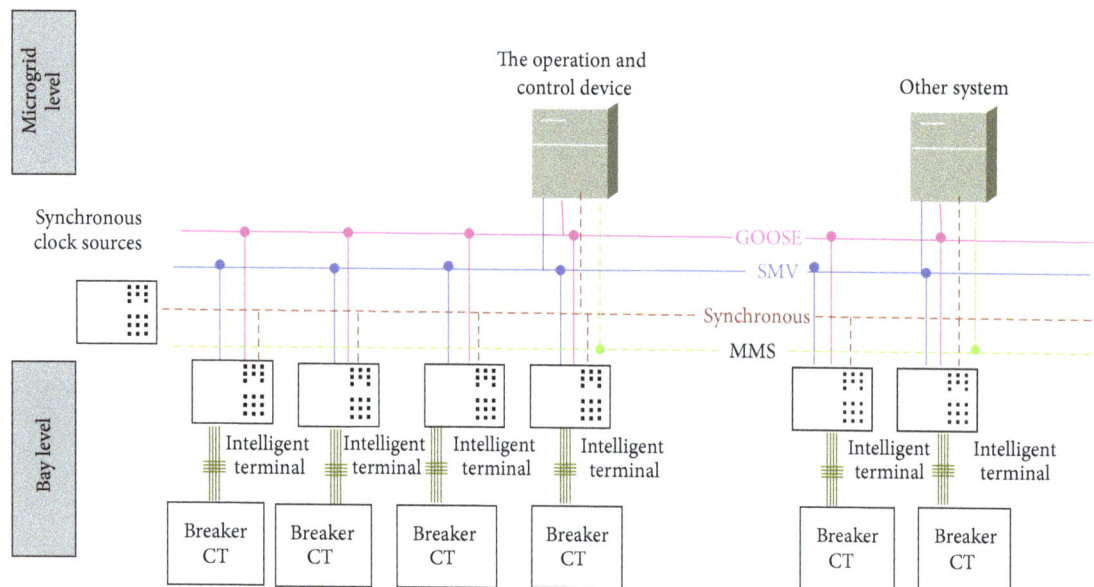

FIGURE 10: Communication architecture of the microgrid based on IEC61850.

5. Conclusions

Microgrid operates under two typical modes. Microgrid mode conversion has been an important part of microgrid research, and the seamless transfer of the microgrid is the control goal. The "master-slave" architecture microgrid which is widely used in engineering is selected as the research focus. The mode conversion function is fulfilled by the mode controller of microgrid and the inverter of the energy storage system. The simulation results show that the mode controller and energy storage system inverter operation mode conversion logic is valid. For further research, characteristic analysis of the equipment in the microgrid can make the microgrid more standardized and more reasonable.

Conflict of Interests

The authors declare that there is no conflict of interests regarding the publication of this paper.

Acknowledgment

This paper was supported by the National Natural Science Foundation of China (no. 51177058 and no. 51277085).

References

[1] E. Sortomme, S. S. Venkata, and J. Mitra, "Microgrid protection using communication-assisted digital relays," *IEEE Transactions on Power Delivery*, vol. 25, no. 4, pp. 2789–2796, 2010.

[2] R. H. Lasseter, J. H. Eto, B. Schenkman et al., "CERTS microgrid laboratory test bed," *IEEE Transactions on Power Delivery*, vol. 26, no. 1, pp. 325–332, 2011.

[3] C. Ramonas and V. Adomavicius, "Research of the reactive power control possibilities in the grid-tied PV power plant," *Electronics and Electrical Engineering*, vol. 19, no. 1, pp. 31–34, 2013.

[4] V. Pilkauskas, R. Plestys, G. Vilutis, and D. Sandonavicius, "Improvement of WMS functionality, aiming to minimize processing time of jobs in grid computing," *Electronics and Electrical Engineering*, vol. 113, no. 7, pp. 111–116, 2011.

[5] G. N. Tsouehnikas, N. L. Soultanis, A. I. Tsouehnikas et al., "Dynamic modeling of microgrids," in *Proceedings of the International Conference on Future Power Systems*, pp. 1–7, Amsterdam, The Netherlands, 2006.

[6] H. R. Lasseter, "Microgrids," in *Proceedings of the Power Engineering Society Winter Meeting*, pp. 305–308, New York, NY, USA, 2002.

[7] K. Sheng, L. Kong, Z.-P. Qi, W. Pei, H. Wu, and P. Xi, "A survey on research of microgrid—a new power system," *Relay*, vol. 35, no. 12, pp. 75–81, 2007.

[8] C.-S. Wang, F. Gao, P. Li, and F. Ding, "Review on the EU research projects of integration of renewable energy sources and distributed generation," *Southern Power System Technology*, vol. 2, no. 6, pp. 1–6, 2008.

[9] C. Li, *Study on micro grid modeling and its connecting operation mode [M.S. dissertation]*, Tianyuan University of Technology, 2010.

[10] Z. Yang, C. Wang, and Y. Che, "A small-scale microgrid system with flexible modes of operation," *Automation of Electric Power Systems*, vol. 33, no. 14, pp. 89–92, 2009.

[11] J. Zeng, *Construction and control of energy storage systems used in renewable energy and micro grid [Ph.D. dissertation]*, Huazhong University of Science and Technology, 2009.

[12] Z. Xiao, *Control and operation characteristic analysis of a micro grid [Ph.D. dissertation]*, Tian University, 2008.

[13] H.-Z. Yang and X.-M. Jin, "Research on grid-connected photovoltaic inverter of maximum power point tracking," *Journal of Northern Jiaotong University*, vol. 28, no. 2, pp. 65–68, 2004.

[14] C. Zhang, M.-Y. Chen, and Z.-C. Wang, "Study on control scheme for smooth transition of micro-grid operation modes," *Power System Protection and Control*, vol. 39, no. 20, pp. 1–10, 2011.

[15] Z. Wang, L. Xiao, Z.-L. Yao, and Y.-G. Yan, "Design and implementation of a high performance utility-interactive inverter," *Proceedings of the Chinese Society of Electrical Engineering*, vol. 27, no. 1, pp. 54–59, 2007.

[16] Z.-L. Yang, C.-S. Wu, and H. Wang, "Design of three-phase inverter system with double mode of grid-connection and stand-alone," *Power Electronics*, vol. 44, no. 1, pp. 14–16, 2010.

[17] J. Jiang, S. Duan, and Z. Chen, "Research on control strategy for three-phase double mode inverter," *Transactions of China Electrotechnical Society*, vol. 27, no. 2, pp. 52–58, 2012.

Permissions

The contributors of this book come from diverse backgrounds, making this book a truly international effort. This book will bring forth new frontiers with its revolutionizing research information and detailed analysis of the nascent developments around the world.

We would like to thank all the contributing authors for lending their expertise to make the book truly unique. They have played a crucial role in the development of this book. Without their invaluable contributions this book wouldn't have been possible. They have made vital efforts to compile up to date information on the varied aspects of this subject to make this book a valuable addition to the collection of many professionals and students.

This book was conceptualized with the vision of imparting up-to-date information and advanced data in this field. To ensure the same, a matchless editorial board was set up. Every individual on the board went through rigorous rounds of assessment to prove their worth. After which they invested a large part of their time researching and compiling the most relevant data for our readers. Conferences and sessions were held from time to time between the editorial board and the contributing authors to present the data in the most comprehensible form. The editorial team has worked tirelessly to provide valuable and valid information to help people across the globe.

Every chapter published in this book has been scrutinized by our experts. Their significance has been extensively debated. The topics covered herein carry significant findings which will fuel the growth of the discipline. They may even be implemented as practical applications or may be referred to as a beginning point for another development. Chapters in this book were first published by Hindawi Publishing Corporation; hereby published with permission under the Creative Commons Attribution License or equivalent.

The editorial board has been involved in producing this book since its inception. They have spent rigorous hours researching and exploring the diverse topics which have resulted in the successful publishing of this book. They have passed on their knowledge of decades through this book. To expedite this challenging task, the publisher supported the team at every step. A small team of assistant editors was also appointed to further simplify the editing procedure and attain best results for the readers.

Our editorial team has been hand-picked from every corner of the world. Their multi-ethnicity adds dynamic inputs to the discussions which result in innovative outcomes. These outcomes are then further discussed with the researchers and contributors who give their valuable feedback and opinion regarding the same. The feedback is then collaborated with the researches and they are edited in a comprehensive manner to aid the understanding of the subject.

Apart from the editorial board, the designing team has also invested a significant amount of their time in understanding the subject and creating the most relevant covers. They scrutinized every image to scout for the most suitable representation of the subject and create an appropriate cover for the book.

The publishing team has been involved in this book since its early stages. They were actively engaged in every process, be it collecting the data, connecting with the contributors or procuring relevant information. The team has been an ardent support to the editorial, designing and production team. Their endless efforts to recruit the best for this project, has resulted in the accomplishment of this book. They are a veteran in the field of academics and their pool of knowledge is as vast as their experience in printing. Their expertise and guidance has proved useful at every step. Their uncompromising quality standards have made this book an exceptional effort. Their encouragement from time to time has been an inspiration for everyone.

The publisher and the editorial board hope that this book will prove to be a valuable piece of knowledge for researchers, students, practitioners and scholars across the globe.

List of Contributors

Xianqing Chen and Lenan Wu
School of Information Science and Engineering, University of Southeast, 2 Sipailou, Nanjing 210096, China

Soobin Lee
Institute for IT Convergence, KAIST, Yuseong-gu, Daejeon 305-701, Republic of Korea

Howon Lee
Department of Electrical, Electronic and Control Engineering, Hankyong National University, Anseong, Gyeonggi 456-749, Republic of Korea

David Petras, Ivan Baronak and Erik Chromy
Institute of Telecommunications, Faculty of Electrical Engineering and Information Technology, Slovak University of Technology in Bratislava, Ilkovicova 3, 812 19 Bratislava, Slovakia

Wei Ding and Hongfa Wang
Zhejiang Water Conservancy and Hydropower College, Hangzhou, Zhejiang 310018, China

Xuerui Wei
Department of Mathematics, Shaoxing University, Shaoxing, Zhejiang 312000, China

Rui Guo, Qiaoyan Wen, Zhengping Jin and Hua Zhang
State Key Laboratory of Networking and Switching Technology, Beijing University of Posts and Telecommunications, Beijing 100876, China

Murat Karakaya
Department of Computer Engineering, Atilim University, Incek, 06836 Ankara, Turkey

Jitendra Mohan
Department of Electronics & Communication Engineering, Jaypee Institute of Information Technology, Noida 201304, India

Sudhanshu Maheshwari
Department of Electronics Engineering, Z.H. College of Engineering and Technology, Aligarh Muslim University, Aligarh 202002, India

Tiefeng Li
Xian Institute of Optics and Precision Mechanics of Chinese Academy of Sciences, Xian 710119, China
The Graduate University of Chinese Academy of Sciences, Beijing 100049, China

Caiwen Ma and Wen Hua Li
Xian Institute of Optics and Precision Mechanics of Chinese Academy of Sciences, Xian 710119, China

Eunsam Kim
Department of Computer Engineering, Hongik University, Seoul 121-791, Republic of Korea

Sangjin Kim
Hyundai Autoever Corporation, 576 Sam, Uiwang, Gyeonggi 437-040, Republic of Korea

Choonhwa Lee
Division of Computer Science and Engineering, Hanyang University, Seoul 133-791, Republic of Korea

Rongzong Kang, Pengwu Tian and Hongyi Yu
Zhengzhou Information Science and Technology Institute, Zhengzhou 450002, China

Ji-Jian Chin and Raphael C.W. Phan
Faculty of Engineering, Multimedia University, 63100 Cyberjaya, Selangor, Malaysia

Syh-Yuan Tan and Swee-Huay Heng
Faculty of Information Science and Technology, Multimedia University, Jalan Ayer Keroh Lama,
75450 Bukit Beruang, Melaka, Malaysia

Kuan-Ta Chen
Institute of information science, Academia Sinica, 128 Academia Road, Section 2, Nankang, Taipei 115, Taiwan

Chin-Laung Lei
Department of Electrical Engineering, National Taiwan University, No. 1, Section 4, Roosevelt Road, Taipei 10617, Taiwan

Fu-Min Wang, Chin-Hua Tsai, San-Chuan Hung, Perng-Hwa Kung and Shou-De Lin
Department of Computer Science and Information Engineering, National Taiwan University, No. 1, Section 4, Roosevelt Road, Taipei 10617, Taiwan

Jing-Kai Lou
Institute of information science, Academia Sinica, 128 Academia Road, Section 2, Nankang, Taipei 115, Taiwan
Department of Electrical Engineering, National Taiwan University, No. 1, Section 4, Roosevelt Road, Taipei 10617, Taiwan

Chun-I Fan, Wei-Zhe Su and Hoi-Tung Hau
Department of Computer Science and Engineering, National Sun Yat-sen University, Kaohsiung 80424, Taiwan

Jinyan Chen
School of Computer Software, Tianjin University, Tianjin 300072, China

Jiansheng Liu
College of Science, Jiangxi University of Science and Technology, Ganzhou 330200, China

Yang Lu and Jiguo Li
College of Computer and Information Engineering, Hohai University, No. 8, Focheng Xi Road, Jiangning District, Nanjing, Jiangsu 211100, China

Xingye Zhu and Jingsha He
School of Software Engineering, Beijing University of Technology, Beijing 100124, China

Jing Xu, Yuqiang Zhang and Ting Zhang
College of Computer Science and Technology, Beijing University of Technology, Beijing 100124, China

Wanqing Fu
Information Center, SINOPEC Research Institute of Petroleum Processing, Beijing 100083, China

Zhenguo Zhao
School of Water Conservancy, North China University of Water Resources and Electric Power, Zhengzhou 450045, China

Wenbo Shi
Department of Electronic Engineering, Northeastern University at Qinhuangdao, Qinhuangdao 066004, China

Tsu-Yang Wu
Shenzhen Graduate School, Harbin Institute of Technology, Shenzhen 518055, China
Shenzhen Key Laboratory of Internet Information Collaboration, Shenzhen 518055, China

Tung-Tso Tsai and Yuh-Min Tseng
Department of Mathematics, National Changhua University of Education, Jin-De Campus, Changhua City 500, Taiwan

Yanxue Zhang
College of Mathematics and Information Science, Hebei Normal University, Shijiazhuang 050000, China

Dongmei Zhao
College of Information Technology, Hebei Normal University, Shijiazhuang 050000, China

Jinxing Liu
The First Aeronautics College of PLAAF, Xinyang 464000, China

Zhaoyun Zhang, Wei Chen and Zhe Zhang
State Key Laboratory of Advanced Electromagnetic Engineering and Technology, Huazhong University of Science & Technology, Wuhan 430074, China